Sheep Farming for Meat & Wool

Editors:
Jane Court, John Webb Ware and Sue Hides

Other contributors:
Noel Campbell, David Champness, Gary Hallam, Leanne Horstman, Linda Hygate, Lachlan King, Trevor Pollard, Nick Rolls, Raquel Waller and Lisa Warn

CSIRO PUBLISHING

National Library of Australia Cataloguing-in-Publication entry

Sheep farming for meat and wool/editors, Jane Court, John Webb Ware and Sue Hides.

9780643092945 (pbk.)

Includes bibliographical references and index.

Sheep – Australia.
Sheep – Breeding – Australia
Wool industry – Australia.
Wool – Australia.
Lamb (Meat) – Australia.
Mutton – Australia.
Meat industry and trade – Australia.

Court, Jane.
Webb Ware, John.
Hides, Sue.

636.30994

Published by
CSIRO Publishing
36 Gardiner Road, Clayton VIC 3168
Private Bag 10, Clayton South VIC 3169
Australia

Telephone: [+613] 9545 8555
Email: csiropublishing@csiro.au
Website: www.publishing.csiro.au

Front cover: EverGraze trial site, Hamilton, Victoria. (Department of Primary Industries)

Set in 10/14 Adobe Palatino and Plantin
Edited by Peter Storer Editorial Services
Cover and text design by James Kelly
Typeset by Desktop Concepts Pty Ltd, Melbourne
Printed by Ingram Lightning Source

CSIRO Publishing publishes and distributes scientific, technical and health science books and journals from Australia to a worldwide audience and conducts these activities autonomously from the research activities of the Commonwealth Scientific and Industrial Research Organisation (CSIRO).

Mar26_RP_ILS

Foreword

This book focuses on the business of sheep farming. Its aim is to play a small part in the DPI's objective to enable transformation in Victoria's primary industries to increase wealth and wellbeing sustainably while protecting and enhancing safety, community, animal welfare and the environment.

My staff in Farm Services Victoria – a division within DPI which focuses on farm businesses – have played an important role in the development of this book. In doing so, they have demonstrated some of DPI's values including an ability to work together and collaborate within and outside DPI to achieve outcomes that make a difference. The collaboration with the Mackinnon Project of the University of Melbourne has been invaluable.

I trust the information in this book will assist primary producers to become even more productive, competitive and sustainable in an era of unprecedented change, including the emerging and profound challenge of climate change, as well as the pressures of competitive and volatile markets and more demanding community attitudes.

Ron Harris
Director, Farm Services Victoria

Arguably, it has never been more important that sheep farming enterprises are run efficiently to achieve the business goals of the owner. A constellation of factors – including an uncertain climate future with impacts on pasture management, patterns of endemic disease and increased risk of emergence of new or exotic diseases, changing rural demographics, increased consumer interest in production methods including animal welfare and wellbeing, changing local and international markers for product and a heightened expectation for food that is safe and nutritious – contribute to increasingly complex challenges to farm management. The complexity of the challenges facing farmers is such that there has never been a time when there was a greater need for expert, impartial advice based on solid science and a comprehensive awareness of the needs of the industry.

This book is both a resource for sheep farmers and an illustration of the high quality of advice that is available to sheep farmers within southern Australia. The collaborative approach to the compilation of this book, the diversity of the backgrounds of the contributors, and the involvement of DPI and the University of Melbourne are examples of the power of collaboration in achieving industry-wide goals. This book underscores the need for cooperation amongst state and federal governments, the tertiary education and research sector, and private industry to address the critical production-level needs of sheep farmers. The University of Melbourne, through the Mackinnon Project, has been a leader in providing this advice for over three decades and, as we enter the second decade of the 21st century, we commit to playing an even greater role in the rural livestock industries.

I congratulate the editors and contributors of this book on compilation of an excellent text on sheep farming in southern Australia. The material is comprehensive, without being excessively dense, and provides farmers with an excellent resource.

Professor Ken Hinchcliff
Dean, Faculty of Veterinary Science
University of Melbourne

Contents

Acknowledgements

Several years ago, Jane Court, Sheep Industry Officer with the Department of Primary Industries instigated a rewrite of *Sheep Farming in East Gippsland*, which was published in 1987 by the Gippsland Sheep Breeders Association. The result is this book: *Sheep Farming for Meat and Wool.*

Sheep Farming in East Gippsland was the result of a collaboration between the then Department of Agriculture and Rural Affairs, the University of Melbourne Mackinnon Project and the Gippsland Sheep Breeders Association. The contributions of Bruce Allworth, Nick Barton, Tony Brightling, Richard Churchley, Chris Ethridge, Ian Grant, Chris Halpin, Leo Hamilton, Laurie Hamlyn, Pat Hedlam, Sue Hides, Leanne Horstman, Peter Jackson, Bob Lambell, Jim Mount, Fred Morley, Mike Phillips, David Sackett, Don Shaw, Brian Shugg, Len Stephens and Tim Watts are gratefully acknowledged because the concept and structure of the renamed, revised and expanded *Sheep Farming for Meat and Wool* are identical to *Sheep Farming in East Gippsland* and parts of the original text are still relevant and have been retained.

Jane's co-editors recognise her persistence, expertise and enthusiasm for *Sheep Farming for Meat and Wool.* The editors also gratefully acknowledge the expertise and dedication of the contributing authors, our colleagues in the Mackinnon Project and the Department of Primary Industries.

In writing this book, we are grateful to the many dedicated scientists and producers whose accumulated experience has helped build a better understanding of sheep production systems.

The contents of this book have evolved from the research and extension projects conducted over many years by the Department of Primary Industries (and its predecessors) and the University of Melbourne.

Many thanks also to Kalinda Wade for word processing and to CSIRO Publishing, especially Ted Hamilton, for his encouragement and patience!

Dedication

Dr Fred HW Morley HDA BVSc PhD DAgSc FTS FACVS FASAP FAIAS FAAABG
14 July 1918 – 14 March 2001

Fred Morley was a remarkable man who made an outstanding contribution to Australian agriculture and the scientific community in general, including the disciplines of veterinary science, animal genetics, plant taxonomy and agricultural systems analysis. Dr Morley was a man with an insatiable appetite for knowledge who demanded, and achieved, the highest level of scientific integrity.

Over a career spanning 50 years at CSIRO and later years at the University of Melbourne, Dr Morley conducted and led research on a huge variety of subjects and published over 150 refereed scientific papers on diverse fields. Some of his research was fundamentally important to the success and profitability of Australian agriculture, including sheep genetic improvement, blowfly control in sheep, the discovery of phyto-oestrogen induced infertility in sheep, and quantifying the production losses associated with internal parasites in sheep and cattle. In addition, he undertook a broad range of research on the taxonomy and genetics of subterranean clover and he initiated work leading to the development of superior varieties of sub clover and conducted research on the relationship of stocking rates, subdivisions and pasture cultivars on animal production. Dr Morley was one of the first researchers to see the potential of computers to aid in the analysis and design of complex farming systems, and developed a group to study agricultural systems.

At the University of Melbourne, he initiated a project to develop methods of delivering a health and production management program to service farmers or corporations engaged in sheep and beef cattle production. The program, known as the Mackinnon Project, began in 1983 and proved to be highly successful. The Mackinnon Project is now conducting a 'whole farm' commercial consultancy sheep and beef farms throughout Victoria and south-east Australia. Over 50 post-graduate students were supervised by Dr Morley during his time at the University and his knowledge and pursuit of excellence have left a legacy that continues to benefit the livestock industries in Australia to this day.

Introduction

> 'Grazing systems are made up of several components. The climate, the soil, plants, animals, parasites and diseases all combine and affect animal production. The farmer, whose objectives are central to any decisions that are to be made, is the driving force behind the system. The profitability of the enterprise he manages depends upon his efficiency of management, the current costs and prices associated directly with the enterprise, and also the state of the economy of his own country and of those countries which trade with it.'
>
> *F.H.W Morley*

This statement remains relevant today, and sheep farmers have embraced practices and philosophies to improve the productivity and efficiency of management of complex grazing systems, to protect the natural environment and to ensure good animal welfare in the face of challenges including commodity price variability and climate unpredictability.

Current significant challenges include the adaptation of sheep farming systems to climate variability and to consumer expectations for the welfare of sheep.

The objective of this book is to compile the wealth of research and knowledge on sheep farming in southern Australia and to provide it in readily accessible form to farmers, farm service providers and students of agriculture. It is a distillation of information on farm business management, marketing, pasture production, nutrition, disease control and prevention, reproductive management, genetic improvement and sheep welfare. It is intended as a first point of reference for a great number of aspects of the sheep farming system and, where relevant, it directs the reader to further sources of information.

'Efficiency of management' is the one factor over which the farmer has direct control. This book provides the background knowledge that can help inform decisions to improve efficiency of management.

CHAPTER 1

Business management

This chapter describes the approach for the development and implementation of a farm business plan to optimise profitability while ensuring sound environmental management, good sheep health and welfare, and meeting personal goals.

Managing a sheep enterprise

There are many facets of a farm business to be considered if the enterprise is to be financially successful, environmentally sustainable and socially acceptable. Managing a sheep flock is demanding and complex. The manager requires a broad set of skills including agronomy, livestock husbandry and financial and personnel management. The manager has to be able to manage the risks of variable climate and volatile commodity prices while ensuring that natural resources are maintained and improved, sheep welfare requirements are met and that the family has an acceptable lifestyle.

Successful farm managers understand and monitor the financial performance of their business, understand the key profit drivers and focus on the key aspects of management to ensure success. They understand pasture production and feed requirements, can manipulate feed supply and feed demand to optimise productivity and profitability. The successful manager is aware of the financial, environmental, animal health and welfare and personal risks that can have an impact on the business, and how to manage them.

Business planning

All producers have different visions and objectives. A key to realising objectives is to develop and constantly update a business plan. The plan must encompass all aspects of the farm business. In a complex system involving physical resources, variable climate and numerous options for pasture and livestock management, having objectives and a vision of what the future will look like is essential to maximise profitability and manage risk. In a constantly changing physical and financial environment, the manager must be motivated to manage change and must have a clear plan.

Plans can be long and short term, strategic or tactical. Long-term strategic plans set direction of the business for the next 5 to 10 years and include specific markets to be targeted. Plans should also be made for immediate or shorter term issues.

Developing a business plan

The **first step** in developing a business plan is to describe the mission, vision, goals and objectives of the farm business.

The **second step** is to describe and review all aspects of the **current** farm business including:

- physical resources
- financial performance
- enterprise mix
- productivity
- environmental management
- human resources
- overall business structures.

The **third step** is to identify the strengths and weaknesses of the farm business and the opportunities for, and threats to, the farm business.

The key issues that must be considered when developing a business plan should include:

- future financial plans
- future enterprise mix
- future production plans
- future pasture development plans
- future marketing plans
- future natural resource management plans
- future human resources.

The plan must contain an implementation schedule and monitoring strategies.

Strategies can then be developed to achieve goals and objectives. Finally, these strategies can be implemented to achieve success. All aspects of the plan should be monitored and reviewed as business conditions and/or personal goals and objectives change.

When developing a farm business plan, the resources and expertise of all people involved in the business should be used to ensure the plans is comprehensive, realistic and achievable.

Setting mission, vision, goals and objectives

An overall mission statement defines the fundamental purpose of the enterprise: describing why it exists and usually the level of performance. The vision defines the desired or intended future state of an enterprise, or what the enterprise wants to be in terms of its objectives: it concentrates on the future and is a source of inspiration.

Objectives are continued statements of intended future outcomes. The objectives set for the business should provide a general direction for all aspects of the business. If numerous objectives are set, it is important to prioritise them.

In contrast, goals are specific, time-bound statements of intended future results. For example, an objective may be to increase pasture productivity and a goal may be to renovate pastures and apply additional fertiliser on a specific area of land within a certain time frame.

Sometimes the terms goal and objective are used interchangeably.

A good goal is clear and specific and can be measured; it is also achievable and realistic and can be completed in a certain time period.

Review of all aspects of the current farm business

There are many aspects of the farm business that should be evaluated. For many producers, analysis of the farm's performance ends with the taxation accounts. For those with financial commitments, more detailed financial reports are required. Some producers are just interested in looking at physical performance such as wool cut per head or lambing percentages. Alone, these offer little insight into enterprise profitability and efficiency. Detailed financial and physical reports offer so much more and, apart from been part of a business plan, provide the foundation for critical analysis of the existing enterprise to identify opportunities for improvement.

The records needed to analyse the farm business are listed below.

- A balance sheet with separate and combined statements of farm and off-farm assets and liabilities (using realistic valuations for all assets). Apart from providing a statement of the farms' financial position, the balance sheet is useful to monitor changes over time.
- A cash flow budget to outline the inflow of cash and outflow of expenses from the business. Cash flow budgets can use actual accounts or budget future cash flow. They are usually conducted on a monthly basis, but can be budgeted long into the future, particularly when major investment decisions are made. Understanding cash flow is critical in a farm business because cash flow budgets allow proper planning for future development and allocation of expenses. There are numerous accounting programs available to assist in development of cash flow budgets.
- A profit and loss statement, which, from a management perspective, is different to taxation profit. The profit and loss statement should take into account the livestock inventory, crop inventory, owner operator wages and depreciation. The profit and loss statement is crucial for benchmarking analysis.
- A stock schedule and crop schedule are essential for an enterprise analysis. An enterprise analysis cannot rely on the numbers sold or value sold alone. The stock schedule should be broken down to each class of livestock and include opening numbers, closing numbers, number sold and numbers purchased. For the crop schedule, all supplementary feeds fed to livestock should be documented, whether produced on farm or purchased.
- Physical performance data including wool (greasy and clean) and meat production (live weight and carcase weight), fertiliser type and amount and land use by area.
- Details of labour use, including family and employed labour and contractors.
- Taxation accounts are a useful starting point for an enterprise analysis. In many cases, taxation accounts are the only information readily available to start an enterprise analysis.

Farm analysis and benchmarking

Farm analyses are undertaken for many reasons and at different levels of the business. To gain a full understanding of a business, a full financial analysis should be conducted. Analyses can be conducted to compare the performance of enterprises in an individual farm between years (within-farm analysis) or to compare performance of different farms in the same year (between-farm analyses). Both strategies are very useful and important.

Within-farm analysis

Within-farm analysis can provide information on performance and a useful comparison of individual enterprises on a farm, especially when conducted over successive seasons. It is important to understand the impact that one enterprise may have on another by comparing them. A livestock enterprise can benefit by periodic grazing of winter wheats or grazing crop stubbles – enabling higher stocking rates compared with a livestock only enterprise – however, for the first few years post cropping, pastures may not be as productive and this may affect livestock profitability. The decision on what enterprises to run is very complex and success is dependent on interactions between enterprises, commodity prices and seasonal conditions. An understanding of relative enterprise performance is very useful when making decisions about enterprise mix.

Between-farm analysis (comparative analysis)

Benchmarking is the process of comparing the cost, production and value of product and financial performance of one farm against others.

Between-farm analysis is a key component of understanding how a business is performing. Although individual farms are very different, valuable information can be obtained by conducting a whole farm production and financial analysis and comparing outcomes (or benchmarking) against other farms. Numerous service providers provide benchmarking services at varying levels of complexity. Engaging professional services will often provide a more detailed interpretation and can assist in setting realistic targets for improvement.

Between-farm analyses can help to identify areas that can be improved either at the whole farm or enterprise level and can help to identify goals. Between-farm analyses also provide valuable information on how productivity and financial performance changes as goals are achieved over a period of years. Interpretation of data must take into account the resource base and management skills available.

Once farm performance has been analysed, the performance is compared with other farms. Ideally, farm performance should be compared with the average and also the top 20% (usually based on return on assets).

Table 1.1 lists the key performance indicators (KPIs) and benchmarks for these indicators for sheep enterprises. These are very useful to better understand farm business performance and to help to identify areas that need improvement. It is important to consider a combination of indicators when evaluating results. Generally, the most profitable farms are not necessarily those with the highest production and lowest cost of production, but have a combination of low cost of production, due to higher production, and lower cost structures. The analysis – particularly when formulated over a period of years (to monitor progress) – provides the framework for developing longer term goals.

Once the overall performance has been evaluated, more detailed analysis of individual enterprises can be conducted. Figure 1.1 provides guidelines for assessing performance at different levels of the farm business.

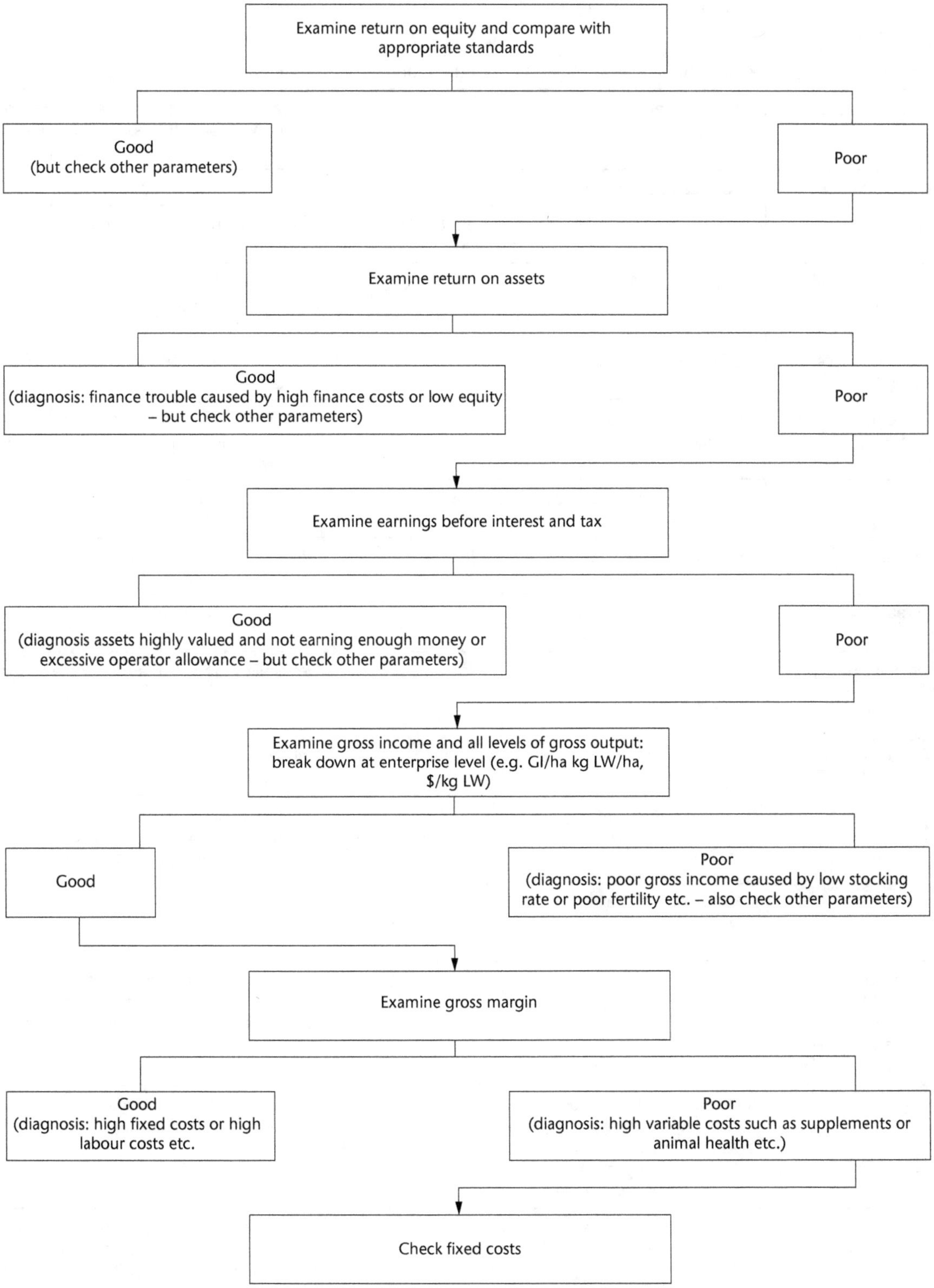

Figure 1.1. Process to identify economic problems at all levels of the farm business.

Table 1.1. Benchmarks and key performance indicators.

Key performance indicator	Calculation	Benchmark[a]	Comments
Return on assets	Operating return/assets managed	>4%	Indicator of profit (be careful interpreting this in regions where asset values are very high)
Return on equity	Business return/equity	>4%	Indicator of profit
Equity	Equity/assets owned	>80–85% usually safe	High asset prices can make this ratio misleading
Operating profit or operating profit/ha		>$100 000	Needs to be large enough to cover finance costs, profit margin, tax, capital investment
Operating profit margin	Percentage of operating profit/gross income	>30%	Indicator of risk – percentage of retained profits
Operating surplus	Operating revenue minus operating expenses	Not applicable	Amount of surplus cash generated which allows investment to improve and sustain the business
Cost structure	Variable and overhead costs/ gross income	<50%	<50% is indicative of a very efficient enterprise
Debt ratio	Interest and lease costs/gross income	<15% safe >20% risky	Indicator of ability to pay financial obligations
Interest cover	Earnings before finance costs and tax/finance costs	>3 good	Ability to service debts
Costs: including variable, pasture, overhead and labour costs $/ha, $/dry sheep equivalent (DSE)		Wide range	Generally, high-profit farms have average costs per DSE but higher costs per hectare, except overhead and labour costs, which are normally lower ($/DSE) in high profit farms
Cost of production	Total costs (may or may not include finance costs)/ production* (for wool including stock trading profit or loss and lamb include sheep trading profit/ loss)	Lamb: <$2.50/kg lamb dressed Wool: <$7.00/kg clean (lower for strong wool flocks; higher for fine wool flocks)	Very useful KPI, but be careful interpreting because the indicator is complex – high cost of production is generally risky and low cost of production usually associated with high profit
Gross margins $ total, $/ha, $/DSE, $/livestock capital. Also consider GM/ ha/100 mm rainfall	Gross income – variable costs	>$40/ha/100 mm	Useful for enterprise comparison – for example, used to compare limiting resources such as land ($/ha) and capital ($/livestock capital)
Production	kg lamb/ha or kg/ha/100 mm rainfall (note live weight or dressed weight). kg wool/ha or kg/ha/100 mm rainfall (note greasy or clean)	>7 kg clean wool/ ha/100 mm >20 kg dressed lamb/ha/100 mm	Useful production indicator and usually correlated with profitability
Product value $/kg lamb, wool, etc.	Note live weight or dressed weight, or greasy or clean for wool		Usually associated with high profit if cost of production is low; also depends on time of sale because commodity price volatility is high

Key performance indicator	Calculation	Benchmark[a]	Comments
Profit per DSE	Subtract all costs/DSE from gross income/DSE		Profit after full cost allocation useful for enterprise comparison
Stocking rate	DSE/ha, DSE/ha/100 mm rainfall or DSE/ha/25 mm rainfall >250 mm	2 DSE/ha/100 mm or 1 DSE/ha/ 25 mm >250 mm	Take care of land class and resource base when making interpretation, but is an important performance indicator. Not applicable in a pastoral zone
Labour efficiency	Total labour (time) DSEs	>10 000 DSE per full time labour unit is very efficient	Labour efficiency indicator
Fertiliser application	kg P/ha, kg P/ha/100 mm, kg P/DSE		Depends on requirement – also consider K, N and S (see Chapter 2)
Weaning percentage	Ewes joined/lambs weaned	>80% in Merinos and >120% in crossbreds	Not most important indicator – more important is production per hectare
Mortality rate	Deaths/total sheep numbers, including weaners	< 3%	Indicator of animal health and welfare standards

[a] based on Victorian South West Farm Monitor Project and Mackinnon Project benchmarking

Identifying opportunities to improve farm profitability

In most farm enterprises there are endless opportunities to invest money and resources to improve profitability to meet long-term goals. The challenge for the farm manager is to identify the strategies that will provide the best return on investment while managing risk, given that money and resources (especially land and people) are usually limiting.

The manager must identify a range of opportunities, often with the assistance of farm analyses. There must be consideration of the time taken to implement a new project or development and the time taken to generate extra revenue. There are many tools that can be used to assist decision making on what strategy to adopt.

Decision making

The key requirements for good decision making include:

- listing and considering all the alternatives
- collecting relevant information
- evaluating the economic impact of the alternatives
- considering intangible factors for which an economic impact can not be defined, such as a personal preference
- making a decision
- carrying it out.

Economic tools that can be used to help decision making include partial budgets, percentage return on capital, gross margins, cash flow budgets and cost–benefit budgets.

Partial budgets

This is a simple, but very important, tool to quantify a new strategy. A simple partial budget should test one alternative and, in the budget, should only consider the changes associated with the new alternative. The new option may result in increased returns, decreased costs, increased costs or decreased returns – all must be considered and compared with the original strategy. Partial budgets are not really appropriate for changes that require capital investment or options with a long time lag before the benefit will accrue. They are appropriate for an alternative with a steady state and do not take the change-over period into account.

Percentage return on capital

The profitability of a project can be assessed by calculating the percentage return on capital. This is defined as change in profit × 100/ change in capital outlay. The percentage return on additional capital is a very useful measure. As a guide, the percentage return on additional capital should exceed 15–20% and will often exceed 30%. Alternatively, the marginal rate of return (per cent) is defined as change in profit × 100/change in costs where capital expenditure is not involved.

Gross margins

The gross margin is defined as the gross income minus the variable costs associated with the enterprise. They provide a simple way to compare enterprises, but can become confused when allocating shared resources, such as equipment, or when one enterprise can have an impact on another, such as benefits of mixing cropping and livestock or interactions between sheep and cattle. It is important to realise that gross margins do not assess the impact of change and do not include overheads and unallocated labour.

Cash flow budgets

Cash flow budgets are an important tool to assess the feasibility of changing one enterprise to another. They are particularly important to consider when making major changes to the business and often should be used once a partial budget has been done. It is important – especially with major investments – to understand any changed financial commitments and when they fall due and any changes to the timing of peak debt.

Cost–benefit budget

Cost–benefit budgets are often more complicated, but are important to estimate the costs and benefits of an option. For projects that take some time to adopt or to yield benefits, a discount factor should be applied at an appropriate interest rate to bring future benefits back to their present value. The future value of a project will diminish rapidly if the interest rate on borrowed money (or opportunity cost) is high. More complicated evaluations can be conducted using net present values, internal rates of return or decision-tree analysis and are usually calculated with the assistance of a consultant.

Once a decision is made to change, it is important to implement the plan in as short a time as possible and to monitor progress to ensure success. Without a plan, no decision to change will succeed, however strong the commitment is.

Managing risk

There are many risks involved with farming that are often a major impediment to change. Farm managers should understand these risks and manage them as much a possible. Ironically, it is often perceived risks that limit the changes that may improve production that expose a farm to substantial financial risks because of low productivity and profitability. This is perhaps the biggest risk that a sheep farm is exposed to: the failure to adopt good management practices, thereby limiting profitability.

There are several broad categories of risk to consider: all of them are important and they all should be managed. The broad categories of risk include:

- financial
- environmental
- animal welfare
- management
- human resource
- personal.

Producers often perceive that financial risk is related only to insolvency and failure to pay financial commitments. Equally as common and important for long term sustainability is the risk of failing to maximise profitability in years when seasonal conditions are favourable and commodity prices are good.

Research shows that, on any given farm with management systems in place, 70% of variation in profit is due to commodity price variation and 25% is associated with climate risk or seasonal variation. Most of the variation in profit associated with climate risk is attributed to higher feed costs and associated lower productivity.

When making decisions on farm, it is imperative that all possible outcomes are considered. For example, when conducting farm budgets, sensitivity analyses should be undertaken to consider all possible outcomes. Sensitivity analysis is used to determine how different values of an independent variable will affect a particular dependent variable under a given set of assumptions.

An example of a sensitivity analysis is outlined in Table 1.2. A farmer is considering running a self-replacing Merino flock or an all wether flock. As part of the process in deciding what enterprise should be run, a gross margin analysis (on a $/DSE basis) is performed to compare a range of production and price outcomes.

Similar resources (labour, infrastructure and land) are available to both livestock enterprises.

In Table 1.2, the gross margin $/DSE of a self-replacing ewe flock with different average fleece weight and different wool prices can be examined. A flock with an average wool production of 5.0 kg greasy has gross margin $/DSE of $20.19 if the micron price guide (MPG) is $10.00. Note that a MPG of $10.00 has been above this price about 25% of the time for a 21-micron flock, 35% of the time for a 20-micron flock and 65% of time for a 19-micron flock over the 10-year period from 1997 to 2007.

In Table 1.3, the gross margin $/DSE of an all wether flock with different average fleece weight and different wool prices can be examined. To achieve a similar gross margin to the self-replacing flock in Table 1.2 when the flock wool production is 5.0 kg/head and the MPG is $10.00, the wethers would need to produce 5.5 kg/head of wool with the same MPG. Other factors the farmer will need

Table 1.2. Self-replacing wool flock gross margin $/DSE with different wool prices (micron price guide) and average greasy fleece weight.

	Micron price guide (MPG) $/kg clean							
kg	**$7.00**	**$8.00**	**$9.00**	**$10.00**	**$11.00**	**$12.00**	**$13.00**	**$14.00**
4.0	$9.83	$11.72	$13.60	$15.48	$17.37	$19.25	$21.14	$23.02
4.5	$11.48	$13.60	$15.72	$17.84	$19.96	$22.08	$24.20	$26.32
5.0	$13.13	$15.48	$17.84	$20.19	$22.55	$24.90	$27.26	$29.61
5.5	$14.78	$17.37	$19.96	$22.55	$25.14	$27.73	$30.32	$32.91
6.0	$16.43	$19.25	$22.08	$24.90	$27.73	$30.55	$33.38	$36.21

Assumptions: ewes 1.5 DSE/head, wool production calculated on average of hoggets, wethers and ewes. Variable cost $12.00/DSE, lamb marking percentage 80%, average sale price of stock $40.15 net. Note: to convert MPG to net greasy price across clip, multiply by 60%.

to consider for the wether enterprise is the availability of wethers of suitable bloodlines at the budgeted price and risk of introducing disease. In addition, a wether enterprise will not fit the pasture growth curve as well as the breeding flock. Other factors that could be budgeted in a sensitivity analysis are the sale price for sheep or lambing percentage in the self-replacing flock or purchase and sale price of wethers in the all wether flock.

The likelihood of achieving the production and price targets should be considered as part of the sensitivity analysis.

There are many strategies that can be implemented on farm that can minimise risk. For example, running a high-profit farm with a low cost of production is an important strategy to minimise the impact of poor prices. Likewise, having large financial reserves can ease the impact of poor seasons and drought. Given the nature of farming is highly variable, debt levels should be lower than what is considered acceptable in many other businesses. Some producers spread the risk by running a number of enterprises, though this must be balanced with compromising enterprise efficiency by reducing scale. Other producers diversify income by investing off farm so that income is not exposed to general rural commodity price downturns or poor seasons. Other strategies include insuring against adverse events such as fire, with the cost of insurance balanced with the impact and likelihood of a negative outcome.

Table 1.3. All wether wool flock gross margin $/DSE with different wool prices (micron price guide) and average greasy fleece weight.

	Micron price guide (MPG) $/kg clean							
Kg	**$7.00**	**$8.00**	**$9.00**	**$10.00**	**$11.00**	**$12.00**	**$13.00**	**$14.00**
5.0	$8.38	$11.32	$14.26	$17.20	$20.14	$23.08	$26.02	$28.96
5.5	$10.44	$13.67	$16.91	$20.14	$23.37	$26.61	$29.84	$33.08
6.0	$12.50	$16.02	$19.55	$23.08	$26.61	$30.14	$33.66	$37.19
6.5	$14.55	$18.38	$22.20	$26.02	$29.84	$33.66	$37.49	$41.31
7.0	$16.61	$20.73	$24.84	$28.96	$33.08	$37.19	$41.31	$45.42

Assumptions: wether 1 DSE/head, wool production calculated on average of all age groups of wethers. Variable cost $11.30/DSE, Wethers are purchased for $45 and sold for an average of $45 net. Wethers are kept for 4 years Note: to convert MPG to net greasy price across clip, multiply by 60%.

Some examples of risk include:

- **Production risk:** Given that productivity is usually an important profit driver, this is an important risk to manage on all farms. Risks can be minimised by adopting management systems that ensure livestock feed supply coincides with feed demand, especially at lambing time and stock sales. In addition, the stocking rate has a very close relationship with productivity. When increasing stocking rate and pasture utilisation, it is imperative to invest adequate resources in growing pasture, especially fertiliser application. Management skill is required to manage the nutritional wellbeing of stock and adequately provide for drought management. Meat production is more exposed to a poor season, because when stock do not meet market specifications, prices often decrease. With wool, the impact is often less because lower production is often associated with lower fibre diameter and higher prices per kilogram, though there is sometimes poorer staple strength associated with lower production due to poor seasons.
- **Commodity price risk:** The biggest cause of variation in farm profit on any given farm is due to variation in commodity prices. Commodity price risk can be managed to some extend by marketing strategies, especially with wool where there are more options for forward sales compared with meat, though strategies cannot cover long-term downward trends in price. Perhaps the best defence against low commodity prices is to ensure the farm business has a low cost of production, so it is in a better position to withstand periods of low prices.
- **Financial risk:** This includes insolvency due high debt levels or the failure of the business or season to generate enough income. In the long term, low income will result in an unsustainable business that can not expand and invest back into improving resources.
- **Climate risk and drought:** Perhaps one of the biggest risks that farm managers face is the risk of drought. The risk in some regions is greater than others because of higher drought frequency and is best managed by ensuring that a drought plan is in place. A key aspect of this is to make informed decisions at specific times if the season fails. There is no doubt that if the farm has either good financial reserves or fodder reserves it will be able to manage drought better. In the longer term, the impact of climate change may be very important, and it is critical that the farm manager can respond to changes in an informed way to ensure profitability is maintained.
- **Environmental risk:** This is a major issue on all farms and must be managed in a rational manner. Environmental risks will be different on different properties with varying soil types, topography, natural resources and climate. All these aspects need to be considered on all farms. Perhaps one of the most important issues is minimising soil erosion associated with inappropriate grazing. As a guide, the risk of soil erosion rapidly escalates if ground cover falls below 80%. Stock should be removed no later than this point and this decision is totally in the control of management. To minimise this risk in drought, stock should be removed to containment areas or sold, depending on the management and budget options. Another example is minimising the risk of nutrient run-off into water ways. Management strategies, such as fencing off waterways, can be adopted to minimise this risk.
- **Animal welfare risk:** As a minimum standard, stock managers should abide by national standards and guidelines or state codes of accepted farming practice for the welfare of livestock. It is in all managers' financial interest to ensure that stock remain healthy and do not fall below minimum nutritional standards. Producers should plan programs that prevent or control important endemic diseases, including gastrointestinal parasites and flystrike.

Prevention programs are highly cost effective and ensure ease of livestock management. The implications of profit must be balanced with risk. For example, in time of high stock prices many producers tend to run a high proportion of ewes and fewer wethers. A potential detrimental risk of this strategy is much higher risk of worms and greater drought risk.

- **Endemic animal disease risk:** Important endemic diseases, including ovine Johne's disease, drench-resistant worms, footrot and lice, are commonly introduced onto farms by purchased, agisted or stray sheep. Often, too little attention is paid to the risks associated with bringing in stock from other properties, and particularly from livestock markets. The greatest disease risk to animals is other animals. Livestock from markets pose the greatest risk because these animals have been mixed with many others of unknown disease status and they may have been stressed by transportation and time off feed. When buying stock: consider the diseases known to occur in the district or region or in areas from which animals will be purchased; know the status of animals that are being purchasing prior to introducing them to a new property; source low-risk animals (vendor declarations and sheep health statements help with animal disease assessment); inspect animals carefully for disease; take appropriate preventative measures; isolate and observe introduced stock closely; and ensure that vehicles used in the transportation of animals are cleaned and disinfected before use.
- **Human resource and personal risks:** As many farm managers and property owners strive to expand and run more efficient businesses, more pressure is placed on themselves, their family and staff. Good human resource management is essential, and good communication is a fundamental part of human resource management. All staff should have clear roles and the opportunity to develop skills. The result of poor staff morale and poor work skills will be inappropriate work practices, higher costs, less efficient implementation of the management calendar and more stress on management and family. All work practices and places must be compliant with occupational health and safety standards.

It is important to understand the likelihood of the risk occurring (such as drought) and the potential impact. Strategies should be in place to prevent or manage all significant risks.

Marketing wool and meat

Managing the marketing of produce is a key aspect of a successful farm business. In a wool flock, approximately 60 to 90% of all income is derived from the sale of wool. The proportion of gross income attributed to wool production is greater with higher wool prices and when more wethers are retained for wool production. In contrast, in prime lamb flocks, between 70 and 90% of income is derived from sale of stock.

Running a profitable enterprise

Long-term enterprise decisions require an understanding of long-term market trends. The South West Victorian Farm Monitor Project has analysed farm performance for over 30 years. Figure 1.2 shows the relative performance of wool, prime lambs and beef cattle since 1971. When selecting an enterprise, long-term trends must be balanced with the market outlook, which is driven by factors such as supply and demand, currency and exchange rate, economic outlook and competing commodity prices.

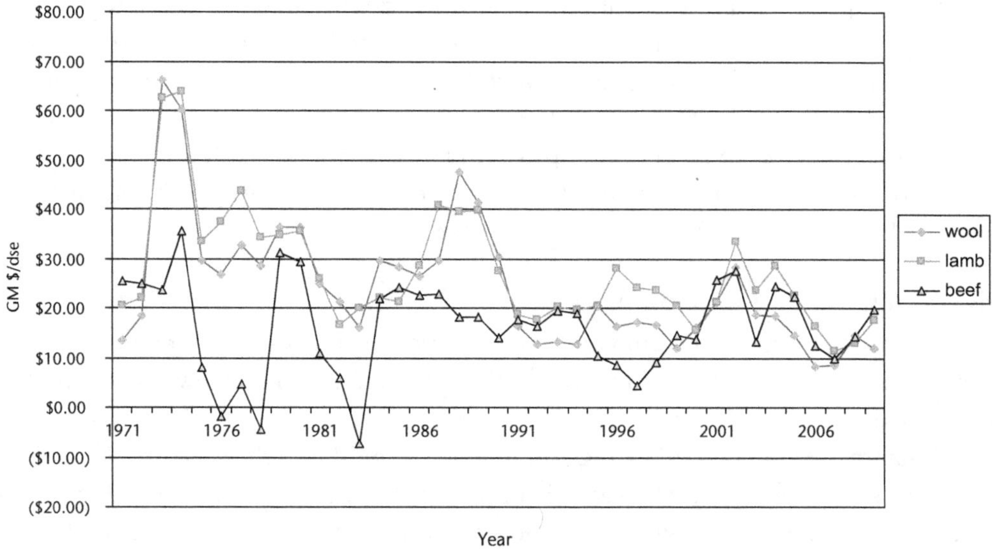

Figure 1.2. Long-term gross margin $/DSE South West Monitor Farm Project.

The relative profitability as measured by the gross margin ($/DSE) has varied over different time periods. Table 1.4 contrasts the gross margin ($/DSE) over different time periods.

Table 1.4 shows the average gross margin ($/DSE). The actual range within enterprises, on individual farms is enormous.

Once the enterprise mix is decided, how the business is run is a critical factor for success.

Low cost of production is a key factor determining the success of a farm business. Knowledge of cost of production enables the farm manager to determine income targets required to achieve profit targets. Without knowledge of cost of production, setting profit targets is difficult. The cost of production is also a benchmark of farm performance.

The drivers of wool price

There are several very important factors that determine wool price. Fibre diameter (micron) and staple strength usually account for over half the variation in wool price. Staple length and the position of break and vegetable matter (per cent) are also important. Factors such as style, staple measurement and marketing factors are of less importance, except in superfine categories.

Adopting management strategies that improve clip quality is very important. Time of shearing can be adjusted to manage staple strength and length and vegetable matter content of wool.

Table 1.4. Average gross margin ($/DSE) of livestock enterprises in the South West Monitor Farm Project over different time periods.

	38 years	10 years	5 years
Beef	$16	$18	$16
Wool	$24	$16	$11
Prime lambs	$27	$20	$16

Producers should review clip specification results and adopt management strategies to improve clip quality. For example, if the vegetable matter content is too high, changing the time of shearing could be considered or at least young sheep (with the finest wool) could be run on paddocks that have a low risk of contamination by vegetable matter.

With the average fibre diameter of the clip being an important determinant of price, it is imperative that producers in wool flocks ensure the genetic merit of the sheep is high, particularly the relationship between fleece weight and fibre diameter.

The drivers of lamb and mutton prices

The major factors that determine lamb and mutton price include carcase weight, fat score, breed, age and skin value. The main factors that determine the skin price include wool length, breed and seed contamination.

Potential market segments – including store lambs, domestic trade, supermarket, export (light and heavy) and sheep meat markets – are also important, especially when deciding on the enterprise structure. For example, export heavy lamb production requires earlier lambing (unless a specialised finishing system is available); the down side of lambing earlier is the mismatch between feed demand and feed supply and the risk of lower lambing percentages in crossbred flocks. Later lambing may result in higher lambing percentages; however, the lambs may not reach the target weights (more lambs of lower value).

Climate, management skill, farm resources and sheep genetics must be taken into consideration when deciding on a specific market segment. Each year may be different in terms of market prices, and managers need to monitor the market very closely. In some years, there may be opportunities to grow lambs to heavier weights if feed costs are low and lamb prices high. In other years, there may be opportunities to sell lambs earlier if feed prices are high and light weight lamb prices are high.

The production system (time of lambing, stocking rate, time of weaning and finishing systems) and genetic selection of sheep should be set to maximise financial returns based on lamb prices and resources available.

Meeting market specifications for the wool clip

Many producers are very skilled at producing wool or growing grass, but forget about managing the wool clip and preparing the wool clip for sale. Research shows that good clip preparation can improve financial return by over 5%. Key factors to consider include:

- ensuring clip preparations meets the code of practice for wool classers
- minimising the risk of dark fibre contamination, especially crutching within 3 months of shearing
- avoiding small lot sizes if possible (small lots increase selling costs): the exception being specialist superfine clips
- avoiding over skirting: ideally the clip should have over 76 to 80% fleece wool
- avoiding over crutching, which will reduce the amount of valuable fleece wool and increase low-value crutchings and locks
- minimising lot building and bulk class lots, which are more expensive to sell

- ensuring heavy bales weights to minimise selling costs
- targeting specialist markets, if cost effective, such as eco-wool and prevent chemical residue contamination.

Managing lamb sales to meet market specifications

A key driver of a profitable prime lamb flock is to meet target weights. Farm managers should learn to fat score and weigh sheep. The specifications for markets are often narrow, and lambs sold outside the weight or fat score range often attract price discounts of more than 10%. For example, lambs sold over the hooks may attract a price premium compared with sale yards within a certain weight or fat score range, but, if they are too fat, the price discounts may mean it is better to sell this category into the yards. Producers should also develop the skill of assessing lamb carcase weight based on live weights. Typically, lambs sold off their ewes will dress out at 43 to 47% with a few hours off feed. Lambs that are lighter will dress out lower and lambs that are left longer off feed will dress out higher. Merino lambs typically dress out about 1.5 to 3% lower than second cross lambs and lambs grazing poor-quality feed will dress out up to 3% less. Understanding the dressing percentage of lambs is important when determining what lambs will be sold within a certain price grid.

Lamb skin values are an important component of the whole lamb price and producers should manage the quality of skins to maximise financial returns. Important considerations include wool length and seed contamination. Avoid over crutching because this will potentially reduce lamb skin value.

All sale stock should not exceed minimum residue levels by ensuring stock sold comply with withholding periods and export slaughter intervals following treatment with or exposure to a veterinary or agricultural chemical.

Selling methods for wool

There are many options available for selling wool. Factors to consider with different options include price, selling costs and flexibility. Comparing prices is often difficult, but with any selling method it is critical that the producer understands the price for each line of wool – particularly as wool is a very volatile commodity. This can be calculated by using tools such as Woolcheque (www.woolcheque.com.au) that enable up to date market valuation of all lines of wool.

Sale options for wool include:

- auction
- electronic sale
- private sale
- forward contracts
- direct sale.

A key factor to consider when selling wool is the timing of the sale. Given that wool can be stored well, there is considerable flexibility of sale timing, with no risk of deterioration of wool quality. This is not possible for meat production. Being able to sell at any time can spread cash flow over a longer period and manage price volatility to some extent. One strategy that can be adopted is to set reserves and sell only if the agreed price is achieved. If sale prices are low, and cash flow is not limiting, then more aggressive reserves can be set or, if income is required, more conservative reserves can be set. One risk with this strategy is that when high reserves are set and wool is passed in, if wool prices then fall and income is required, the wool may need to be sold at lower prices. Decisions on the timing of sale, requires a clear understanding of financial commitments and wool price outlook. A useful strategy is to compare the current price of wool with long-term prices and compare on a percentile basis when deciding on the time of sale

Selling methods for lamb and mutton

The timing of sale of prime lambs is more reliant on the production cycle and time of year than wool. There is less ability to hold back stock if prices are not favourable. Even so, it is important to assess different selling options with gross prices and selling costs, including transport and agent commission. Often live weight price ($/head) must be compared with carcase prices ($/kg dressed and skin price).

Common sale options that should be considered include:

- auction (on-farm or public sale yards)
- private sale
- direct consignment (to abattoir) either at current price or forward sale 'over the hooks'
- Auction Plus (internet based sale by description).

Risk management

Commodity price variation is a major risk on wool and prime lamb properties. There are numerous strategies that can be adopted to manage price volatility, including spreading sales over the year, forward selling, the use of derivatives and fixed price contracts and other tools available through agents, brokers and customers of the farm's produce. Although setting future prices can reduce downside risk, it can also reduce upside price premiums. It is important to understand the cost of production and the impact of setting certain price targets on profitability. Seek specialist advice when considering adopting forward selling and marketing strategies.

Setting up a farm management calendar

A farm management calendar is a fundamental business requirement of a sheep enterprise. A farm management calendar ensures efficient use of labour and other resources and timing of events to maximise livestock productivity and profitability. In a sheep enterprise, the two major decisions that drive virtually all events on the management calendar are the time of lambing and the time of shearing. The timing of sale of livestock is also critical, as is nutritional management and disease prevention. The management calendar will vary according to enterprise mix and flock structure.

A properly designed management calendar should be displayed for all staff involved in the farm business for their understanding and implementation.

Example management calendars for:

- autumn lambing Merino ewes joined to Border Leicester rams
- spring lambing wool producing flock

- winter lambing crossbred ewes joined to terminal sires

can be found in Appendix 1. They are accompanied by general monthly management reminders that are relevant to all sheep farming enterprises.

Time of lambing

Deciding on the time of lambing is a critical aspect of determining farm profitability. The key factors that determine the time of lambing include the following.

- **Type of enterprise:** The optimal time (from a financial perspective) to lamb wool flocks tends to be 3 to 4 months before the end of the growing season. In contrast, with meat flocks the optimal time of lambing is earlier at 4 to 5 months before the end of the growing season. Running more stock with a later lambing must be traded off against ensuring lambs are heavy enough to meet market specifications.
- **Pasture growth patterns:** A key objective should be to fit lambing time with the pasture growth curve. This should ensure more stock can be run and production is greater. As a guide, about 0.5 DSE extra (per lambing ewe changed to later lambing), should be run in winter with a spring lambing flock compared with an autumn lambing flock. This factor alone makes spring lambing more profitable in most wool growing flocks.
- **Cost of supplementary feeding:** Early (autumn) lambing will result in a greater risk of supplementary feeding for ewes and later (spring) lambing may mean more feed is required for lambs to ensure survival in summer or for prime lambs to ensure they meet market specifications.
- **Disease and nutrition:** Generally, later (spring) lambing helps ewe nutrition and reduces winter worm problems, but sometimes may increase the risk of weaner ill-thrift in summer on poor-quality pastures. This is more of an issue in Mediterranean climates when feed quality is poor by late summer.
- **Labour efficiency:** Early (autumn) lambing increases staff work load in autumn because of a higher likelihood of feeding ewes and late lambing may increase the need for weaner management skills in summer.
- **Reproductive efficiency:** Reproductive performance of Merinos does not vary greatly with season. More lambs are born with later (spring) lambing, but lamb losses associated with cold, wet, windy weather are greater. British breeds of sheep are seasonal breeders, with later (spring) lambing usually resulting in much higher lambing percentages.
- **Weaner management:** Early (autumn) lambing flocks have less weaner ill-thrift in summer because lambs are older and bigger; however, in autumn-lambing flocks there is a risk of ewe nutritional stress and high worm challenge in late winter.

All of the above factors must be considered when setting the time of lambing. There are many flow-on effects of lambing time on the management calendar. For example, the time of weaning should be set about 12 to 14 weeks after the start of lambing in wool flocks and 14 to 16 weeks after the start of lambing in prime lamb flocks. Time of weaning may have implications on the number of drenches required for lambs and the need for jetting. In addition, the time of lambing will have an impact on the time of sales of surplus stock and time of shearing.

Time of shearing

The time of shearing is the second major decision for the management calendar because it has major implications for the timing of many events during the year.

The key factors that determine time of shearing include the following.

- **Sale of surplus sheep:** Shearing should coincide with the end of the spring flush so that surplus sheep can be sold in good condition. Later shearing may result in sheep attracting higher prices per kilogram, but they will have eaten valuable feed and maybe lighter and therefore not necessarily more valuable.
- **Flystrike:** Shearing sheep just before major flystrike periods is an important factor for many producers, because it results in less reliance on chemical applications for fly control. This is not a major consideration for producers with a low fly risk.
- **Seed set:** Shearing sheep before seed set will reduce contamination of wool with vegetable matter. This is an important consideration in regions with lots of annual pastures.
- **Shearer availability:** In some regions, this in an important consideration.
- **Staple strength:** This is the second most important determinant of wool value after fibre diameter. The weakest point in the wool staple usually occurs in autumn – particularly in Mediterranean climates – so producers try to shear at this time. However, the time of lambing is another consideration because a wool break may occur at the point of lambing.
- **Labour resources on farm:** With limited labour on many farms, shearing at a time that fits labour availability is an important consideration.
- **Pre-lamb shearing:** This is commonly adopted in some regions, though it is generally not encouraged because it increases ewe feed demand when feed is limiting and increases the risk of hypocalcaemia and pregnancy toxaemia.
- **Managing dags in sheep:** In some regions where scouring (and dags) is a common problem, shearing time is set to minimise the amount of crutching required.

The time of shearing must be balanced between these factors and other management events on farm. Design a calendar that enables high production with less risk (such as reducing the amount of feeding or reducing disease risk), labour efficiency and the maximisation of profit.

CHAPTER 2

Pasture growth and management

Pasture is the cheapest source of feed for livestock. The key to optimising the profitability of a grazing enterprise is to optimise the amount of pasture grown and consumed, with minimal supplementary feeding and minimal impact on the environment.

This chapter deals with optimising pasture growth and the persistence of the desirable pasture species. Making the best use of pasture, and assessing the quality and quantity of pastures to meet animal needs, is dealt with in Chapter 3.

Pasture growth

The amount of pasture grown is determined by the rainfall, temperature, length of the growing season, soil type, nutrients and plant species.

Plants use sunlight, water and carbon dioxide to produce carbohydrates through photosynthesis. Carbohydrates are used immediately for growth or stored as water-soluble carbohydrates (WSC). The rate of plant growth is strongly influenced by both the leaf area (to capture light) and the WSC reserves in the tiller bases (if leaf area is insufficient). Although rainfall, temperature and soil type cannot be controlled, pasture growth can be optimised by ensuring nutrients are not limiting, plant species are appropriate and leaf area is sufficient.

Pasture growth curves

Figures 2.1–2.6 illustrate some indicative monthly pasture growth rates for a number of sites in south-eastern Australia. Each figure illustrates the growth rates for an improved perennial pasture with good fertility (adequate nutrients) and an annual pasture with lower soil fertility. Pasture growth is measured as kilograms of dry matter per hectare per day (kg DM/ha/day).

These pasture growth curves illustrate the range in growth rates that can occur within the same soil type and environment. Therefore they give an indication of the potential for maximising growth by addressing plant species and nutrients, given the same amount of rainfall. Note that an 'average' year is depicted, so there will be monthly variation around each pasture growth curve

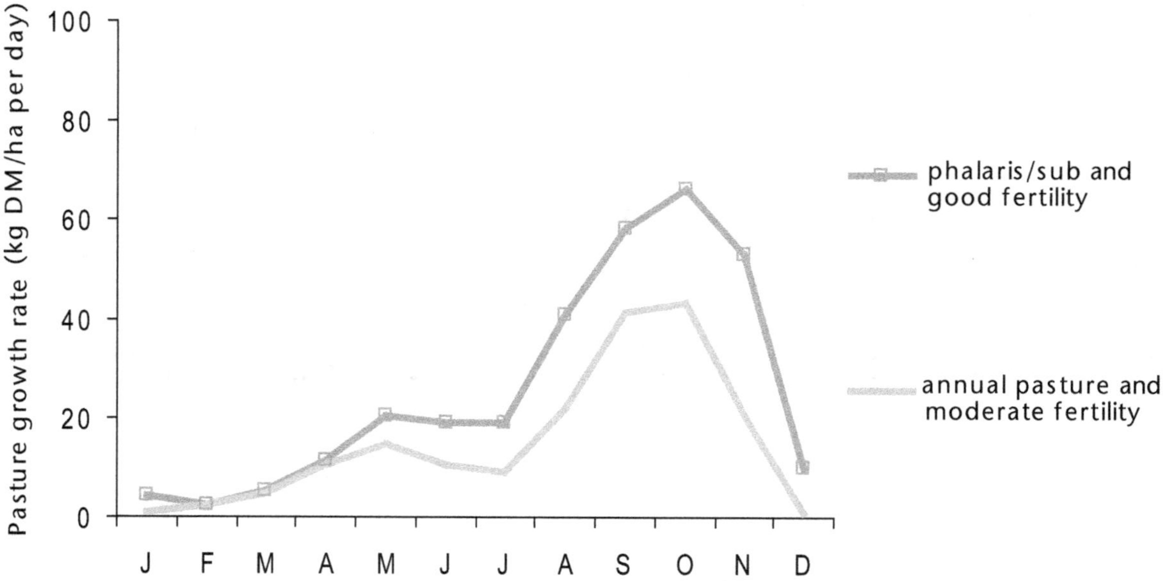

Figure 2.1. Average pasture growth rates for an annual (sub clover/barley grass dominant) pasture and a sub clover/phalaris pasture at Bendigo, Victoria. (GrassGro™)

each year. However, the average year curve gives a guide to the length and magnitude of the growing season. This is essential information to help make strategic management decisions such as the most appropriate enterprise mix, flock or herd structure, time of lambing or calving, turn-off dates and stocking rates.

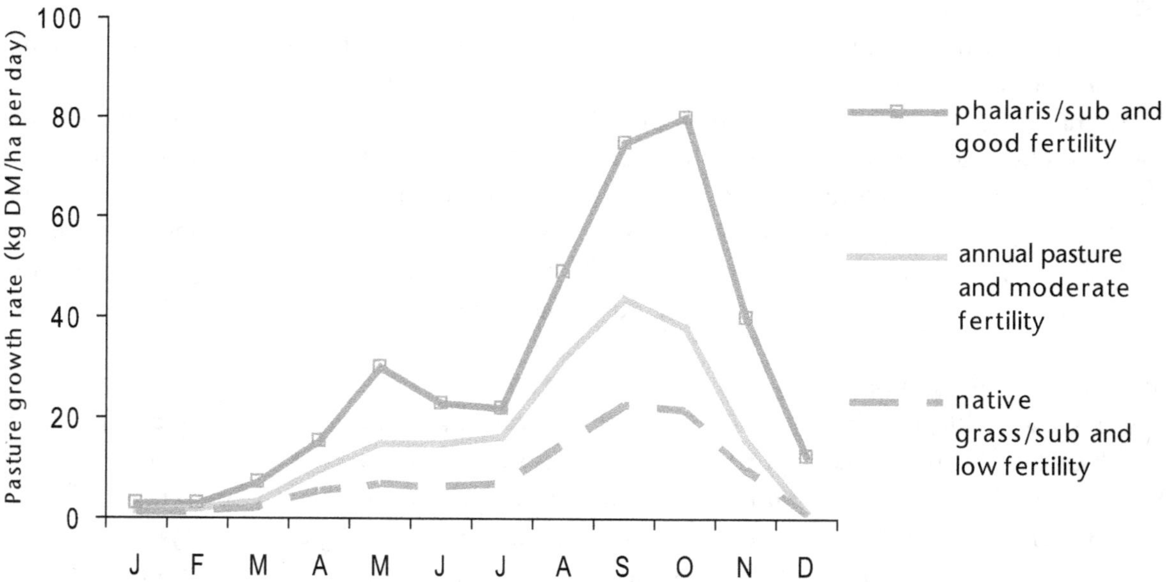

Figure 2.2. Average pasture growth rates for a sub clover/phalaris pasture, an annual (sub clover/capeweed dominant) pasture and a native grass/sub clover pasture at Seymour, Victoria. (GrassGro™)

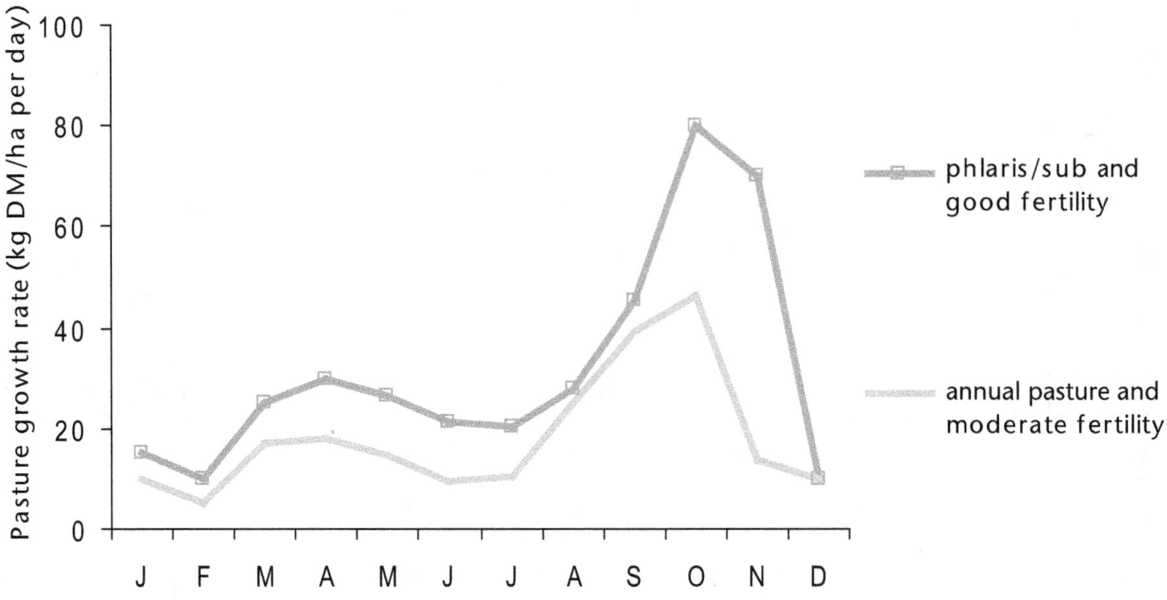

Figure 2.3. Average pasture growth rates for a sub clover/phalaris pasture and an annual (sub clover/barley grass dominant) pasture at Bairnsdale, Victoria. (GrassGro™)

The pasture curves illustrate the significant differences between regions in total growth and in the length of the growing season. These ultimately determine or constrain stocking rates and production systems.

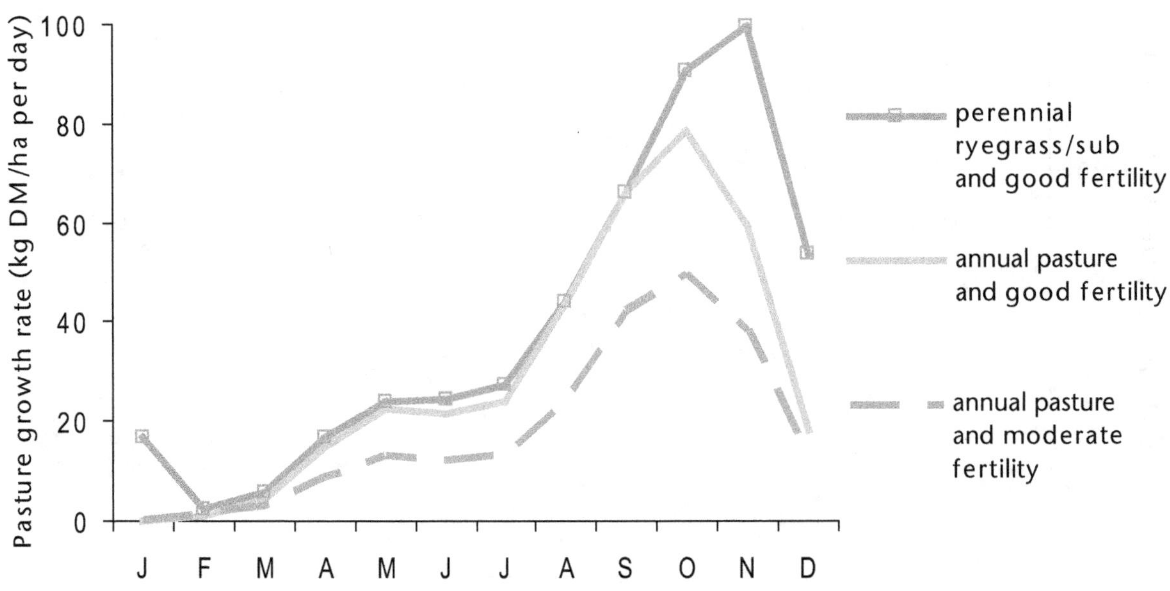

Figure 2.4. Average pasture growth rates for an annual (sub clover/capeweed dominant) pasture and a sub clover/perennial ryegrass pasture at Hamilton, Victoria. Two levels of soil fertility are shown for the annual pasture. (GrassGro™)

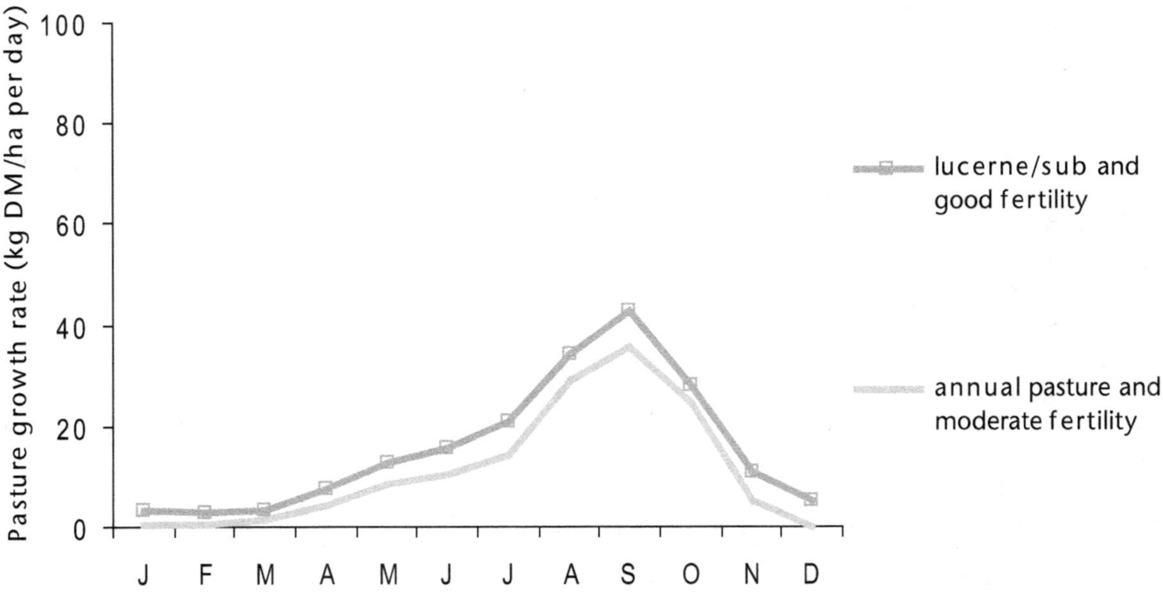

Figure 2.5 Average pasture growth rates for a lucerne/sub clover pasture and an annual (sub clover/barley grass) pasture at Walpeup, Victoria. (GrassGro™)

GrassGro™ is a decision-support tool that can model pasture growth, animal production and economics for a range of environments in southern Australia. The tool was developed by CSIRO Plant Industry.

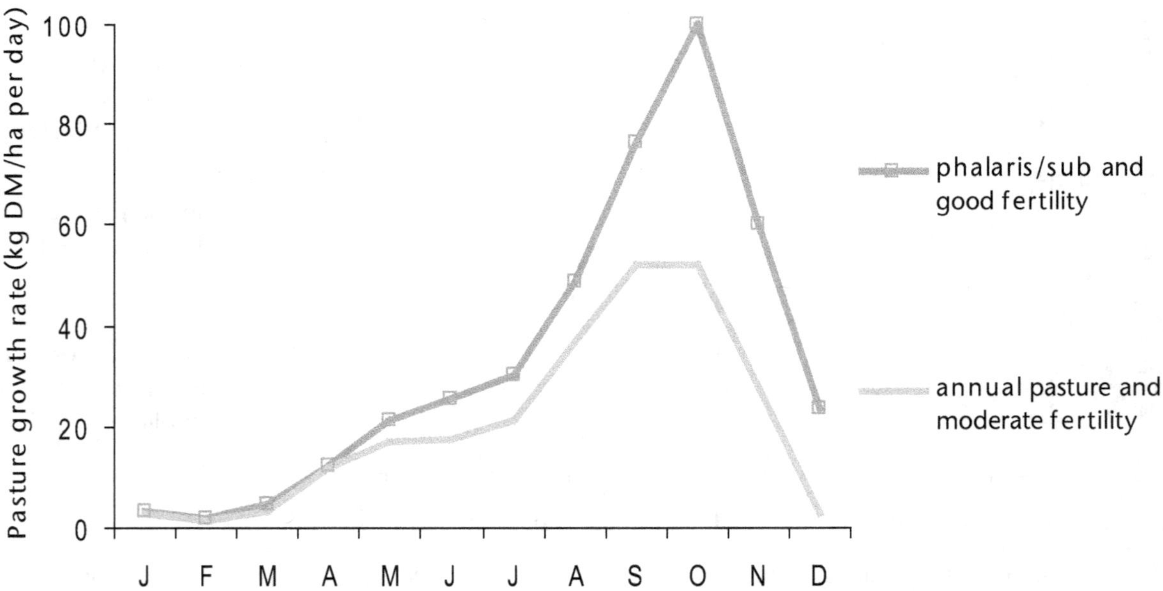

Figure 2.6 Average pasture growth rates for a sub clover/phalaris pasture and an annual (sub clover/annual ryegrass) pasture at Naracoorte, South Australia. (GrassGro™)

Key messages – pasture growth

- Plant growth is affected by temperature, rainfall and soil type, which cannot generally be controlled.
- Plant growth is also affected by species, nutrition (fertility) and leaf area, which can be managed. Reducing limitations from these factors increase both the amount of pasture grown and the length of the growing season.
- The amount of pasture grown and length of the growing season are key determinants of critical issues such as stocking rate, lambing times and enterprise mix.

Plant and soil nutrition

The major or macro elements required for plant growth are carbon (C), hydrogen (H), oxygen (O), nitrogen (N), phosphorus (P), potassium (K), calcium (Ca), magnesium (Mg) and sulphur (S).

Carbon, hydrogen and oxygen are obtained from the atmosphere and from water. The remaining elements are absorbed from the soil through the roots, and deficiencies can be detected through soil tests. Legumes, such as clover, medics and lucerne, are able to use nitrogen from the atmosphere, which is released and made available to other plants when legume root nodules break down.

Trace elements are those nutrients required in very small amounts They include copper (Cu), zinc (Zn), boron (B), manganese (Mn), iron (Fe), chlorine (Cl), molybdenum (Mo), sodium (Na) and cobalt (Co).

The main nutrient limitations to pasture growth in Victoria, in order of importance, are phosphorus, potassium, nitrogen and sulphur. In acid soils, the trace element molybdenum is often limiting. In many coastal, sandy soils in Victoria, both molybdenum and copper are deficient.

Phosphorus (P)

Many soils in Australia are naturally deficient in phosphorus. Phosphorus is an essential element for plant and animal growth, being a component of nucleic acids, cell membranes and the energy-carrying adenosine triphosphate (ATP).

When phosphorus fertiliser, such as superphosphate, is applied to soil and dissolves, only around 10% of the phosphorous stays in solution and is readily available to plants (Figure 2.7). The rest either becomes adsorbed onto the soil particles or is 'fixed' and becomes unavailable. However, the adsorbed and fixed phosphorus will be recycled and become available in the longer term. The amount of nutrient cycling of the various forms of phosphorus is an important process that dictates pasture growth.

Most phosphorus will remain in the top 10 cm of the soil and does not leach down the profile, unlike more mobile nutrients such as nitrogen and sulphur. To preserve phosphorus in the soil, it is important to maintain ground cover to protect the topsoil from wind or water erosion.

Figure 2.7 shows the fate of applying 10 kg of phosphorus per hectare. In this example, ewes are run at 15 sheep per hectare, 10 kg P/ha is applied and soil phosphorus levels are maintained. 1 kg P/ha is sold off the farm in animal product and about 3 kg is 'lost' through excretion in sheep camps and a very minor (<0.04 kg) amount is lost in erosion or run-off. The rest is returned to the soil pool and so is 'fixed'. As 24 kg P/ha is used to grow the required amount of pasture, the phosphorus in the soil's reserve pools is also used.

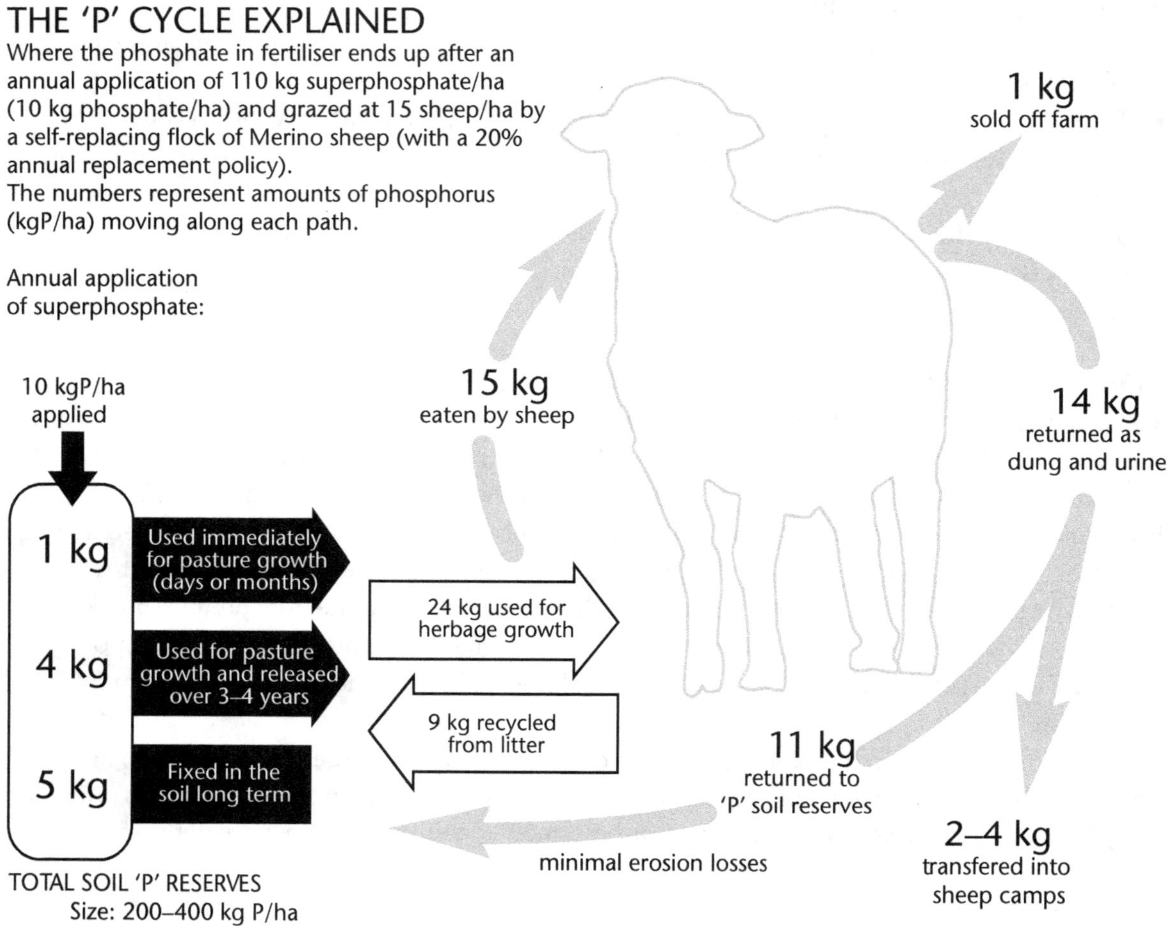

Figure 2.7. The phosphorus cycle in a self-replacing flock of Merino sheep (Source: Richard Simpson CSIRO and used with permission MLA 2008).

There are target levels of phosphorus in the soil under which plant growth will be limited. The soil nutrient concentration at which 95% of potential pasture growth occurs is called the critical soil-test value. In other words, improving soil levels above these targets is unlikely to give large incremental increases in pasture growth, although it does give the option of omitting fertiliser in years where costs of applying fertilisers are high. The target, or critical soil-test, value for phosphorus is shown in Table 2.1. Available soil phosphorus can be measured using the Olsen or Colwell soil-test method, but the critical soil-test value is different for each test.

The ability of soils to adsorb phosphorus is measured as the phosphorus buffering index (PBI). Soils with low phosphorus buffering index (such as sands) will require less phosphorus fertiliser than soils with a high PBI (such as ferrosols) because less phosphorus will be adsorbed onto the soil particles and hence more will be available for plant growth. The critical soil-test value for Colwell phosphorus varies with PBI, but the critical value when using the Olsen phosphorus test does not vary (Table 2.1).

In southern Australia, it has been extremely cost effective to apply phosphorus as a fertiliser to reach target levels as described above, providing there are plant species that will respond. Phosphorus can be supplied from a range of sources, but a common fertiliser is one fortified with sulphur (such as 'single super'). There are also a range of high-analysis products available that contain higher amounts of phosphorus and less sulphur (such as 'double or triple super') and are suitable for use where sulphur is not limiting. Phosphorus may also be supplied in a range of manures from intensive industries, but the cost, concentration and risk of contaminants needs to be assessed. Reactive phosphate rock (RPR) can also be used to supply phosphorus but this product is not as soluble as previous products mentioned. In southern Australia, RPR is an option in environments with high rainfall (greater than 700 mm for loams or clays; greater than 800 mm for sandy loams) and acid soils (pH water less than 6.0 or pH $CaCl_2$ less than 5.2) because it will dissolve more readily under these conditions. In these environments, it generally takes 3 to 5 years of RPR application before RPR performance is equal to more the more soluble forms of phosphorus fertiliser.

Potassium (K)

Potassium is required by all plants for photosynthesis, translocation, cellulose formation, enzyme activities and stomatal control.

Potassium deficiencies have long been known to occur in the southern coastal regions of Victoria and the sandy soils in the south-east of South Australia. However, potassium responses have also been identified in other areas and soil types, particularly where high production grazing systems are being run or where hay is cut frequently.

The critical value for potassium in soil, as identified by a Colwell soil test, varies with soil texture (see Table 2.1). Potassium deficiencies can be corrected by application of potash fertiliser or various blended products that contain potassium and other nutrients.

Sulphur (S)

Sulphur is required for the formation of several amino acids, proteins and vitamins and for chlorophyll production.

Sulphur is a relatively mobile nutrient that can readily be leached down the soil profile. Hence, it is common to find sulphur deficiencies on light textured, sandy or granitic soils. In the past, the widespread use of superphosphate (a phosphorus and sulphur blend) has masked potential sulphur responses.

Sulphur can be measured using the CPC or KCl-40 test. The critical values for sulphur in soil, as identified by a soil test, are shown in Table 2.1. Sulphur deficiencies are commonly corrected by application of phosphorus fertilisers fortified with sulphur. Some forms of sulphur are more soluble (as sulphate), and hence available more quickly to plants, so, if sulphur is deficient, the more soluble form should be used. For example, superphosphate contains sulphur in this more soluble form. There are products that have sulphur in a less soluble form (such as elemental sulphur) that release sulphur much more slowly and so can be used when sulphur levels are adequate and only maintenance is required. The high-analysis phosphorus products often contain more sulphur in the elemental form. If phosphorus levels are adequate, gypsum (calcium sulphate) may provide an alternative source of sulphur depending on availability and price.

Nitrogen (N)

Nitrogen is an essential plant nutrient because it is a building block for making protein. Nitrogen is very mobile and so can also leach through the soil. Pasture legumes have the ability to capture nitrogen from the air and make it available to other pasture plants as plant legume matter is broken down. Therefore nitrogen is derived largely from soil and plant litter. Responses to nitrogen fertiliser can still be significant at times when nitrogen availability from legumes is low (for example, in late autumn, winter) or when legumes are not part of the pasture or cropping system, but species have high nitrogen requirements.

Due to the dynamic nature of the supply of nitrogen during the year, a soil test taken at one point in time may not successfully predict the supply of plant available nitrogen. Hence, for pastures, nitrogen levels in soil are not routinely determined. In cropping systems it is more common to test soil nitrate levels before sowing and sub-soil nitrate is often monitored during the growing season.

In perennial and annual pastures, the companion legume (usually clover) is relied on as the primary supply of nitrogen for the grasses. However, temporary deficiencies can occur during the season, such as in cooler weather because clover cannot supply sufficient nitrogen at soil temperatures below 10°C for maximum grass growth, but productive grasses can respond to nitrogen at temperatures above 4°C.

Nitrogen fertiliser – such as urea, calcium ammonium nitrate (CAN) or liquid nitrogen fertilisers – can be applied tactically to overcome deficiencies and boost dry matter production. Unlike phosphorus, potash and sulphur, which have a residual effect, nitrogen responses usually last only for a couple of months.

Molybdenum (Mo)

Molybdenum is required by pasture legumes because it is plays a role in fixing nitrogen in their root nodules. Deficiencies are characterised by poor legume growth or poor nodulation and subsequent nitrogen deficiency.

Molybdenum is often limiting in acid soils (that is soil with pH (H_2O) less than 5.5 or pH ($CaCl_2$) less than 4.8 – see the section on soil pH later in this chapter). The recommendation is to apply molybdenum in fertiliser once every 5 to 7 years in these areas. Applying lime can increase the availability of molybdenum in acid soils. A plant tissue test can be done to check if molybdenum is limiting. Alternatively, a series of fertiliser test strips, including a phosphorus fertiliser and a phosphorus/molybdenum blend, can be set up to observe a response (see the section on 'Evaluating products' later in this chapter).

Applying too much molybdenum can induce copper deficiencies, in both plants and livestock, so both molybdenum and lime should not be applied to a pasture in any one year. In areas where copper is also deficient, copper should be applied with the molybdenum.

Sourcing nutrients – alternative products

There are numerous commercial fertiliser products that should be able to supply most combinations of limiting nutrients, but there are alternative products that may be used, such as animal manures, composts or liquid fertilisers.

Table 2.1. Critical values or ranges for soil-test parameters.

Soil test (ASPAC test)[a]	Target/desirable range		
Olsen Phosphorus	15 mg/kg		
Colwell Phosphorus	PBI category[b]		Critical value
	15–35	Extremely low	26
	36–70	Very low	29
	71–140	Low	34
	141–280	Moderate	40
	281–840	High	55
Colwell Potassium	Sand	126	
	Sandy loam	139	
	Sandy clay loam	143	
	Clay loam	161	
Sulphur (CPC) Sulphur (KCl-40)	3 mg/kg 8 mg/kg		
pH (water)[c]	5.3–6.0 (>5.8 for lucerne)		
pH ($CaCl_2$)[c]	4.6–5.3		
Aluminium (0.01M $CaCl_2$) Aluminium (percentage of cation exchange)[c,d]	<8 mg/kg (<2 mg/kg for lucerne) <20% (perennial ryegrass, fescue and cocksfoot) <10% (phalaris) <5% (lucerne)		
Organic carbon[e]	2–3% low rainfall 3–6% high rainfall		
Salt[f] (EC 1:5)	<0.15 dS/m (sand) <0.3 dS/m (clay)		
Exchangeable sodium (Na+)[d,f]	<6%		

[a] ASPAC – Australian Soil and Plant Analysis Council
[b] PBI – phosphorus buffering index
[c] see section on acidity later in this chapter
[d] see section on soil structure later in this chapter
[e] see section on organic matter later in this chapter
[f] see section on salinity later in this chapter

The following is a checklist for consideration when comparing any fertiliser source to ensure the right nutrients are being applied at the lowest cost.

- What is in it – will the product supply the nutrients needed?
- What is the cost per unit of nutrient required? Some products may supply a range of key nutrients (nitrogen, phosphorus, sulphur and potassium) but if one nutrient (say phosphorus) is needed the most, it is important to compare the costs of supplying the required amount of

phosphorus. If a large amount of product is required because the product contains a lot of water or low concentrations of nutrients, it may be very expensive. The cost to cart and spread the product should be included in the cost comparison, not just the product price.

- Are there good data to support the claims of products promoted to reduce the need for fertilisers (such as those products used to stimulate soil biological activity)? If there are not, but it is considered worth trying, some easy method of evaluating the response should be used. Seasonal variations make evaluating products very difficult. Simple fertiliser strips can be run out across a paddock (see the section on 'Evaluating products' later in this chapter).

Comparing the cost of nutrients

To calculate the price in dollars per kilogram of nutrient spread (Table 2.2) use the following formula:

$$\$/\text{kg of nutrient} = \frac{\$/\text{tonne product (ex works plus delivery and spreading costs)}}{\text{Nutrient concentration divided by 10}}$$

Example – phosphorus is required in a product which costs $400/tonne on farm and there is 8.8% phosphorus in the product

$$\text{Cost of phosphorus} = \frac{\$400 \div 8.8}{10}$$
$$= \$4.50/\text{kg}$$

Table 2.2 compares the cost per nutrient for a range of products with several available nutrients. The costs are examples only and will vary enormously over time. However, by updating calculations and making such a comparison when prices change, opportunities can be identified to save money.

Evaluating products

The market for nutrient and fertiliser products is large and while many are useful and cost-effective products, many may not be. It is extremely difficult to evaluate products in agriculture because there are so many things that can affect the response. The main ones are the season and paddock variation. For example, if a product is spread on a paddock and there is no area left untreated (nothing to compare the response with) and there is a fantastic season, it is impossible to know how much growth or pasture composition change is due to the season and how much is due to the product.

There is huge variation between and within paddocks in things such as soil type, pasture composition, waterlogging, and so on. If a product is spread on the best part of the paddock and the poorer half is left as a comparison, it is not possible to attribute any difference to the product because any or some of the difference seen may have occurred anyway.

Simple strips can be set up and run along a fence to make some fairly crude, but useful, nutrient and product comparisons. Each strip does not need to be very large, but it is very important to know the size so that the appropriate amount of fertiliser is spread. For example, if a strip is 10 m × 3 m (30 m^2) then this is 0.003 of a hectare (30/10 000 m^2). If the spreading rate is 125 kg/ha then 375 g needs to be spread in this plot (125 × 0.003 = 0.375 kg = 375 g). Because this is a very small amount, and it is difficult to spread evenly, it can be mixed in a bucket with sand or sawdust to

Table 2.2. Example of nutrient contents and comparative prices for a range of fertiliser products.

Product	Price $/t spread	N $/kg N	P $/kg P	K $/kg K	S $/kg S
Single superphosphate/ 'superfect' (8.8% P, 11% S)	$250/t $400/t	– –	$2.80 $4.60	– –	$2.30 $3.60
Double superphosphate (16.8% P, 4% S)	$400 $760	–	$2.40 $4.50	–	$10.00 $19.00
Reactive phosphorus rock (RPR) (12.5% P, 1.7% S)	$480	–	$3.84	–	$28.23
DAP (18% N, 20% P, 1.6% S)	$700 $1200	$3.90 $6.70	$3.50 $6.00	– –	$44.00 $75.50
Super Potash 4&1 (7% P, 10%K, 8.8%S)	$480	–	$6.90	$4.80	$5.50
Muriate of potash (50% K)	$705	–	–	$1.41	–
Poultry litter[a] – broiler shed (2.6% N, 1.5% P, 1.5% K, 0.6% S)	$70/t	$2.70	$4.70	$4.70	$11.70
Poultry manure – layer shed (5% N, 3% P, 3% K, 1% S)	$90/t	$1.80	$3.00	$3.00	$9.00

[a]Note: the composition of animal manures is highly variable, as are cartage and spreading costs

increase the bulk. Walk over the strip four times; that is, spread one-quarter of the product with one pass and repeat this three times to ensure even coverage. The ideal is to repeat all applications in more than one part of the paddock to take into account the paddock variation. If this is too hard, pick an area with no visual variation. Strips along a fence line are practical because they are easy to fence to exclude stock. Figure 2.8 gives an example of a trial looking at responses to three forms of phosphorus, as well as potash and molybdenum.

In evaluating the response, it is important to consider composition differences as well as the total amount grown. Animal production from pastures with more legumes will be greatly improved. Estimating amount of pasture and effects of pasture quality are outlined in Chapter 3.

Testing for nutrient deficiencies

Soil tests should be taken every 3 to 5 years to assess the status of the macro nutrients such as P, K and S in order to develop appropriate maintenance or capital fertiliser programs. Soil tests also identify other soil problems such as acidity (pH), salinity or sodicity.

Plant tissue tests (leaf analysis) should be done to assess the need for trace elements required for plant nutrition. Some laboratories may provide measures of trace elements from a soil sample, but soil tests are not reliable indicators of most trace element deficiencies in plants or animals. Some of these nutrients (for example, molybdenum) are present in such small amounts in the soil that accurately measuring their availability with a soil test is difficult.

Similarly, trace element deficiencies in livestock are not well correlated to plant deficiencies and so, if trace element deficiencies are suspected in animals, the animals themselves need to be tested (often using a blood sample). Different plant species take up nutrients differently: sheep selectively graze and eat soil, so it is not easy to determine what trace elements they are ingesting.

Nutrient	P only	P + moly	K only	High P	RPR	Control	P and more
Amount	11 kgP/ha	11 kgP/ha	0 P	22 kgP/ha	22 kgP/ha	0	11 kgP/ha
Product amount	125 kg/ha super-phosphate	125 kg/ha super and moly 0.05%	63 kg/ha potash	250 kg/ha super	176 kg/ha RPR		863 kg/ha poultry litter

Figure 2.8. Example of a simple farm nutrient and fertiliser product trial.

Guidelines for soil testing

- Sample each soil type or management unit separately. Having information on every paddock would be ideal, but, if several paddocks are on a similar soil type, and have been managed in a similar way, a test from one paddock can be used as a guide for the other paddocks. Take separate samples from within the same paddock if there are vastly different soil types or if one area of the paddock has noticeably poorer pasture growth. Avoid sheep camps.
- Samples should be taken at the same time of year. Spring is a good time for sampling because the ground is softer and there is sufficient time to get results and plan for autumn fertiliser needs.
- For each soil type, take 30 soil cores to a depth of 10 cm. This depth is critical because the interpretation of the soil test results is based on samples of this depth.
- The samples should be sent to an ASPAC (Australian Soil and Plant Analysis Council) accredited laboratory (see website listing at the end of this chapter). Tests are accredited separately so is important to ensure the selected laboratory is accredited for all tests required.

Guidelines for fertiliser application – maintaining fertility

Maintenance applications of fertiliser are aimed at replacing the amounts of nutrients exported off the farm in produce (meat, wool, hay or silage) or tied up in the soil. Maintenance applications are designed to maintain the target soil levels for each nutrient and maintain the current level of productivity. Maintenance applications usually involve phosphorus and sulphur, but may involve potassium in paddocks cut for hay or silage.

Maintenance rates of phosphorus are expressed in terms of kilograms of phosphorus per dry sheep equivalent (kgP/DSE). They are calculated using a combination of the rainfall, the phosphorus buffering index (PBI) of the soil, the topography of the paddock, the grazing system and the pasture type as shown in Tables 2.3 to 2.5.

In Table 2.3 an animal loss factor is used to estimate losses from different grazing systems and topographies.

The soil loss factor is a measure of the phosphorus buffering capacity (soils P fixing capacity). Table 2.4 shows the soil types and the severity of soil phosphorus loss.

The amount of kgP/DSE is then multiplied by the stocking rate (DSE/ha) to give the required phosphorus per hectare to be applied. Maintenance rates for sulphur and potassium are best determined by reviewing changes in these elements on individual soil tests.

Another approach to calculating maintenance fertiliser requirements – more commonly used in the dairy industry – is to construct a full nutrient audit or budget. This is a more complex method

Table 2.3. Animal loss factors.

Loss factor	Topography	Grazing
Very low	Flat or nearly flat paddocks (mostly less than 10°)	Rotationally grazed.
Low	Hills less than 25° Flat or nearly flat paddocks	Rotationally grazed Set stocked
Medium	Steep hills (a third of the area over 35°) Hills less than 25°	Rotationally grazed Set stocked
High	Steep hills	Set stocked

and is more thorough in that it looks at nutrient inputs against nutrients exported and adjusts for phosphorus buffering index (PBI). Potassium and sulphur requirements can also be estimated. Amounts of nutrients removed from pasture in animal and plant products are outlined in Table 2.6. For the less-intensive grazing industries, such as sheep and beef cattle, this approach may over estimate some nutrient requirements. Whichever method is used, monitoring of soil nutrient levels over time is the only sure way of assessing changing levels of nutrients. Soil nutrient levels will vary from year to year owing to seasonal conditions (for example, they will tend to be higher after a drought). Sampling and testing procedures will result in some variation; therefore looking at trends in changes over time (that is, more than 2 years of data) is more useful than two individual tests.

Nutrient removal from hay (Table 2.6) can be used to estimate nutrient requirements after hay has been cut from a paddock. It should also be used a part of the calculation in estimating the true cost of hay or silage production.

For example, if 5 tonnes/ha of dry matter are cut from a paddock, then nutrients required to maintain levels will be:

- Phosphorus – 15 kgP/ha (5 tonnes × 3 kg P)
- Potassium – 75 kgK/ha (5 tonnes × 15 kg K)
- Sulphur – 15 kgS/ha (5 tonnes × 3 kg S)

This could be provided by using 300 kg/ha of 'super potash 1 + 1' that has 4.4% P; 25% K and 5.5% S. This cost should be included in estimating the total cost of hay or silage.

Audit method for calculating maintenance fertiliser requirements

Generally, in sheep and beef enterprises, the annual stocking rate expressed in DSE/ha and soil type are used to calculate maintenance fertiliser requirements for P.

In the dairy industry, a more common approach is to complete a full nutrient audit for P, K and S, where the following items are accounted for:

Table 2.4. Soil loss factor.

Soil type	Phosphorus buffering index (PBI)	Loss factor
Recent alluvial soils, loams, sandy loams	Less than 100	Low
Podzols, clay loams, redzinas	100–300	Medium
Krasnozems, other clays, organic soils	Greater than 300	High
Acid sands (leaching leads to high loss factor).	Less than 100	High

Table 2.5. Recommended phosphorus rate (kg P/DSE) required to maintain soil phosphorus, according to loss factors, pasture type and rainfall (Cayley and Quigley 2005).

Soil loss factor	Animal loss factor	Poor pasture Annual rainfall (mm)			Improved pasture Annual rainfall (mm)		
		400	600	800	400	600	800
Low	V. low	0.42	0.45	0.48	0.43	0.48	0.53
	Low	0.54	0.58	0.62	0.55	0.62	0.68
	Med	0.65	0.70	0.75	0.67	0.75	0.83
	High	0.77	0.83	0.89	0.80	0.89	0.98
Medium	V. low	0.61	0.65	0.70	0.63	0.70	0.77
	Low	0.72	0.78	0.84	0.75	0.83	0.92
	Med	0.84	0.91	0.97	0.87	0.97	1.07
	High	0.96	1.03	1.11	0.99	1.11	1.22
High	V. low	0.80	0.86	0.92	0.82	0.92	1.01
	Low	0.91	0.98	1.05	0.94	1.05	1.16
	Med	1.01	1.11	1.19	1.06	1.19	1.31
	High	1.15	1.24	1.32	1.18	1.32	1.46

- nutrients exported in animal product, grain or fodder
- nutrients imported in grain or fodder
- the soil maintenance factor (that is nutrients required to maintain soil at target level).

The amount of bought-in feed is much higher on a dairy farm, so it makes sense to allow for the large amounts of nutrients imported in hay, silage or grain.

The soil factors used for P,K,S calculations are shown in Table 2.7. An example of how to undertake a nutrient audit is shown in Table 2.8.

Guidelines for fertiliser application – increasing fertility

Capital application of fertiliser aims to optimise growth and persistence of a particular pasture species by lifting soil nutrient levels (see Table 2.1). This involves making one or more

Table 2.6. Amount of nutrient removed from pasture in animal or plant products.

Product	Amount of nutrient removed (kg) Phosphorus	Potassium	Sulphur
Hay (1 tonne dry matter)	3	15	3
Grain (1 tonne dry matter)	3	4	2.5
Wool (1 kg greasy)	0.004	0.02	0.04
Meat (1 kg live weight)	0.008	0.002	0.008
Milk (1 L)	0.001	0.0014	0.0006

Table 2.7. P, K and S soil factors for maintenance fertiliser applications.

Soil type	Phosphorus[a] (with Olsen P 14-17 mg/kg)	Potassium	Sulphur
Sand	9	25	12
Sandy loam	18	25	12
Clay loam	23	15	12
Clay/red soil	28	15	12
Peats	31	15	12

[a] The soil factor for P varies with Olsen P level

A worked example is shown in Table 2.8 below

Assumptions:

- a farm produces 100 kg/ha greasy wool and 300 kg/ha live weight in sheep meat
- 0.6 t/ha grain (0.5 t dry matter/ha) is purchased; no grain or fodder is sold
- nutrients exported/imported in product are found in Table 2.6
- soil type is a clay loam; soil factors are found in Table 2.7
- soil does not require capital applications of P, K or S.

applications over and above the maintenance levels outlined above. Capital application usually refers to phosphorus fertilisers or limestone because of their availability over time and lower mobility through the soil. Expenditure on capital application must be regarded as an investment of financial capital. To get a positive return on capital invested, the increased pasture grown must be converted into increased animal product per hectare and hence profit. The speed at which nutrients should be built up will depend on cash flow, access to capital and the ability to quickly increase stock numbers (purchase or breed) to use the additional pasture grown. The extra capital required to invest in stock may be much larger than that invested in fertiliser. The higher the marginal return per unit increase in turn-off or stocking rate, the higher the return from capital applications.

The amount of phosphorus needed to increase soil phosphorus by one unit varies with soil type as discussed in relation to phosphorus buffering capacity. Heavier textured soils (clays) require more phosphorus than lighter, sandier soils. Depending on the soil's phosphorus

Table 2.8. An example of a nutrient audit/budget.

Nutrients exported (a)	Phosphorus	Potassium	Sulphur
Meat 300 kg/ha LWT	300 × 0.008 = 2.4 kg P/ha	300 × 0.002 = 0.6 kg K/ha	300 × 0.008 = 2.4 kg S/ha
Wool 100 kg/ha greasy	100 × 0.004 = 0.4 kg P/ha	100 × 0.02 = 2.0 kg K/ha	100 × 0.04 = 4.0 kg S/ha
TOTAL EXPORTED Kg/ha	2.8 P	2.6 K	6.4 S
Nutrients imported (b)			
Grain 0.5 t/ha	0.5 × 3 = 1.5 kg P/ha	0.5 × 4 = 2.0 kg K/ha	0.5 × 2.5 = 1.3 kg S/ha
Soil factor (c)	23	15	12
Nutrient required (kg/ha) **a – b + c**	2.8 – 1.5 + 23 = 24.3 kg P/ha	2.6 – 2.0 + 15 = 15.6 kg K/ha	6.4 – 1.3 +12 = 17.1 kg S/ha

buffering index (PBI), it may take 5–15 kg P/ha to increase Olsen P by 1 unit, or 2–4 kg P to shift Colwell P by 1 unit.

The amount of potassium needed to increase soil Colwell K by 1 unit also varies with soil type. For example, a clay soil will need 2 kg K/ha, while a clay loam or a sandy loam will need 1 kg K/ha, to lift Colwell K by 1 unit.

There are no rules of thumb for capital sulphur application and so rates need to be based on soil tests.

Pasture type should be taken into consideration when determining the type and amount of nutrient to be used. Priority should be given to newly sown pastures that have low levels of soil fertility because the persistence of the sown species improves with increasing soil fertility. Pasture establishment will not be a profitable exercise unless sown species persist. Improved species of grasses and legumes have a higher requirement for phosphorus than native grasses. Fertilising improved pastures will give a better response compared with native pastures. Native grass pastures containing some legume will respond to some fertiliser, but many native species will decline under competition from legumes, annual grasses or broadleaf weeds when phosphorus levels are raised.

Correct all nutrient deficiencies where economic responses are indicated. If phosphorus, sulphur and potassium are all identified to be below critical levels, then a fertiliser blend containing all three nutrients should be applied. Correcting only one nutrient will limit the response in pasture growth.

Timing of application

Most fertilisers can be applied in late summer or early autumn. This ensures that vehicles can easily get onto paddocks and that the performance of early germinating plants is not limited by lack of nutrients. It may be cost effective to have fertiliser application split between autumn and spring for highly productive perennial pastures, such as white clover/perennial ryegrass. A spring application may provide the best response for summer-active perennial pastures (such as lucerne).

Phosphorus should not be applied to waterlogged soils or at the coldest time of the year because most of it will be rapidly converted into a form that is not available to plants. To avoid nutrient contamination of streams, fertiliser should not be spread near waterways, on waterlogged soils or on steep slopes or when storm events are likely to increase risk of nutrient run-off.

Fertiliser and/or legume seed are not recommended for native grass pastures in high conservation areas.

Guidelines for applying nitrogen fertilisers

Nitrogen is a more mobile nutrient with a short-term response and is best applied when pastures are actively growing. Do not apply nitrogen to waterlogged paddocks or if it is too cold or too dry.

The best responses to nitrogen occur when soil fertility is high and there is a high proportion of improved grass species. Annual ryegrass is highly responsive, followed by short rotation ryegrass then perennial ryegrass. In a trial at Hamilton, responses per kg of nitrogen applied ranged from 7 to 13 kg of additional dry matter. The better responses were achieved on the higher fertility sites. Pastures were perennial ryegrass–sub clover-based.

Nitrogen fertiliser, such as urea, can be applied tactically during the growing season to overcome deficiencies and boost dry matter production. Unlike phosphorus, potash and sulphur – which have a residual effect – the nitrogen response usually only lasts for a couple of months.

Inefficient application of nitrogen can lead to large losses through a number of processes, including the formation of nitrous oxide. Nitrous oxide is of concern because it is a greenhouse gas. Nitrous oxide is primarily lost from agricultural soils as a result of cultivation, legumes, application of nitrogen fertilisers and animal excreta. Nitrous oxide is primarily formed through denitrification: a process maximised in warm, anaerobic (wet) soil conditions with large amounts of nitrate and available carbon present.

Pasture growth responses can be optimised, and nitrous oxide losses from nitrogen fertiliser can be minimised, by following the guidelines below.

- **Rate:** do not apply above 50 to 60 kg N/ha in any single application and do not apply N closer than 21 to 28 days apart, because this will increase N losses exponentially.
- **Source:** use ammonia-based N fertilisers (for example, urea or diammonium phosphate (DAP)) rather than a nitrate-containing fertiliser.
- **Timing:** only apply fertiliser when pasture is actively growing and can use the N.
- **Limiting factors:** ensure other nutrients (P,K and S) are not limiting pasture growth.
- **Slow release and inhibitors:** there are a number of slow release N fertiliser products, as well as coated fertiliser products, that reduce nitrous oxide losses from fertiliser. A spray is also available to prevent urine from forming nitrous oxide, or from leaching as nitrate. Under a greenhouse gas emissions trading regime, these products may become more cost effective.
- **Grazing:** do not graze pasture for 7 to 14 days after nitrogen application, because nitrate levels are at their highest and can be toxic to stock. Nitrate levels in perennial ryegrass seldom reach toxic levels, but other plants in the pasture, such as capeweed, may reach toxic levels.

Reducing nutrient loss

The guidelines for fertiliser use, outlined above, provide tips on minimising the loss of applied nutrients to the environment. Loss of nutrients can be costly to the producer and detrimental to the environment. Loss of nitrogen can be as high as 150 kg N/ha/year and loss of phosphorus can be up to 11 kg P/ha/year. Small nutrient losses can cause significant nutrient pollution of waterways and groundwater and add to greenhouse gases. Landscape features, climatic conditions, pasture and stock management practices all affect the risk of losing phosphorus and nitrogen to the environment.

There are four pathways for loss of nutrients:

- run-off across the soil surface (run-off)
- drainage past the root zone (deep drainage)
- lateral flow within the root-zone of the soil profile (subsurface lateral flow)
- emission of nitrous oxide, which is a powerful greenhouse gas (gaseous emission) (Figure 2.9).

A 'farm nutrient loss index' can be estimated using easily quantifiable inputs such as landscape features, climatic conditions and pasture and stock management practices to calculate the risk of nitrogen and phosphorous loss at the paddock scale and to evaluate the effects of different management

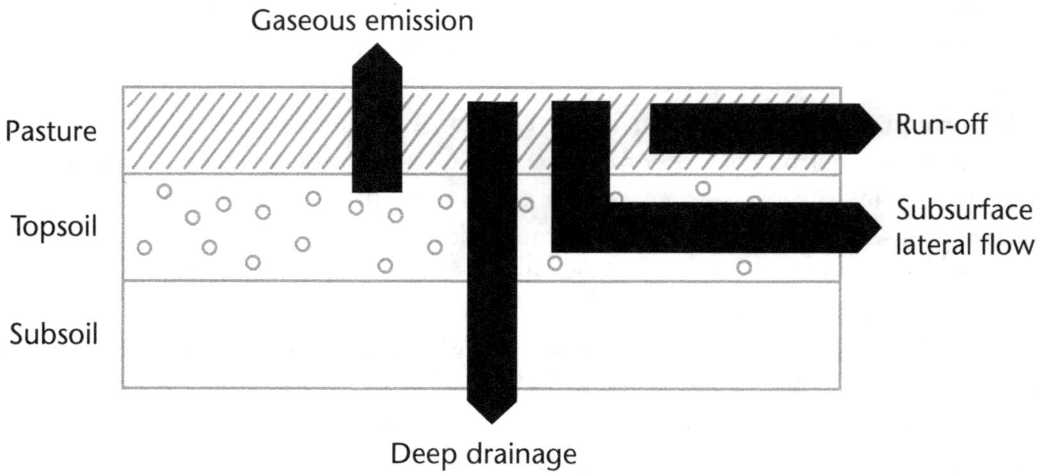

Figure 2.9. Pathways of nutrient loss (Gourley *et al.* 2007. Making Better Fertiliser Decisions for Grazed Pastures in Australia).

practices. The index identifies the risk of N and P loss from the paddock to the wider environment via the four nutrient loss pathways, but is not designed to estimate actual loads of nutrients lost from farms. The model can be accessed and used from the CSIRO website (see list at end of this book).

Key messages – plant and soil nutrition

- The main nutrient limitations to pasture growth in Victoria are, in order of importance: phosphorus, potassium, nitrogen and sulphur. In acid soils, the trace element molybdenum is often limiting.
- Phosphorus, potassium and sulphur can be assessed by regular soil tests.
- Most trace elements need to be assessed by plant tissues tests or, if animal deficiencies are suspected, by animal testing.
- Targets for all the nutrients, which aim to optimise plant growth, have been developed.
- Decide on the type and quantity of nutrient to use by determining what nutrients are limiting and what product will provide the most cost effective response.
- Follow all recommendations to reduce the risk of losing nutrients to the environment.

Other soil factors that affect pasture growth

Acidity

The acidification of soil occurs naturally as soils age. Many Australian soils are naturally acidic, but many agricultural practices that promote nutrient removal and leaching have accelerated the acidification process.

The number of hydrogen ions (H^+) determines how acidic the soil is. A pH scale is used to express acidity. The pH scale (as measured in water) goes from 1 to 14: pH below 7 is considered acidic and pH above 7 is alkaline; a pH of 7 is neutral. Most living systems operate at a stable pH.

Causes of soil acidity

Rainfall – soils in high-rainfall areas are naturally more acid than soils in low-rainfall areas. In high-rainfall areas, more leaching of nutrients occurs, which eventually leads to lower soil pH. There is little risk of soil acidification where the annual rainfall is less than 500 mm.

Soil properties – Soil texture and inherent soil pH affect how much the soil pH will decline. Clay soils have a much greater ability to resist (or buffer against) pH change. Therefore larger pH drops are measured more often in sandy soils than in clay soils under agriculture over the same period of time. Strong chemical reactions prevent the soil pH dropping below about 4. Soils with a low inherent pH (e.g. 4.3) will tend to drop more slowly than neutral or high pH soils.

Soils that are very acidic may not show a drop in pH, but may exhibit other types of soil degradation, such as increased aluminium levels. As pH decreases, aluminium becomes more available. Aluminium can be toxic to many plant species, and this toxicity may have a higher impact on plant production than the low pH itself.

Leaching of nitrogen – legumes provide nitrogen to other plants. Nitrogen is important for plant growth. During the nitrogen cycle, organic nitrogen (mostly plant protein in organic matter), is converted to an ammonium ion ($NH4^+$), which is then converted to nitrate (NO^{3-}). This results in a hydrogen ion being released for each NO^{3-} ion generated. If the nitrate is taken up by plants, then the soil remains neutral. However, if the nitrate is leached through the soil, it can take other cations – such as calcium, magnesium and potassium – with it. This results in the accumulation of free hydrogen ions in the soil. The more hydrogen ions, the more acid the soil becomes.

Nitrogen is available in organic matter, which is released as the organic matter breaks down. Green leafy material is high in protein and so produces more ammonia than plant residues such as straw, which is rich in carbohydrates but low in protein.

Some nitrogenous fertilisers can also be acidifying to varying degrees, as discussed below.

Relative acidification caused by fertilisers

Here is a list of fertilisers from the most acidifying to the least acidifying:

- mono-ammonium phosphate (MAP)
- sulphate of ammonia
- di-ammonium phosphate (DAP)
- anhydrous ammonia
- urea
- ammonium nitrate
- superphosphate
- potassium phosphate
- potassium chloride.

Removal of soil alkalinity by farm products

Plants consume a greater proportion of cations (such as calcium and potassium) than anions. The removal of produce containing these cations results in an excess of hydrogen ions in the soil and hence a decrease in soil pH. Table 2.9 gives an indication of the lime (calcium carbonate) equivalent required to compensate for the removal of some crop and animal products.

Table 2.9. Soil alkalinity removed by some farm products (kilograms).

Product	Yield	Lime equivalent per year (kg)
Lucerne hay	3–7 tonnes per ha	360–840
Clover/grass hay	3–7 tonnes per ha	270–630
Lupins	0.6–1.3 tonnes per ha	12–26
Wheat	1.2–2.1 tonnes per ha	11–19
Lambs	10 per ha	6
Wool	30 kg per ha	4

Effects of soil acidity

As the soil becomes more acid, certain nutrients are made less available to plants, while others can become quite toxic. With increasing acidity, nutrients attached to clay or organic matter surfaces will be displaced by hydrogen ions. Therefore the availability of many major nutrients (such as N, P, K, Ca, Mg and Mo) is likely to be depleted in an acid environment. However, improving soil pH will not guarantee increased availability of these nutrients. In field observations, short-term pasture responses to liming have been diagnosed as molybdenum responses (molybdenum is released with increased pH). No such measured responses have been reported due to increased availability of phosphorus.

The significant impact of low pH is the increased availability of aluminium in the soil. In acid soils, aluminium is released from clay particles and becomes increasingly available to plants. Aluminium is toxic to plants and to most micro-organisms. Low pH can reduce the activity of soil biota, including rhizobia (bacteria that allow clovers to fix nitrogen) and earthworms. The availability of aluminium reduces root tip growth and can prevent plants from taking up nutrients and water. Manganese can also become more available in acid soils.

Some plants, such as lucerne, will be affected more than others by pH. As the soil pH drops, the availability of toxic elements, particularly aluminium, restricts growth more than the acidity does.

Soil pH

Soil pH can be determined by a soil test. There are two common measures of soil acidity or alkalinity: pH measured in water (pHw) and pH measured in $CaCl_2$ (pH $CaCl_2$). The soil pH as measured in water is considered to be closer to the pH that the plant roots are exposed to in the soil, but varies within a paddock and over the year owing to changes in soil moisture. Seasonal variation as high as 0.6 units of pH has been recorded in a single year. Soil pH measured in $CaCl_2$ is less affected by the seasons. Soil pH in water may be 0.6 to 1.2 units higher than soil pH in $CaCl_2$ in low salinity soils and 0.1 to 0.5 units higher in high salinity soils.

For most plant species, the desirable pH (water) range is between 5.5 and 7.0. Table 2.10 shows species that are specifically susceptible to low pH and high aluminium. If soils have a pH below the desirable level and have high aluminium levels (see Table 2.10), the application of lime will raise pH and reduce aluminium availability. The amount of lime required will depend on how much the pH needs to be lifted to reach the target pH, for a particular crop or pasture. The soil texture also affects how much lime is required. For example, a heavier textured soil such as clay needs around 4.0 t/ha of lime to shift the pH by 1 unit, whereas a sandy loam will only need 1.8 t/ha.

Lime can vary considerably in ability to reduce pH. Lime contains calcium carbonate and the higher the calcium carbonate content and the finer the lime, the more effective the lime is in reducing pH (or the higher its effective neutralising value). Dolomite is calcium magnesium carbonate ($CaMg(CO_3)_2$) and can be described as a type of lime that also contains magnesium. It can be used in situations where pH needs to be lifted and a magnesium deficiency corrected.

Soil structure – cation exchange capacity

Cation exchange capacity is a measure of the soil's capacity to absorb and hold cations (positively charged ions). This measurement provides an indication of the amount of nutrients available in the soil, their ratios and the soil's ability to hold nutrients. The higher the cation exchange capacity, the more nutrients a soil can hold. Cation exchange capacity can indicate if there is a physical or structural problem with the soil. Calcium and potassium aid soil structure, but sodium and magnesium harm it. To a large extent, cation exchange is dependent on soil texture, with light (sandy) soils having lower cation exchange capacity than heavy (clay) soils.

The major cations are exchangeable calcium (Ca), magnesium (Mg), sodium (Na), potassium (K) and aluminium (Al). They are measured from a soil test in milliequivalents per 100 grams and then expressed as a percentage of the total. This is because the proportions of the exchangeable cations are often more relevant than the actual levels.

Exchangeable calcium (Ca) should make up the highest proportion of the cations and the desirable range is 65 to 80%. Calcium is the dominant cation in many soils and if low, can be increased by adding lime.

The ratio of exchangeable calcium to magnesium provides a guide to a soil's structure and potential problems influencing soil drainage, root development and hence plant growth.

A calcium-to-magnesium ratio of 2:1 or less indicates reduced soil stability in heavier soil types (clays and clay loams), but is not as important for lighter soils. However, if the exchangeable sodium is less than 6% then soil structure may not be affected.

A calcium-to-magnesium ratio of more than 10:1 indicates potential magnesium deficiency.

Exchangeable magnesium (Mg) is the second most important cation and the level should be between 10 and 20% of the total cations. The balance of magnesium and potassium may have an impact on animal mineral balance and health (such as in grass tetany in cattle), although soil test results are not the sole indicators of a problem.

Exchangeable potassium should be between 3 and 8% of the total cations.

Exchangeable aluminium should be the lowest of the cations and less than 1% of the total cations. Aluminium was discussed earlier under acidity because it becomes more available (and toxic) with high acidity.

Exchangeable sodium should be less than 6% of the total cations. If the levels are greater than 6% then soils may be considered to be sodic.

Sodicity

Calcium holds soil particles together creating a good soil structure allowing free movement of water and nutrients. On the other hand, sodium can be detrimental to soil structure. Sodium causes swelling and dispersion of clay; clay particles may then move through the soil, clogging pores, and reducing infiltration and drainage. Sodic soils are characterised by waterlogging and surface crusting, and are associated with erosion and can cause cultivation problems.

Sodicity is measured by the exchangeable sodium percentage. Soils with an exchangeable sodium percentage greater than 6% are classed as sodic and are potentially dispersive. Gypsum (hydrated calcium sulphate $CaSO_42H_2O$) can be used to prevent soil dispersion because the calcium from gypsum displaces the sodium, which can then be leached deeper into the soil.

Sodicity has no direct effect on plant growth, but may have an indirect effect through poor water penetration of the soil profile. When sowing crops or pastures into dispersive soils, a surface 'crust' will hamper the establishment of seedlings.

Organic matter

Organic matter in the soil is an important source of nutrients for the soil and its population of micro-organisms. Soils will inherently differ in their organic matter levels, but organic matter can be built up by growing more pasture or crop.

Organic matter can be assessed directly but can be estimated from a measure of organic carbon, which is measured in a standard soil test. Low organic carbon levels in soils indicate that the soil is low in organic matter and target levels are included in Table 2.1. The minimum target for organic carbon is around 3%, but this does vary with soil type. Organic matter is usually calculated as 1.72 times the organic carbon level to give an estimate of organic matter, though this may vary with the type of organic matter.

Salinity

A salt-affected soil is one in which growth of useful plants is inhibited by salts in the soil solution, usually sodium chloride.

Dryland salinity is caused primarily by the clearing of deep-rooted native vegetation, such as trees, and replacing them with pastures or crops that use less water. Removal of deep-rooted plants allows more water to soak into the soil, raising the level of the groundwater. During and after rainfall, water can either run off the land, be used by the trees or plants, or soak into the ground. Water that soaks into the ground is eventually stopped by layers of rock or clay. This water then begins to move up towards the surface. The upper surface of the groundwater is called the watertable. When the watertable rises to within 2 metres of the surface, the land is at risk because the salts dissolved in the groundwater can move up to the surface through pores in the soil. Salts that exist naturally in the soil are brought to the surface and when this water evaporates these salts become more concentrated, affecting the vegetation. Areas in which the surface water from rainfall, irrigation or streams filters through the soil and is added to the groundwater are known as recharge areas. Rocky hilltops or ridge areas with course, porous soils are recharge areas. Areas affected by salt are called discharge areas. This is where groundwater is discharged from the soil surface and so usually occurs on the lower slopes or in drainage lines. Bare areas caused by salinity are prone to erosion. Indicators of saline soils are the presence of plants that can tolerate salt such as sea barley grass and spiny rush.

To manage areas that are salt affected (i.e. discharge areas), species that are tolerant to salt can be planted to increase groundcover and hence reduce erosion, and also to improve agricultural production. Grasses that are most tolerant to salt are tall wheat grass, *Puccinella,* some of the fescues, and legumes such as *Balansa* clover and red clover. These areas should be fenced off from the rest of the paddock to control grazing pressure. However, to reduce the problem of increasing salinity, increased planting and managing of deep-rooted perennials to increase water use in recharge areas is recommended. Deep-rooted perennials can include trees and shrubs and, for agricultural production, can include species such as lucerne and phalaris.

Soil salinity is assessed by measuring the electrical conductivity (EC) of the soil. The most commonly used method is ECw – electrical conductivity measured in soil/water solution. The interpretation of this test varies with the soil type (see Table 2.1).

Soil health

'Soil health' is a frequently used term, but it can have a wide range of definitions. 'Healthy agricultural soils are able to balance a range of functions to meet the needs of both farmers and the community. Healthy soils function to sustain soil biota and plant life, store and cycle water and nutrients, decompose organic matter, inactivate toxic compounds, suppress pathogens and protect water quality' (Slavich 2001).

All the issues discussed above in terms of nutrients and soil factors that aim to produce a productive pasture, must logically support a healthy soil. Understanding issues that affect soil factors will assist in managing a productive pasture and vice versa. Soil chemical, physical and biological properties all interact, and are all important in determining whether a soil is 'healthy'. The soil tests and targets discussed in the previous sections relate to soil chemical and physical properties. Tests and targets for healthy soil biology are not so clear cut. However, the biological component of the soil system has a high dependence on the chemical and physical soil components.

Soil biology

Soil organisms range from bacteria to giant tunnelling earthworms. They all play an essential role in decomposing organic matter, cycling nutrients and fertilising the soil.

There are four levels of soil organisms: microflora, microfauna, mesofauna and macrofauna. The microflora (bacteria and fungi) make up 75 to 90% of the soil living biomass and are the primary decomposers of organic matter (plant residues and roots). They transform organic molecules into mineral nutrients (such as nitrate, ammonium and phosphate) that are then available for uptake by plants. The microfauna (protozoa and nematodes) prey on the microflora. The mesofauna (springtails and mites) also prey on the microflora. The macrofauna include earthworms, beetles, ants and termites.

There are tests available to measure the soil biology. They include:

- **population analysis** (such as nematode counts, earthworm counts and food web analysis, which identifies/counts species of microflora and microfauna)
- **biological activity tests** (such as enzyme assays, microbial respiration measured by amount of carbon dioxide produced, and microbial activity measured using cotton strips buried in soil to assess decay rate)
- **indirect indicators** (such as soil organic carbon levels or ground cover/litter; these indirect indicators simply provide an indication of healthy soil biology, by giving a measure of the amount of food present for organisms).

Apart from the cost of the direct tests, there are other problems associated with measuring soil biology and interpreting the results. The sheer number of organisms and species in a soil sample make this a complex task. Soil organisms change rapidly in response to season, rainfall, crops, soil management and even the time of day. A 'high' or 'low' result may simply be a reflection of the conditions at time of sampling. However, such information could be used to compare paddocks, and the impact of management on soil organisms in a relative sense, rather than expecting absolute benchmarks. Because microbial activity is sensitive to many factors (soil pH, organic matter supply, salinity, temperature and moisture), it is difficult to use this information to assess soil problems and make specific recommendations.

Encouraging soil organisms

Generally, the more fertile and better structured the soil is, the more pasture (or crop) can be grown, the root mass will be greater, and more food will be available to support a greater population of soil organisms.

It can be difficult to measure soil organisms and activity, but, by understanding what conditions they like, soils can be managed to suit them. The concept is the same with managing weeds: if the environment is right for a weed and the weed is sprayed, the weed will die but, if the environment is not changed, the weed will return.

Almost all soil organisms need the same things to survive: food, water and oxygen. They need a carbon-based food source that provides all their nutrients, including phosphorus and nitrogen. They require a moist habitat, with access to oxygen in the spaces in the soil. Hence 75% of the soil organisms are found in the top 5 cm of soil.

To encourage soil organisms:

- **minimise soil disturbance** (Cultivation destroys the habitat for soil organisms and compaction from machinery reduces the spaces in the soil, reducing air and water movement. Minimal tillage/direct drilling can reduce these problems. In a permanent pasture, soil disturbance is not an issue. A standard soil test will indicate if there are structural problems such as dispersion (responsive to gypsum) or slaking. Slaking is when the soil aggregates fall apart when wet. This can be rectified by increasing organic matter.)
- **check the pH** (Add lime if necessary: the ideal range for most organisms is pH (water) 5– 8. Strongly acidic soils discourage nitrogen-fixing bacteria and earthworms. In a pasture situation, with consideration of the economics of liming, the aim is to lift the pH (water) to around 5.4–5.5, whereby aluminium toxicity (for the plants) is alleviated. A standard soil test will measure pH and aluminium levels and indicate if lime is required.)
- **maintain ground cover/ litter** (Over the summer/autumn period there should be more than 70% ground cover at least 800–1000 kg DM/ha dead pasture/litter left by the time of the autumn break. Litter is important for several reasons: it provides insulation to protect soil biota; provides a carbon source for soil biota; allows more rain to infiltrate/less evaporation; and protects soil from wind/ water erosion.)

Grazing management has a major impact on the clover, grass and weed content in a pasture, which then affects how much litter is left on the paddock by late summer/autumn. Rotational grazing can help maintain a desirable grass content and maintain more even cover across paddocks (see 'Grazing management' later in this chapter).

Practices that build soil organic matter will help to build the organic pool of nutrients such as phosphorus and nitrogen.

Adding organisms to the soil

The most common and successful example of addition of micro-organisms to the soil is the addition of the *Rhizobium* bacterium via inoculation of clover. The *Rhizobium* bacteria help legumes to fix atmospheric nitrogen and convert it to useable forms. Developing a technique to add *Rhizobium* bacteria took a huge amount of research and development. There have been very few successes in incorporating other micro-organism in the soil through direct addition.

The general rule of thumb for treating any symptom or problem, whether animal or soil health, is to identify the problem first and then address it. If pasture productivity is felt to be limiting because the response to fertiliser or nutrients is poor or plant composition is deteriorating, then the following areas should be checked:

- a soil test will rule out any macro nutrient deficiencies, acidity, salinity or structural problems
- plant leaf analysis will rule out trace element deficiencies
- other measures, such as organic carbon or groundcover/litter, can indicate if there is problem with soil biology.

Products containing biological preparations or ingredients that claim to improve soil health or production should be assessed just like any other product. There should be proven scientific data to back-up claims and, if there is not, they should either be avoided or at least evaluated reliably for their performance and cost. The section on 'Evaluating products' above provides a guide for setting up on farm comparisons.

Greenhouse gases – the role of soils in storing carbon

Agriculture contributed 15.7% of Australia's total net greenhouse gas emissions in the 2005 National Greenhouse Gas Inventory.

The main greenhouse gases emitted from agriculture are methane (from livestock) and nitrous oxide (from soils). Both are powerful greenhouse gases with respectively 23 and 298 times the global warming potential of carbon dioxide.

The Australian government signed the Kyoto protocol in 2007. The Kyoto protocol is an international legally binding plan that set targets for reducing greenhouse gases and established an international emissions trading scheme. A carbon emission trading scheme can be a cap-and-trade scheme, whereby the amount of carbon dioxide allowed to be emitted will have an upper limit imposed, and polluters have to reduce their emissions or buy carbon credits from others. One way to generate carbon credits or offsets is to store it (sequester) in the soil or by growing trees.

The amount of organic matter in soil results from the balance of inputs (plant residues) and outputs (mineralisation/decomposition). Carbon occurs in different forms in different soils and under different management practices. A high proportion of carbon in pasture soils is in a form that can be rapidly decomposed, so, if inputs of organic matter are reduced, the soil carbon levels can rapidly drop.

Studies show that increasing soil organic carbon is a very slow process, and is correlated with the amount of dry matter grown. For example, to double soil carbon from 2 to 4% under pasture, it has been estimated the dry matter production has to be doubled – and this level of production increase may have to be maintained for 200 years. Larger increases in dry matter production would allow carbon to build up more quickly.

The amount of soil organic carbon that can be stored is also affected by soil texture. Clay soils can hold onto more carbon and protect it from decomposing, compared with sandy soils.

Key messages – other soil factors

- Factors that affect the nutrient value and structure of the soil include acidity, salinity, exchangeable cations and organic matter.
- Growing perennial plants can reduce salinity (by using more water), reduce acidification (by using nitrogen) and increase organic matter (by producing more).
- The proportions of exchangeable cations (Ca, Mg, K, Na and Al) in the soil provide some insights into soil structure problems, as well as potential deficiency or toxicity issues. Products such as lime, gypsum and dolomite can be used to address specific problems.
- A system that maintains a productive and healthy pasture system will be sustained by a 'healthy' soil. Many of the factors that positively influence plant growth also create a favourable environment for the soil population of micro-organisms. These include nutrients supplied by organic matter and good soil structure.
- Managing a system that provides a favourable environment for desirable plants to grow, and for soil micro-organisms to thrive, will ultimately sustain itself.

Pasture species

Pastures in Victoria are often described under the following classifications.

- **Native (or unimproved) pastures** have a high proportion of native perennial grass species present such as kangaroo grass, wallaby grass, weeping grass or spear grass. They may also contain some sub clover and volunteer annual grasses or broadleaf weeds: the amount dependent on fertiliser and grazing history.
- **Annual pastures** consist of annual grasses, clovers and weeds. Some of these pastures may have originally been sown down to a mix like annual ryegrass and sub clover, but many may have simply been fertilised and had sub clover seed broadcast, with the annual species eventually out competing the native species.
  Common volunteer (not sown) annual species (weeds) found in pastures include silver grass (*Vulpia spp*), barley grass, the brome grasses (several species), annual (Wimmera) ryegrass, capeweed and *Erodium* spp.
- **Improved perennial pastures** have undergone a renovation program and been sown to perennial grass (such as phalaris, cocksfoot, fescue or ryegrass) and legumes (such as white clover, lucerne or sub clover).

As a guide, a good pasture composition consists of 50 to 60% desirable grasses (preferably perennial, but annual ryegrass is quite productive and very high feed quality) and 40 to 50% legume and a minimal amount of weed species.

In the higher rainfall areas, it is also common to find volunteer perennial species (weeds) in pastures, such as bent grass, fog grass and sorrel.

Perennials – native and introduced

Improved perennial species (grasses and lucerne) and native perennial grasses have the following advantages over annual species:

- they have deeper roots and can access more water stored in the soil profile
- they have a longer growing season
- they can respond quickly to rainfall and provide feed if summer rain occurs
- they provide a more stable pasture composition from year to year, which helps keep weeds out
- they provide better ground cover over summer/early autumn (depending on grazing intensity), protecting the soil from wind or water erosion.

There are also many undesirable perennial species such as bent grass, sorrel and the noxious weeds: serrated tussock and Chilean needle grass. Good fertiliser and grazing management is the key to keeping undesirable perennial or annual species out of improved perennial pastures, thereby reducing reliance on herbicides.

Native grasses

The native species are acclimatised to our soils and environment and hence have low requirements for phosphorus and nitrogen. Introduction of exotic perennial species generally requires higher fertility levels and therefore respond with higher pasture production.

It is difficult to get good comparative pasture production figures for different species, given their different environmental requirements (for example, comparing a perennial ryegrass, with its high requirements for water, phosphorus and nitrogen, with *Microlaena* spp. (weeping grass) in a low rainfall and low fertility environment is not entirely relevant. However, Figure 2.10 gives an indication of the growth habit of some of the species.

The native grasses have different growth patterns. Some are more winter-active (C3 category) and some, such as tropical grasses, are more summer-active (C4 category). The C3 group includes wallaby grass (*Austrodanthonia species*) and weeping grass (*Microlaena species*). They grow in low fertile conditions, are frost hardy and need some rotation or rest period to persist. The C4 groups includes kangaroo grass (*Themeda trianda*) and red grass (*Bothriochloa macra*) and, while they grow over summer, they are less palatable than C3 grasses and will not persist with heavy grazing or set stocking.

Therefore, in a whole farm system, pastures that are to be maintained as predominantly native pastures will require management to avoid overgrazing over summer and invasion of annual grasses and clover. Application of low rates of fertiliser (for example, 100 kg/ha every 3 years) will ensure that sub clover can grow in order to improve feed quality and quantity. However, higher or more frequent applications of fertiliser will encourage more competition from volunteer annual species.

Introduced grasses

Figure 2.11 shows the potential improvement, in both the amount of feed grown and the length of the growing season, by improving the fertility and/or the grass species with an introduced perennial (in this case perennial ryegrass).

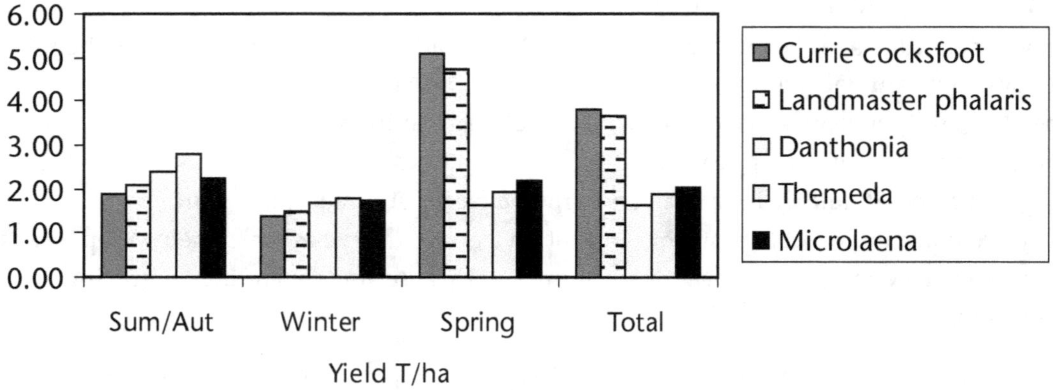

Figure 2.10. Growth habit of some improved perennial and native grasses at Springhurst (north-east Victoria) in low fertility environment.

In situations where there are no, or very few, desirable perennial grasses present in a pasture, pasture renovation may be warranted. For degraded or run down pastures where there are still more than five to seven plants per square metre of phalaris or cocksfoot or eight to 10 plants per square metre of perennial ryegrass or fescue, trying to improve the growth and size of the current pasture plants with grazing and fertiliser is an option. Often this can be achieved over 2 to 3 years and is cheaper than having to re-sow the paddock.

Pasture renovation can give a very good return on investment, if the following conditions apply:

- pasture establishment is successful the first time to contain capital costs (pasture renovation can be extremely costly and will depend on the inputs required and the amount of time the paddock is not available for grazing)
- stocking rate needs to be increased to turn the extra feed into product and dollars (the greater the lift in production, the quicker it takes to break even)
- for permanent pastures, fertiliser and grazing management needs to be appropriate to ensure that productivity and persistence is optimised
- for short-term special purpose pastures to be worthwhile, they either have to play an integral role in the paddock preparation (weed control or levelling a rough paddock) or they have to provide high-quantity/quality feed at a lower cost than supplementary feed.

Feed gaps

Pasture growth curves can assist in making strategic decisions about enterprises and management, and if there is a need to increase seasonal pasture growth to meet the animal feed demand. The various times of the year when the pasture supply does not meet animal demands are called 'feed gaps'. Fertiliser (N, P, K and S) can be used to increase pasture supply during the growing season, but there may also be opportunities to introduce new pasture species to increase feed supply at critical times of the year.

There is a wide range of pasture or forage species available that can be used to establish long-term perennial pastures or short-term, special purpose pastures, depending on the need. For example, species that can help fill summer feed gaps are lucerne, tall fescue (continental types), kikuyu, chicory and brassicas. Species that can provide a bulk of good quality winter feed – provided they are sown early in autumn – are the Italian and hybrid ryegrasses.

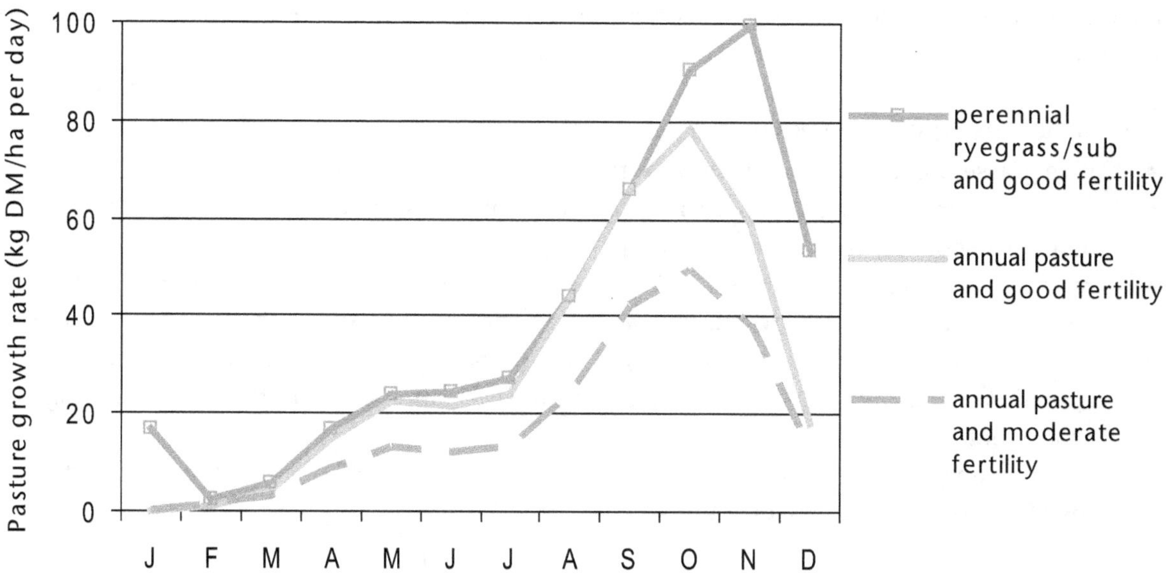

Figure 2.11. Average pasture growth rates for an annual (sub clover/capeweed dominant) pasture and a sub clover/perennial ryegrass pasture at Hamilton, Victoria. Two levels of soil fertility are shown for the annual pasture. (GrassGro™)

The key questions to be addressed before sowing new species are as follows:

- When is/are the feed gap/s?
- Will the proposed species/varieties (or mixture) fill an identified feed gap at a lower cost than other options (grain, hay or nitrogen fertiliser)?
- Is the environment right (rainfall or irrigation, soil type and adequate soil moisture over summer) for the species to grow and persist for the desired length of time?
- Do the species require special grazing management to persist?

Pastures/fodder crops that only last 1 to 2 years have to supply more feed or higher quality feed than a longer-term pasture to be profitable; hence, there is more risk associated with growing short-term pastures/fodder crops.

Table 2.10 summarises a range of improved perennial and annual species and the environments to which they are suited. Examples of cultivars are included, but these will be continually updated and improved over time.

Cereal crops to fill feed gaps

In mixed farming systems, cereal crops can be used to help fill feed gaps. The most common ways of using cereal grain crops are to either graze the stubbles over summer/early autumn, or graze during the vegetative phase in winter. Grazing cereal crops (wheat, barley or oats) at the vegetative growth stages (up to the onset of stem elongation) has been shown to have little impact on grain yield, and can provide valuable winter feed.

During winter and early spring when grazing crops is practised, cereal forage quality is extremely high. The timing of grazing will generally be determined by the trade-off between maximising herbage mass and avoiding grain yield losses. Because wheat forage at this time is commonly deficient in sodium and, often, magnesium, supplementation of livestock with these elements is an inexpensive and effective insurance policy.

Key messages – pasture species

- Native grasses have low requirements for phosphorus and nitrogen.
- Introduced perennial grasses can increase the amount of pasture grown and length of the growing season, but have higher requirements for phosphorus and nitrogen.
- Perennial pastures (native or introduced) are beneficial because they have deeper roots and can access water, provide better ground cover and many are able to respond to out-of-season rains.
- Establishing new pastures can be costly, so it is important to undertake good paddock preparation, use the appropriate species, and follow with appropriate nutrition and grazing management.
- Feed gaps can be filled by using special purpose pastures or cereal crops, but the cost and value of short-term pastures must be assessed carefully.

Grazing management

Grazing management can have an impact on both the growth and the utilisation of pastures, but can also have beneficial impacts on weed control, environmental management and animal health, especially worm control. The impacts could be the reverse with poor grazing management. In developing a grazing management plan, there needs to be a clear idea of what the objectives are.

The objectives may be:

- to grow more grass at a particular time of year
- to improve the persistence of perennial pasture species
- to control the animals' intake
- to match animal requirements to pasture availability
- to manage ground cover
- to manage pasture composition
- to control weeds.

This section deals with the role of grazing method in optimising the growth and persistence of desirable species and controlling weeds.

How plants respond to grazing

The growth habit of a pasture species gives a guide as to what grazing method might best suit it. Understanding the growth habit of the existing, or desired, species is the first step to deciding what grazing method/s are appropriate.

In general, pasture species that have an upright growth habit will produce more dry matter and persist better if they are rested after grazing rather than subjected to continuous, hard grazing. If eaten off close to the ground, they cannot maintain adequate leaf area to photosynthesise and have to rely on any stored carbohydrate to regrow. Depleting carbohydrate reserves reduces the growth rate, because the leaf size is smaller, reduces root growth and ultimately reduces plant survival. Resting the plant after grazing the leaves allows the plant to initially draw on water soluble carbohydrate (WSC) reserves, but, once some leaves have grown, the plant can then photosynthesise and divert excess WSCs back into storage (Figure 2.12). Setting up a grazing system that allows the plants to be rested at critical times during the year is superior to a system of continuous grazing, particularly if the pasture is heavily stocked.

Species that grow flatter to the ground such as many broadleaf weeds (capeweed, erodium and Paterson's curse) and some grasses such as bent, couch and kikuyu, or even sub clover, can tolerate continuous grazing. This is because they can still maintain some leaf area for photosynthesis and replenishment of WSCs, even when grazed low to the ground.

Hence, rotational grazing systems tend to increase the content and productivity of upright growing perennial and annual grasses (phalaris, perennial or annual ryegrass) or legumes such as lucerne, and set stocking tends to increase the content of capeweed and sub clover. Improved fertility will favour improved grasses, clover, capeweed and barley grass, but reduce native grasses and onion grass.

If there are no desirable species present, or unlikely to volunteer into the pasture over time with improvements in grazing management, then pasture renovation may be necessary.

Grazing systems

Grazing method falls into two broad categories: pastures are either continuously grazed (set stocked) or they are rested after grazing (rotationally grazed). In practice, most producers would do a bit of both at different times of the year or on different parts of the farm, depending on their enterprise and environmental objectives. A typical example might be to graze perennial pastures rotationally during autumn and winter in order to improve pasture growth rates, but to graze continuously just before lambing through to marking (for example August to October), to reduce the risk of mismothering. Pasture growth rates are increasing in spring, so there is less need to grow more pasture. The rotation might start again after marking (or weaning) to make sure the pasture that has built up in spring is used evenly over summer and to prevent some areas being over grazed. In this example, the focus changes from optimising pasture growth in autumn and winter to managing animal grazing behaviour to improve utilisation in late spring and summer.

A number of terms or labels have been created to describe different ways of rotationally grazing a pasture, which can create confusion. For example:

- **simple rotation** – usually means a small number of paddocks (perhaps three to six) are used, or the decision on how long to rest pastures/move stock is based on rigid time periods rather than pasture monitoring.
- **intensive rotation** – usually means a larger number of paddocks are used to reduce the grazing time in each paddock, and the decision on how long to rest a paddock may be associated with monitoring the plants.
- **'cell grazing'** – essentially the same as an intensive rotation. A 'cell' refers to a number of small paddocks around which one mob might be allocated to graze.
- **'Technograzing'** – a system originally developed for bull beef enterprises with a focus on controlling daily pasture intake and bull growth rates. Paddocks are electric fenced into a series of laneways to separate bulls into smaller, more manageable mobs. A single electric wire is run across the laneways in front of the bulls, and another wire behind them, to give them a

Table 2.10. Improved grass, legume and special purpose pasture species and their preferred soil conditions.

Species	Rainfall zone	Features	Preferred soil conditions	Cultivars available
Perennial grasses				
Phalaris	> 450 mm	Deep-rooted grass that tolerates a wide range of environments. More drought tolerant than ryegrass and fescue. The older variety, Australian, has persisted well even with overgrazing. Winter-active varieties will persist better with rotational grazing. Can cause phalaris staggers or phalaris toxicity/sudden death.	Tolerates heavy soils and moderate waterlogging and salinity. Moderately sensitive to aluminium – prefers pH of >5.3 (water) and aluminium levels <8 mg/kg ($CaCl_2$).	Winter active: Sirosa Holdfast, Landmaster, Atlas Less winter active: Australian, Uneta
Cocksfoot	> 500 mm	Hardy perennial grass, useful component of perennial pastures, especially on light-textured, acid soils in medium-high rainfall areas.	Tolerates acid soils and light textures/sandy soils.	Currie, Grasslands Kara, Grasslands Tekapo, Grasslands Wana, Kasbah, Porto, Cambria, Grasslands Vision, Oxen, Yarek
Fescue	> 600 mm	A productive perennial grass, with superior tolerance of low soil moisture and other soil limitations such as aluminium and waterlogging, and high temperatures compared with perennial ryegrass	Prefers heavy to medium, soils, with high fertility. Mediterranean types (winter active) more drought tolerant than continental types (summer active). Continental types require a minimum of 650 mm rainfall. Tolerant of moderately saline soils: more so than ryegrass and phalaris. Can tolerate elevated levels of aluminium.	***Winter active (Mediterranean):*** Fraydo, Grasslands Flecha, Grasslands Flecha MaxP (endophyte for improved insect tolerance), Resolute, Prosper, Lunibelle, Origin. ***Summer active (Continental):*** Grasslands Advance, Quantum (replaces AuTriumph), Demeter, Dovey, Ceres Torpedo, Cajun, Jesup, Jesup MaxP (endophyte), Grasslands Advance MaxP (endophyte), Typhoon, Vulcan II, Carmine.
Perennial ryegrass	> 650 mm + rainfall or irrigation.	A productive, high-quality perennial grass. Quicker to establish than other perennial grasses owing to good seedling vigour. Presence of wild endophyte can cause ryegrass staggers. New cultivars can be purchased with nil or safe (novel) endophyte. Presence of endophyte enhances persistence.	Prefers heavy to medium, soils with high fertility, and prefers environments with mild summers or irrigation.	Bronsyn, Fitzroy, Grasslands Lincoln, Grasslands Samson, Ceres Kingston, Meridian, Roper, Victorian, Camel, Matilda, Extreme, Avalon, Embassy Ceres Marathon, Quartet,Ceres Cannon, Matrix, Prolong, Revolution, Cordoba, Sierra

Species	Rainfall zone	Features	Preferred soil conditions	Cultivars available
Hybrid ryegrass	> 650 mm + rainfall or irrigation.	Hybrid ryegrasses are crosses between perennial ryegrass and Italian ryegrass. They are productive, short-lived grasses inheriting some perenniality from the perennial ryegrass parent, combined with higher growth potential of the Italian ryegrass parent. Can live from 2–4 years.	Prefers heavy to medium, soils. Require good soil moisture to perform to potential, and not tolerant of dry conditions. Require high fertility (Olsen P 20+ mg/kg)	Grasslands Greenstone, Grasslands Supreme, Ceres Horizon, Ceres Geyser, Marsden, Grasslands Impact, Maverick Gold, Banquet
Italian ryegrass Includes true annuals that grow for only 1 year (for example, the 'Westerwolds' type), and biennials where some plants may survive into a second winter.	>650 mm rainfall or irrigation.	Highly productive annual or biennial grass Can provide a bulk of high-quality winter and spring feed, and can be used like a special purpose fodder crop for these times of the year. Also valuable for high-quality silage and hay production.	Prefers heavy to medium, soils. Require good soil moisture to perform to potential and is not tolerant of dry conditions. Require high fertility (Olsen P 20+ mg/kg)	*Annual* New Tetila, Winter Star, Surrey, Progrow, Biennial, Roberta, Ceres Crusader, Concord, Conquest, Flanker.
Tall wheat grass	>400 mm	A hardy, salt-tolerant perennial with a specialist role in the reclamation of salt-affected land, stabilisation of eroding land and for dealing with other soil limitations.	Tolerates saline conditions better than all other common species, and will grow well in moderately saline areas with an electrical conductivity reading in the top 10 cm of soil of less than 5 dS/m. It also establishes in areas with far more severe salinity problems (EC in top 10 cm of up to 40 dS/m). Very tolerant of waterlogged soils and 'heavy' soil conditions Tolerant of soils with moderate aluminium levels (2–4 ug/g).	Tyrrell, Dundas.

Species	Rainfall zone	Features	Preferred soil conditions	Cultivars available
Kikuyu	>600 mm or irrigation	Summer-growing, C4 perennial; native of the highlands of East Africa. Adapted to high rainfall sub-tropical and coastal temperate areas, or inland areas with mild winters and irrigation. Has an established role for dairy production in NSW and Queensland, and is finding a role for sheep/beef production in south-west Western Australia where deep roots and active summer growth give good control of groundwater recharge, watertable height, waterlogging and salinity. Must be heavily grazed during summer to prevent build-up of a dense, stemmy mat, which can choke out companion species.	Prefers light to medium textured, free-draining soils. Moderately tolerant of acid soil with slightly elevated aluminium levels. Moderately tolerant of saline soils.	Common Kikuyu, Whittet.
Legumes				
Sub clover	375–700 mm	The 'standard' legume for grazing systems in the medium–high rainfall zones of southern Australia. Easy to establish, adapted to a wide range of soil types and climatic regions, and tolerant of a range of grazing managements.	Moderately tolerant of saline soils, acidity and some cultivars are very tolerant of waterlogging. Requires moderate fertility (Olsen P 12+), and will respond strongly to phosphorus fertiliser.	Clare, Nuba, Rosedale, Daliak, Dalkeith, Denmark, Enfield, Goulburn, Karridale, Leura, Seaton Park, Woogenellup, York, Gosse, Larisa, Riverina, Trikkala, Antas, Losa, Campeda, Limbara, Coolamon, Urana, Dalsa, Koala
Balansa clover	>450 mm	A source of high-quality, special purpose winter and spring feed, either when sown alone or in mixture with species such as Italian ryegrass.	Adapted to waterlogging and to saline soils, and grows well on a wide range of soil types.	Bolta, Frontier, Paradana, Enduromax.
Barrel medic	250–500 mm	Winter-growing annual. Used in pasture phase of crop rotations in areas with a short growing season.	Suited to alkaline soils. Grows best on clay loams and sands. Moderately tolerant of salinity.	Caliph, Mogul, Parabinga, Paraggio.
Strand medic	250–500 mm	Winter-growing annual. More winter growth than Barrel medic. Used in pasture phase of crop rotations in areas with a short growing season.	Suited to alkaline soils. Prefers lighter soils. Moderately tolerant of salinity.	Harbinger, Herald.
White clover	>700 mm	A valuable perennial legume, best suited to areas with high rainfall/long growing seasons and/or irrigation. High-quality feed for animal production.	Moderately tolerant of waterlogging and soil acidity. Not tolerant to salinity.	Aran, Grasslands Demand, Grasslands Huia, Grasslands Pitau, Grasslands Prestige, Grasslands Sustain, Grasslands Tahora, Haifa, Irrigation, Mink, Grasslands NuSiral, Osceola, Tamar, Waverley, Will Ladino, Grasslands Kopu II, Tribute, Quest, Trifol Sweet.

Species	Rainfall zone	Features	Preferred soil conditions	Cultivars available
Strawberry clover	>550 mm	Deep rooted, drought-tolerant perennial, able to cope with a range of soil limitations.	Copes better with waterlogging, heavy soils and salinity, than white clover. Prefers neutral/alkaline soils.	O'Connors, Grasslands Onward, Palestine.
Lucerne	400–900 mm rainfall or irrigation	An important perennial legume, renowned for its drought tolerance and high-quality as an animal feed. Winter-dormant types are generally more persistent than winter-active types under higher stocking rates.	Requires well-drained, fertile soils, and is best adapted to neutral or alkaline soil conditions. Not tolerant of saline soils.	Nova, Pioneer L55, Aurora, Genesis, Hunterfield, Quadrella, Trifecta, WL 414, WL 525HQ, CUF 101, Rippa, Sequel, Siriver, Cimaron, Prime, Sceptre, Pioneer L90, Salado, Sardi Seven, Aquarius, Kaituna, Pioneer 54Q53, Pioneer L56, Pioneer L69 , Pioneer 57Q75, Stamina GT6.
Special purpose pastures/fodder crops				
Plantain	>550 mm	A perennial forage herb producing high-quality animal feed from spring to summer. Poor winter growth. More persistent than Chicory. Leaves are rich in mineral elements, and the species is reputed to have a number of other benefits for animal production, including reduced incidence of scouring and dags, and reduced internal parasite burdens.	Tolerates low fertility and mild waterlogging. Not salt tolerant.	Ceres Tonic, Grasslands Lancelot.
Chicory	>550 mm	A moderately persistent, leafy herb with a deep tap root, providing excellent quality feed from spring to summer in warm, dry regions. Should be regarded and managed as a perennial summer crop, with an effective life of 3–4 years.	Best suited to fertile, free-draining soil. Deep tap root allows chicory to access water at depth, but growth will stop when the soil dries out. Well-suited to regions with summer rainfall.	Grasslands Puna, Ceres Grouse, Chico, Choice, Grasslands Puna II.
Brassicas Includes turnips, rapes, kales and swedes.	>600 mm Summer moisture	Generally short lived and sown as part of a pasture renovation program to fill feed gaps. Provide good quality feed for stock in summer, autumn and early winter.	Prefer well drained soils, with pH 5.5 or above. Require soils with good moisture content in profile or summer rainfall.	Wide range.

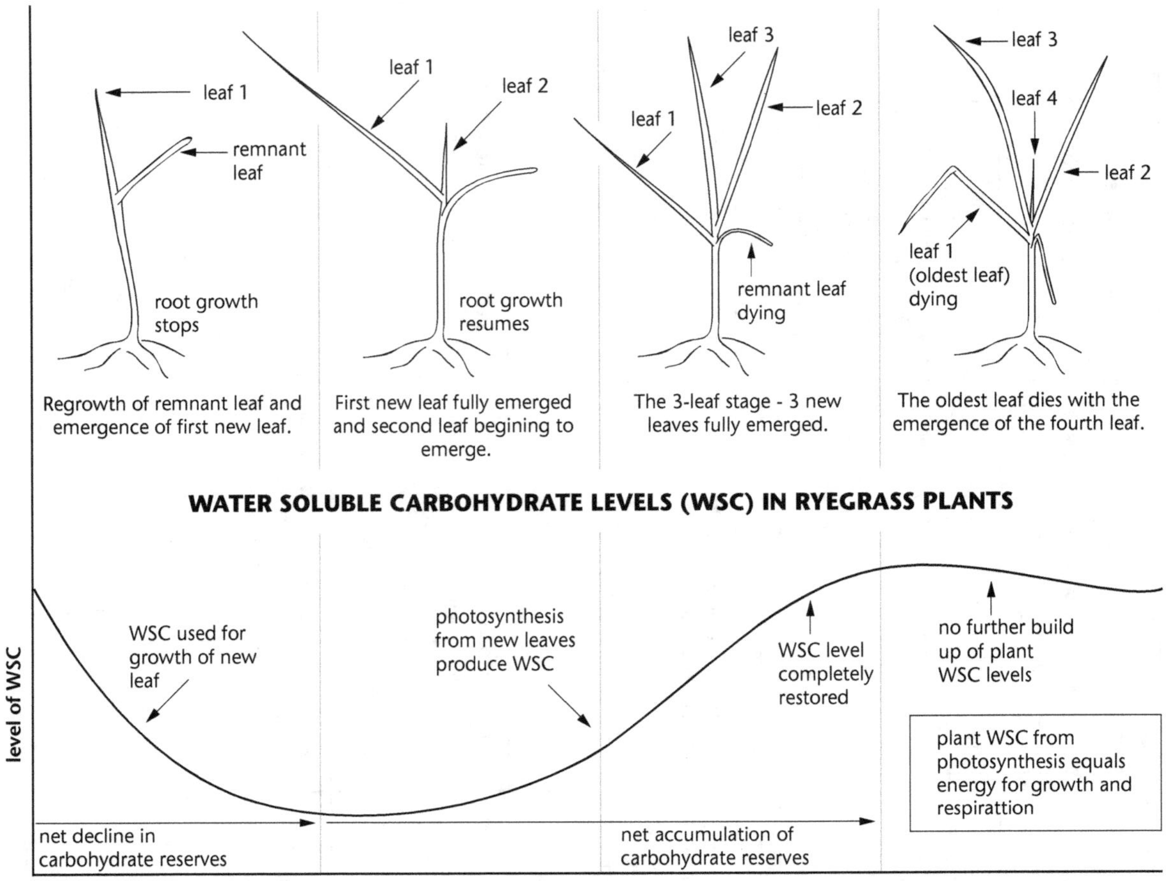

Figure 2.12. Regrowth of a perennial ryegrass tiller and accompanying change in stubble water-soluble carbohydrate (WSC) levels. (Source: D.Donaghy, TIAR, Tasmania)

certain allocation of pasture. All mobs can be shifted along the laneways by moving the front and back single wires.

The name of the system is not important. What is important is knowing what the objectives are, and using a system that will achieve these.

There are a number of decisions to be made about how to rotationally graze a pasture. The three most important factors to consider are:

- **rest:** how long to rest the pasture/paddock
- **grazing time:** how long stock will be left in a paddock to graze
- **grazing height:** how hard to graze a pasture down.

The criteria set for these three factors can vary from season to season, and the focus can change from a plant objective to an animal objective or to an environmental objective (e.g. ground cover). The criteria used for all three factors will dictate how many paddocks are needed in the rotation. This, in turn, will dictate paddock size. Work through developing criteria for each factor before undertaking any paddock sub-division. In many cases, producers have used existing infrastructure (paddocks and water supply) and simply boxed up mobs to get a rotation going.

These are all factors that should be tailored to the specific producer and property: there are no hard and fast rules.

Rest period

For plants that have an upright growth habit, some rest is better than none. However, if the aim is to optimise plant growth and persistence, information about how plants grow should be used to work out the optimum rest time for a species.

Grasses have a maximum number of live leaves they can maintain per tiller. After grazing, once the full complement of leaves have regrown, the oldest leaf will die off and there is no further build up in WSCs in tiller bases. The time it takes for the full number of leaves to regrow therefore becomes the rest period. Each species has its own leaf number; for example, perennial ryegrass and fescue have three leaves per tiller, while phalaris has four and cocksfoot has five (see Figure 2.13).

Rest times can be worked out by monitoring the leaf appearance rates for the relevant species.

Some typical rest times for phalaris are:

- autumn–winter: 30 to 40 days
- spring: 20 days
- summer: 70 plus days (less if rain occurs).

Grazing time

When a plant is grazed and all leaf material is removed, the plant relies on WSCs in tiller bases to produce its first new leaf. Usually this takes 2 to 3 days to happen. So, ideally, to reduce the risk of grazing the first new leaf, stock should not graze paddock longer than 3 days. If feed availability is low and stocking density is high (intensive rotation, high stock numbers in small area, e.g. >200 DSE/ha) the risk is high, and growth rates will be reduced. If pasture mass is high or stocking density is low, the risk is lower, and leaving stock in the paddock for longer will have less impact.

It has been estimated that in an intensive grazing situation (dairy), leaving stock in a paddock for 6 days rather than 3 days can reduce pasture growth rates by around 14%.

Having a shorter grazing time in a paddock means more paddocks in the rotation and a higher stocking density in a paddock. This will reduce the ability for animals to graze selectively. It also reduces the formation of stock camps and leads to more even distribution of nutrients, particularly in hill country.

In many situations, it is not practical to move stock every 1 to 3 days. A compromise might to move them every 1 or 2 weeks to get some of the growth benefits. This would result in fewer paddocks in the rotation.

Grazing height

Plants store their WSCs in their tiller bases (leaf stems and crowns), so grazing hard into the base of plants can remove these reserves and, in some species, can remove the growing points and lead to tiller or plant death.

Leaving some leaf stem material behind when stock are moved out of the paddock will aid plant recovery and growth rates.

It has been estimated that in an intensive grazing situation (dairy) with perennial ryegrass, leaving 5 cm of stem material behind after grazing can increase pasture growth rates by at least 10%

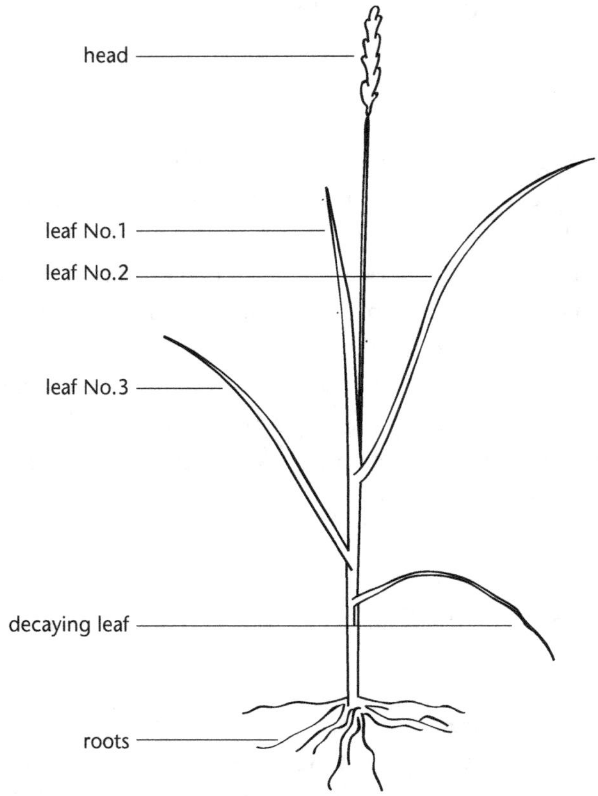

Figure 2.13. The number of live leaves per tiller for perennial ryegrass.

compared with leaving 2 cm. (Note: these heights do not directly correspond with all other perennial grass species or the structure of grasses when they are grazed by sheep.)

In most sheep pastures, feed is limiting in autumn and early winter, so it can be difficult to leave a target amount of pasture behind. If autumn and early winter rotation management is solely focused on grazing height, the rest period and time in paddock would be compromised.

Results from a grazing trial in north-east Victoria

The impacts that three grazing systems had on both stocking rates and pasture composition are well illustrated in the results of from a grazing trial at Broadford in north-east Victoria. The trial was on Sirosa phalaris pastures and both low (maintenance) and high (capital and maintenance) levels of fertiliser were compared, as well as set stocking, simple and intensive rotations. At the start of the trial the pasture only had seven phalaris plants per square metre.

The following systems were evaluated:

- **Set stocked/continuous grazing**
- **Simple rotation** – four paddocks, shifts based on rigid times. Stock in paddocks for 3 weeks during summer (70+ days rest), 2 weeks during autumn and winter (42 days rest) and 1 week during spring (21 days rest). This system ran up to 10% more stock than set stocked paddocks and requires less management skill than the more intensive rotations.

- **Intensive rotation** – 20 paddocks, stock moved every 3 days, shifts based on time for phalaris to regrow to four-leaf stage. This system ran up to 20% more stock than the set stocked paddocks.

Figure 2.14 shows the change in stocking rates over time under all six treatments. Note that the grazing management treatments did not start until 1997. Figure 2.14 highlights that the response to phosphorus fertiliser was almost immediate, and the stocking rates on plots with two different fertiliser rates diverged quickly. After 1997, the grazing treatments took 2 to 3 years to show an effect and after 3 to 4 years the stocking rate in the intensive system with low phosphorus (maintained Olsen P of 8) was similar to the high phosphorus/set stocked treatments. This was because, over time, the phalaris content increased dramatically, increasing total pasture yields, and the nutrient distribution and recycling improved under the rotation. Therefore, for degraded perennial pastures, fertiliser can give quick responses and allow stocking rates to be quickly increased, but it may take 2 to 3 years for the improved grazing to increase the perennial grass content, and get further lifts in stocking rate.

Effect on pasture composition

Set stocking resulted in:

- more sub clover and capeweed than grasses
- small phalaris plants
- large sheep camps
- large bare areas in late summer and autumn.

This change in composition indicates that these pastures were overstocked (despite good condition of stock) and it should be noted that these paddocks were carrying well over the local stocking rates.

Simple rotation resulted in:

- larger phalaris plants but capeweed and stock camps were still obvious.

Intensive rotation resulted in:

- large phalaris plants
- much more ground cover
- less capeweed
- stock camps not obvious.

The root depth of phalaris was greater in the intensive rotation and the nutrient spread was more even across the paddock, with no obvious stock camps. The increased grass content, better root systems, higher pasture yields and more even spread of nutrients all contributed to reducing the maintenance requirement for phosphorus. The maintenance rate of P was estimated to go from 0.6 kg P/DSE under set stocking to 0.4 kg P/DSE under the intensive rotation.

Putting it together

At different times of the year it can be difficult to manage a grazing rotation to optimise all three factors (rest, grazing time and grazing height) in order to optimise pasture growth rates, and

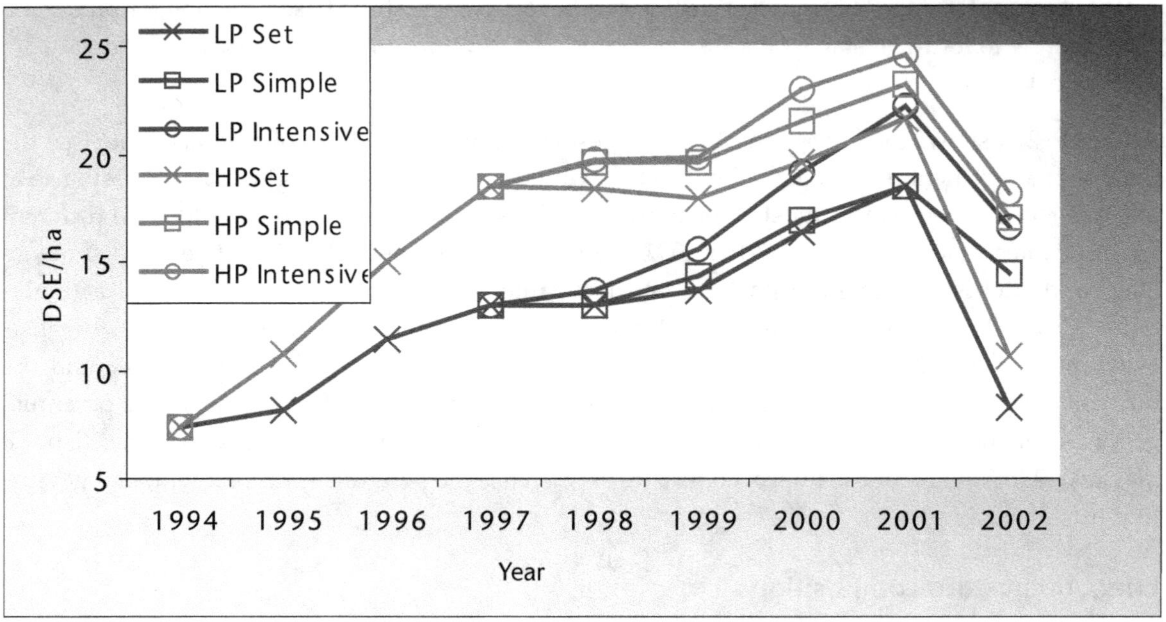

Figure 2.14. Stocking rate changes at Broadford for three grazing systems with two fertiliser treatments.

compromises have to be made. However, it is more important to aim to get the rest period right, and make compromises on the grazing time and grazing height. In seasons when ground cover drops below target levels, such as in drought, the stock should be placed in containment areas to protect the pasture and soil (see Chapter 4).

The other major consideration is livestock performance targets. Flock structure (the proportion of wethers, ewes and weaners), time of lambing and markets (wool and meat) will also determine different target pasture requirements for different times of the year (see Chapter 3). All these factors need to be considered when making decisions about grazing times and grazing heights.

Important considerations for grazing management

Managing how animals graze pastures has the potential to increase pasture growth by optimising leaf area and ensuring plant carbohydrate reserves are not depleted. The feed requirements of the animals are paramount, and flexibility and common sense must be used. The keys to success are as follows:

- If using a rotation system, the needs of the animals must be balanced with the needs of the pasture.
- There is little financial benefit in confining stock to a small area of the farm and feeding expensive supplements simply to allow plants to reach the 'optimum' leaf stage or to establish a 'wedge' in years with a poor autumn break. This practice is known as 'autumn saving' of pasture and research has shown that it is not economic. The exceptions are:
  - after drought, when stock may be in a stock containment area or drought-lot (delaying release of stock from drought-lot can help drought affected pastures recover)
  - on mixed farms where stubbles can be grazed and pastures allowed to get away in autumn.
- Grasses can dominate clover under any rotational grazing system; speeding up the rotation in autumn or set stocking can improve clover content.

- Closing up a thick germination of silver grass and barley grass for 6 weeks in autumn will mean that these species will dominate for the next 12 months.
- In spring, animals should not be moved into the next paddock in the rotation, simply because it 'is next', if another paddock needs grazing more. Poor quality silver grass, barley grass, fog and bent grass paddocks should be eaten out while they still have some nutritional value and the improved grasses/clover paddocks left for later.
- The spring flush can increase by up to 50 to 100% under rotational grazing, if grasses become more dominant, particularly in a good season. So, strategies are needed to use it, or quality will decline rapidly and next year's growth will be reduced.
- Phalaris toxicity (or sudden death) can be a problem if rotationally grazing. There may be more phalaris in the pasture, it may have accumulated more toxins, stock may be hungry after eating out the last paddock and they may not be used to green grass or phalaris. Phalaris toxicity can occur in autumn/winter and seems to be associated with periods of moisture stress (for example, a poor autumn break with patchy rainfall events). If moving stock onto potentially toxic phalaris pastures, stock should be moved in the afternoon and fed hay if hungry. Alternatively, a test mob of 30 sheep should be put in the paddock and observed over the next 2 to 3 days, to determine if it is safe, before the rest of the mob is put in the paddock. (Note: Phalaris toxicity is different to phalaris staggers. The risk of phalaris staggers can be reduced by using cobalt pellets in sheep. The risk of phalaris staggers is greater with the older varieties of phalaris and in districts with cobalt deficiencies.)
- Rotationally grazing while lambing is not generally recommended because an unacceptable level of mismothering can occur. Ewes can be set stocked 2 to 3 weeks before lambing.
- Any rotational grazing system will improve the persistence of winter-active phalaris cultivars in marginal soils. In good basalt country, in the higher rainfall areas with a long growing season, where phalaris thrives, there is less benefit, and rotational grazing may exacerbate phalaris dominance problems and reduce clover content.
- There is local evidence to support New Zealand observations that white clover and fescue will out-compete bent grass when rotationally grazed.
- The more intensive the rotation, the less sub clover will be in the pasture (although white clover may increase). If sub clover is very dominant, this may not matter, and may even be an advantage. Grasses may become dominant in any rotational system and the build up of grass weed species may create feed-quality problems if there are inadequate desirable perennial grasses to benefit from the rotation. Grass that has been allowed to become rank will also decline in quality and animal performance may well be reduced.

Key messages – grazing management

- Managing how animals graze pastures has the potential to increase pasture growth by optimising leaf area and ensuring plant carbohydrate reserves are not depleted.
- Managing the grazing of pastures, by knowing the growth habits and characteristics of both desirable and undesirable species, also has the potential to manipulate and improve pasture composition. Ground cover over summer can also be increased.
- Pastures should not be managed with only the pastures in mind because the requirements of the animals are paramount.

- Rotational grazing requires good animal management skills as there can be risks associated with wool strength, pasture composition, phalaris toxicity and reduced per head production if not done properly.
- Simple rotations are an easy way to start and can improve pasture utilisation, productivity and profitability.

CHAPTER 3

Nutrition and feed management

The most crucial aspect of animal production systems is the management of feed. Animal requirements need to be matched with the available feed supply from pastures and crops, and conserved and purchased feedstuffs.

The principles outlined in this chapter aim to improve the decision making when assessing feed requirements of a sheep flock. Chapter 4 outlines the particular feed and management issues associated with drought.

The nutritional requirements of livestock

Rumen structure and function

Sheep are ruminants, which means that they have a forestomach designed to use cellulose, the largest source of energy available in plants. The forestomach consists of the reticulum, rumen and omasum.

The rumen is the largest compartment of the forestomach and is lined by absorptive papillae and is partly subdivided by folds. It can hold up to 7 litres of food and fluid. It connects with the reticulum and it covers most of the floor of the abdomen on the left side. The reticulum, also called the 'honeycomb' because of the hexagonal pattern of its mucous membrane, connects with the oesophagus (the gullet) and to the omasum. The omasum is the smallest compartment of the forestomach and connects with the abomasum. The omasum is also known as the 'bible' because of its many leaves. The abomasum, sometimes referred to as the true or fourth stomach, is an elongated sac, which is comparable in structure and function to the stomach of non-ruminants.

The rumen acts as a fermentation vat and provides an anaerobic environment (no oxygen), and operates at constant temperature and pH. The pH of the rumen is 6 to 7 when animals are on a pasture, and 5.5 to 6 when on a grain diet. The forestomach has specialised contraction patterns that serve to mix and regurgitate food. Rumination consists of regurgitation of 'cud' (a bolus of food) for re-chewing and re-swallowing. This allows for further physical breakdown of feed, with the addition of large quantities of saliva, which also acts as a buffer. Time devoted to rumination is determined by the nature of the diet.

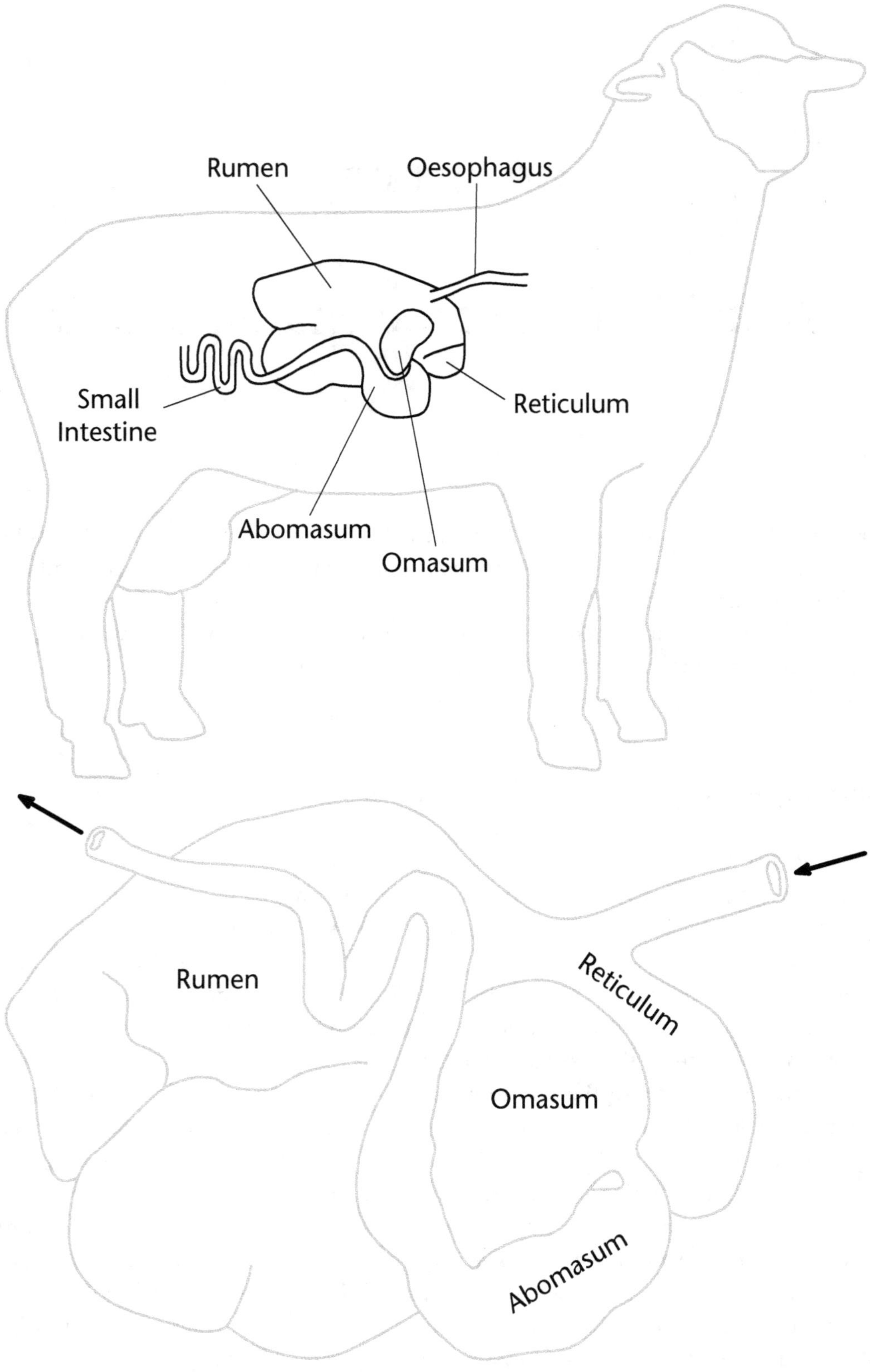

Figure 3.1. The stomachs of a sheep.

Cellulose is the main structural carbohydrate in plants, but it cannot be digested by most animals. The forestomach of ruminants is inhabited by a large population of micro-organisms, which maintain a stable system and break down, via fermentation, the plant materials eaten. The products of fermentation are then available to both the animal and the micro-organisms. Rumen micro-organisms are also able to use nitrogen to make new protein that is then available to the animal and also to synthesise many vitamins required by the animal. Ingested material continues to be broken down in the omasum and both feed and microbial nutrients are further broken down and absorbed in the abomasum and small intestine. Undigested material passes into the large intestine and eventually is excreted as faeces. Fermentation in the rumen also produces carbon dioxide and methane, which are both greenhouse gases.

The value of this system is that sheep are able to use relatively poor quality feed and access the abundant energy available in pastures. The downside is that the symbiotic relationship between the micro-organisms and the animal requires that the rumen needs time to adjust to changes in feed types. Therefore, when considering the nutritional requirements of the sheep, both the animal and the microbial population of its rumen must be taken into account.

Nutritional requirements can be broadly classified into:

- energy
- protein
- minerals
- vitamins
- fibre
- water.

Nutritional requirements vary with live weight, level of production, physiological status (dry, pregnant or lactating), climatic conditions and body condition.

Energy

Energy is the most important requirement for stock and is the most common limitation to animal production in southern Australia. An animal's requirement for energy is expressed as metabolisable energy (ME) and is measured in megajoules (MJ). One megajoule is equivalent to 1000 kilojoules.

To estimate the nutritional requirements of different types and sizes of animals approximately and easily, metabolisable energy can be expressed in 'dry sheep equivalents' (DSE). A DSE is the energy required to maintain the body weight of a 2-year-old, non-lactating sheep in condition score 3. The weight of the 'standard' dry sheep is debatable and not always consistent. Traditionally, one DSE referred to a 45 kg sheep, with a metabolisable energy requirement for maintenance of 7.4 MJ per day. A 'standard' dry sheep weight of 50 kg is now commonly used, at 8.5 MJ/day.

The actual amount of energy that a dry sheep requires for maintenance can vary because it is affected by factors such as feed type and paddock size (or the need to exercise). For example, sheep need more energy to break down the fibrous material in the feed source, so sheep on low-quality feeds (containing more fibrous material) have a higher energy requirement than sheep on high-quality diets (containing less fibrous material). Sheep that need to walk long distances for feed and water will have a higher energy requirement compared with those that do not.

Table 3.1. Dry sheep equivalents for different classes of stock on a daily feed requirement basis.

Class of stock	DSE at specified live weights		
Weaner lambs	**20 kg**	**25 kg**	**35 kg**
gaining 50 g/day	0.6	0.8	0.9
gaining 100 g/day	0.8	1.0	1.2
gaining 200 g/day	1.2	1.4	1.6
Mature sheep	**Medium Merino (50 kg)**	**Large Merino or crossbred (60 kg)**	
Dry ewes[a] and wethers			
maintaining weight	1.0	1.2	
gaining 100 g/day	1.7	2	
Lactating and pregnant ewes			
Late pregnant ewes (singles)[b]	1.2	1.4	
Late pregnant ewes (twins)[b]	1.4	1.6	
Ewes with lambs at foot (singles – twins)	2.5–3.4	3–4.1	
Dry cows and steers	**400 kg**	**500 kg**	
maintaining weight	8	9	

[a]pregnant ewe energy requirements do not change significantly in early pregnancy, so they are generally classed as dry until the last 6 weeks of pregnancy

[b]average requirement during the last 6 weeks of pregnancy

A general guide, taking into consideration traditional and current 'standard' dry sheep weights, one DSE is equivalent to between 7 and 8.5 MJ of ME per day.

Common uses of the DSE system include:

- to compare the stocking rate between properties with different enterprises (because DSE ratings can be applied to other livestock)
- to compare and estimate the feed requirements of different classes of stock
- to describe the carrying capacity or potential productivity of grazing land.

Table 3.1 provides estimates of DSE ratings for different classes of stock and can be used to estimate their energy requirements. Energy requirements increase for larger animals and for animals that need to grow or rear lambs. For example, a medium Merino ewe when dry is 1 DSE (requiring about 8 MJ ME/day to maintain weight), but will increase to approximately 3 DSE (requiring up to 24 MJ ME/day) when feeding twin lambs.

Weather effects

Cold temperatures, especially combined with wind and/or rain, can significantly increase energy requirements. Bare shorn and thin animals will have a higher requirement for energy at lower temperatures. Cold temperatures can increase energy requirements by as much as 100% and so this should be taken into account when setting pasture targets or supplementary feed rations. In one trial, ewes that were 5 weeks off shears, increased their feed intake by 44% when the temperature fell to 8°C. It has been estimated that this impact could reduce winter carrying capacity by 18 to 40% if shearing occurs in winter in cold climates. Fat reserves and wool length will help reduce the need for extra energy to generate warmth. Newborn lambs are particularly prone to chilling owing to their small size and relatively large surface area. Cold storms in the warmer months of the year pose a great risk to shorn sheep because sheep are not metabolically adapted to cold temperatures during the warmer months.

High temperatures are thought to reduce appetite, and therefore intake; however, indoor experiments with lot-fed sheep have not measured the reduction in feed intake associated with hot weather as seen in beef cattle. However, there is variation between sheep (some sheep do reduce feed intake in hot weather while others do not).

Sources of energy

An understanding of the different sources of energy is useful, because the rumen micro-organisms need to adapt to different forms of energy. A sudden change in some energy sources can lead to severe problems.

Carbohydrates: Carbohydrates are the main component of the dry matter of plants and consist of water-soluble carbohydrates and sugars, starch, cellulose and lignin. The water-soluble sugars are the most digestible and the lignin component is largely indigestible. Other forms of energy are provided in the form of fats or oils.

Water-soluble carbohydrates are the total soluble sugars present in forages and include glucose, fructose, sucrose and complex sugars, which are all almost completely digestible. Pastures contain only small amounts of these sugars. As plants mature, the proportion of lignin, cellulose and water soluble carbohydrates change and this affects their nutritive value. The water-soluble carbohydrate component is slightly higher in spring than in autumn and reduces as plants mature. Sugars are not commonly used to supplement the diets of sheep, but, if used, a period of adaptation is required.

Cellulose is the most common source of energy supplied by pastures, grass, hay and stubbles. The micro-organisms in the rumen enable sheep to digest cellulose.

Lignin is a structural plant carbohydrate that is indigestible. As plants mature, the lignin component increases.

Starch is the main source of energy in cereal grains. However, the micro-organisms that process starch are different to those that process cellulose. Hence, the introduction of cereals must be slow to enable the rumen micro-organisms to adapt.

Fats/oils: Fats or oils are not common sources of energy for ruminants. They generally have an energy value about 2.25 times that of carbohydrates. However, fat levels at or above 10% will affect microbial fermentation and feed intake will be reduced. Some of the oilseeds, and their by-products, can have high oil levels.

Protein

In addition to energy, sheep require protein. Dietary protein is measured as 'crude protein' (CP), which is the percentage of protein (by weight) in the feed. The digestion of protein by rumen

Table 3.2. Protein requirements for growing lambs and ewes.

Stock	Weight gain (g/day)	Dry sheep equivalent (DSE)	Approximate protein requirement CP (%)
Wether (45 kg)	0	1	7–8
	100	1.7	9–11
Weaner		0.8	16
20 kg	200	1.2	18
25 kg	0	0.7	9–11
25 kg	100	1.0	12–14
35 kg	0	0.9	8–10
35 kg	100	1.1	12–14
Finishing lamb (35 kg)	>200	2.6	15–18
Ewe – late pregnancy (50 kg)	0	1.4	7–9
Ewe – with lamb (50 kg)	0	3	11–14

micro-organisms is much faster than the digestion of cellulose and, as a result, high protein levels in pasture are more likely to be used by rumen micro-organisms and not directly by the sheep itself. For maintenance, mature, dry stock need crude protein to comprise 7 to 8% of the diet. Most pasture and commonly used feeds usually contain adequate protein. If crude protein is below 7% of the diet (such as in poor quality oats, hay, straw and mature pasture with no green pick), feed intake will be reduced and the animal may lose weight. The requirements for protein increase with the need to grow or to produce milk, as shown in Table 3.2. Requirements for energy must be considered and met first. High protein diets can be unnecessarily costly and wasteful.

Fibre

Fibre, or roughage, is not a nutrient as such but has physiological benefits and is required for a stable digestive system. Fibre promotes saliva production, which helps to maintain pH, rumen activity ('motility') and stable rumen fermentation, resulting in better digestion of nutrients.

Fibre includes cellulose, hemicellulose and lignin. Ruminants can digest cellulose and hemicellulose, but not lignin. Generally, as plants mature the fibre content increases but the digestibility of the fibre decreases. The best measure of fibre for ruminants is neutral detergent fibre (NDF), which is a measure of all of the components of fibre.

When sheep are grazing pasture or crop stubble, a fibre supplement is unlikely to be of any benefit. Cereal grains are very low in fibre and a source of fibre is desirable in a high-grain diet and will reduce the likelihood of acidosis. Where sheep are confined or shedded (as with Sharlae sheep), and are therefore unable to graze, the use of additional roughage has reduced the incidence of wool biting and therefore may also contribute to other behavioural requirements.

Minerals

Sheep require certain minerals in their diet. These are classified as either macro-minerals or trace elements. In most cases, the mineral requirements are obtained from the pasture and soil, except in areas that have known trace element deficiencies (see Chapter 6). It has been estimated that mineral and vitamin deficiencies probably cause less than 2% of overall loss of production in the sheep industry. Pregnant and lactating animals are the most susceptible to inadequate macro mineral nutrition. Young growing lambs are more susceptible to trace element deficiencies, and also calcium deficiency.

Macro-minerals

Calcium is needed for bone growth, muscle contraction, growth of the unborn lamb and milk production. Grazing sheep normally receive a plentiful supply of calcium from pasture. Legumes are particularly rich in calcium. Cereal grains are generally deficient in calcium. Low blood concentrations of calcium cause milk fever (hypocalcaemia), which most commonly occurs when pregnant or lactating ewes are deprived of food (see Chapter 6). Young sheep with poor calcium intake can suffer from poor growth and bone deformities.

Magnesium is needed for muscle function and milk production. Magnesium is usually sufficient in grazed pastures. Deficiency may cause grass tetany, which is mainly a problem in cattle rather than sheep. Grass tetany may occur on short, young shallow-rooted pastures with high potash and nitrogen levels.

Phosphorus is needed by sheep in balance with calcium. Phosphorus deficiencies are seldom seen in sheep in southern Australia, but a phosphorus–calcium imbalance, where there is too much phosphorus for the amount of calcium available, can lead to milk fever or other diseases.

Sodium is seldom limiting except in grain diets and can easily be supplied as salt.

Sulphur is required for wool growth, but intake in southern Australia is usually adequate because responses to sulphur supplementation are seldom seen.

The other macro-minerals are chlorine and potassium.

Trace elements

The trace elements are iron, zinc, manganese, copper, cobalt, iodine, molybdenum and selenium. They are required in very small quantities: hence the term 'trace' element. Iron and zinc are unlikely to be deficient in sheep diets. In some areas, copper, cobalt, selenium and iodine can be deficient and these are discussed more fully in Chapter 6.

Mineral nutrition will vary depending on soil types and can also vary throughout the year, depending on mineral uptake and seasonal growth patterns of pastures, how much soil the stock ingest and the mineral content of the water supply.

Vitamins

Most vitamins required by the sheep are made by the rumen micro-organisms. Vitamin A and Vitamin E are sourced from green plants, so deficiencies have been reported during prolonged drought, especially in young sheep, and supplementation is recommended for sheep in feedlots. Vitamin D is produced by the action of sunlight on the skin and so Vitamin D is needed when sheep are shedded for long periods. Hay, rather than green grass, contains vitamin D (owing to the action of the sunlight during the curing).

Water

Water is not a nutrient source (although it may supply some minerals), but nevertheless it is an essential requirement. Sheep will drink about 3 to 5 litres per head per day in hot weather, but up

Table 3.3. Recommended minimum mineral concentrations in pasture dry matter (DM) for grazing livestock.

Mineral	Minimum concentration required for sheep	Comments and toxicity
Major elements	**g/kg DM**	
Calcium	1.4–4	Low in grains. Decreases as pastures mature, but increases as pastures dry and die. Deficiencies on pasture are not common; most seen in lactating ewes
Phosphorus	0.9–3.0	Need to consider P and Ca together Should be in proportion 1:1 (Ca:P)
Magnesium	0.9–1.2	
Sodium	0.7–1.0	
Potassium	5.0	
Sulphur	2.0	Should not exceed 4% DM from all sources
Trace elements	**mg/kg DM**	
Copper	5.0 (4–14)	Copper can be affected adversely by concentrations of molybdenum (>2 mg/kg DM); by sulphur at >3 g/kg DM; by iron >500 mg/kg DM; by zinc >100 mg/kg DM; or by cadmium >5 mg/kg DM. Liver damage caused by pyrrolizidines in weeds such as heliotrope, ragwort and Paterson's curse can result in copper toxicity. Toxic: 20–50 mg/kg DM for lambs or 130 mg/kg for adults
Cobalt	0.11	Doses in excess of 40 mg/kg may cause death
Selenium	0.05	Toxic at levels greater than 5 mg/kg DM
Iodine	0.5	2.0 in presence of goitrogens Can be toxic (do not give >8 mg/kg DM)
Zinc	20–30 (9–20)	
Manganese	20–25	Levels over 2000 mg/kg DM are toxic
Iron	40	Levels over 2000 mg/kg DM are toxic

Note: These amounts represent the average requirements for growth, pregnancy and lactation. When a range is given, the higher value is for rapidly growing or lactating animals and the lower values for maintenance and a low level of production.

to 9 litres in extreme weather. In winter, dry stock may get all of their water requirements from grass, because lush green grass contains 70 to 90% water. Sheep have quite good tolerances of salt in water, but it is necessary to test new sources of water, particularly bore water. When testing water for stock, the main considerations are pH, salinity and chloride levels. The pH of water should be in the range of 6.5 to 8.5.

Salinity actually refers to all mineral salts present in the water, including sodium, calcium, magnesium, chloride, sulphate and carbonate. Salinity is expressed as electrical conductivity (EC) in units of microsiemens per centimetre (µS/cm). Tolerance to salinity depends on the breed, age and production level of stock and the water and mineral content of the feed. Magnesium content will reduce tolerance at certain salinity levels, as shown in Table 3.4.

Excess chloride will reduce intake through a decrease in microbial population in the rumen. Excess sodium chloride can lead to dehydration, kidney failure, nervous system dysfunction and death. The maximum acceptable level of chloride in water for ewes and lambs is 2400 mg/litre and for adult dry sheep 5600 mg/litre.

Table 3.4. Maximum acceptable salinity levels for different classes of livestock.

Electrical conductivity (µS/cm)	Magnesium (ppm or mg/L)	Suitable for:
<8500	<400	All classes of sheep (production declines for lactating ewes/young stock >6000 µS/cm)
8500–16 500	<600	Adult sheep (production declines >9300 µS/cm). Caution required for lactating ewes; unsuitable for young stock
21 800	<600	Upper limit for adult dry sheep

Blue green algae in water can be a problem in dry warm summers where water flow is restricted. Many farm dams are conducive to algal blooms, but not all blooms are toxic to stock.

Key messages – sheep nutritional requirements

- Sheep are ruminants, which gives them the ability to digest the main carbohydrate in grass (cellulose).
- Energy and protein are the main nutrient requirements of sheep.
- Energy is measured in megajoules of metabolisable energy (MJ ME).
- Protein is measured as the percentage of crude protein in the diet (CP per cent).
- A sheep's energy and protein requirements depend on age, body weight and whether or not the animal needs to grow, or produce and feed a lamb.
- A dry sheep equivalent (DSE) is the energy required to maintain the body weight of a 2-year-old, non-lactating sheep in condition score 3 and is approximately 7–8 MJ ME/day. A lactating ewe needs a daily energy intake equivalent to two-and-a-half to four dry sheep.
- A diet with 7 to 8% crude protein is adequate for dry sheep.
- Fibre and vitamins are rarely limiting in pasture based diets. Mineral deficiencies can occur on some soil types.

Assessing the feed supply

For most of the year, sheep in southern Australia source their feed requirements from pasture. Both the quantity and quality of the pasture available is important for assessing whether the supply will be adequate for the animals' needs. The ability to accurately assess pasture quantity and quality can be easily learnt. The annual growth pattern of pasture will be an important determinant of the potential numbers of stock that can be carried without the need for significant supplementary feeding.

Pasture supply

The quantity of feed grown and used is the key driver of production and profit in any grazing business. Feed grown varies widely over individual farms, but broadly depends on environmental factors such as rainfall, soil, fertility and pasture species.

Pasture quantity or supply is measured in kilograms of dry matter per hectare (kg DM/ha). It is also referred to as 'pasture (or feed) availability' or 'feed on offer' (FOO). This is the average amount of feed, with the water removed, present in each hectare of pasture. Pasture growth rates are

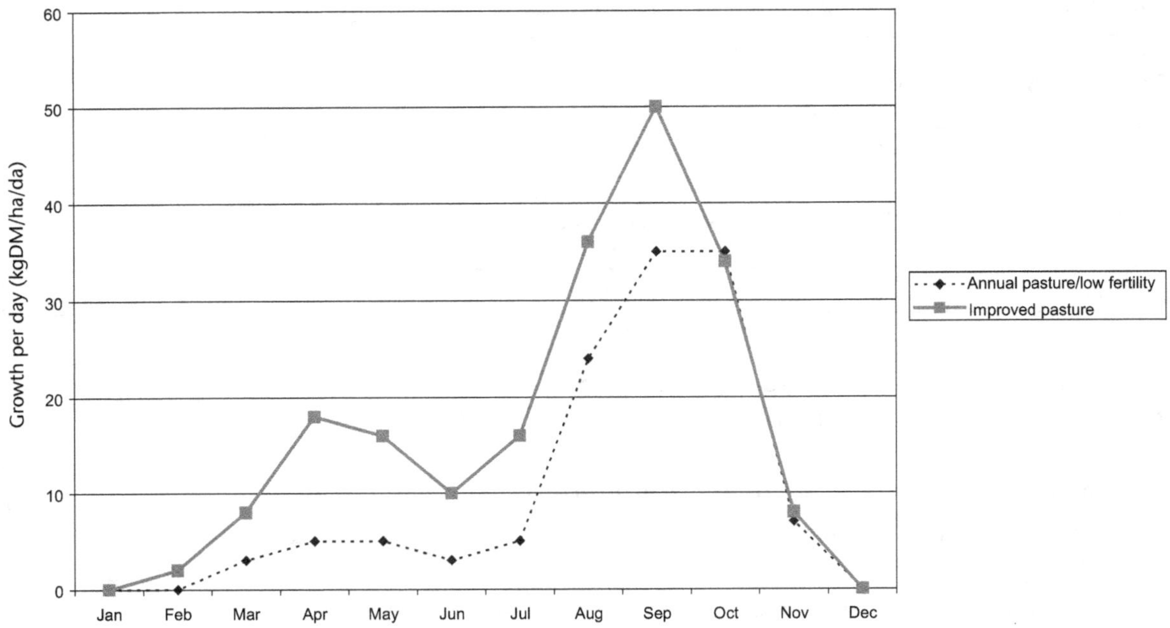

Figure 3.2. Example of pasture growth rates for low fertility annual and high fertility perennial pastures in Central Victoria.

measured in kilograms of dry matter grown per hectare per day. Rates after the autumn break commonly vary from 10 kg DM/ha/day in winter to 70–100 kg DM/ha/day in spring. Figure 3.2 illustrates an example of annual pasture growth for a low-fertility annual pasture compared with a highly fertile perennial pasture in one environment. Examples for a range of environments are given in Chapter 2.

As Figure 3.2 illustrates, the predominance of growth occurs over 2 to 3 months of the year during spring, but the length of the growing season and height of growth can vary quite significantly. With green pastures, there is a reasonably reliable relationship between pasture quantity and average pasture height, so that available feed can be assessed by measuring the average pasture height. Height is measured in centimetres, with the first half centimetre being assumed to be unavailable to stock. Figure 3.3 illustrates how to measure pasture height using a simple stick. To measure the average pasture height of the paddock, 50 random height measurements are made, ensuring that representative areas of the entire paddock are sampled. The stick is thrown in front of the measurer and the base of the thumb is run down the stick until it touches the first green leaf (not stalk or dead leaf). If the stick lands in a bare area, a zero is recorded. Similarly, plants that stock obviously do not eat, such as thistles and onion weed, should be ignored and a zero recorded for that measurement. Appendix 2 provides an example of a recording sheet used to estimate this. Although this may seem a laborious task, with practice, average pasture height can be assessed visually without measurement and immediately related to feed availability.

The average height of the green feed can then be converted to the amount, in kg DM/ha using Table 3.5. This is also referred to as food on offer (FOO). A simple guide for a moderately dense sheep pasture is that the first centimetre is equivalent to 500 kg DM/ha and every centimetre

Figure 3.3. Estimating the height of green pasture using a pasture stick.

thereafter is equal to 300 kg DM/ha (until the density drops). Using this rule of thumb, a pasture of 4 cm in height would be equivalent to 1400 kg DM/ha (1 × 500 kg DM/ha plus 3 × 300 kg DM/ha). However, there will be some variance in the density of pastures because rotationally grazed pastures tend to be less dense, so some compensation should be made, as shown in Table 3.5. This method will also be inaccurate for very sparse pastures.

Table 3.5. Feed availability (in kgDM/ha) estimated by height of green pasture and pasture density.

Height (cm)	Dense pasture (kg DM/ha)	Moderate density pasture (kg DM/ha)	Clover dominant (kg DM/ha)
1	400	500	700
2	700	800	1000
3	1000	1100	1500
4	1200	1400	2000
5	1400	1700	
6	1600	2000	
7	1750	2300	
8	1900	2600	
9	2000	2800	
10	2100	3000	

Table 3.6. Feed availability (kg DM/ha) required for maintenance of different classes of sheep grazing good quality green pasture.

	kg DM/ha good quality (75% digestibility)	Pasture height (cm)
Dry sheep	400	1
Pregnant ewes (last month)	800	2
Lactating ewes: single twin	 1000 1500	 3–5

As pastures mature and go to seed, or if they contain significant amounts of dry feed, this measure will no longer be accurate. The only way of getting a true estimate of available feed is then to cut some small samples, and dry and weigh them. This method is outlined in Appendix 3 because it can be handy for fine-tuning estimates of production for hay or crop.

Sheep can graze for a maximum of 13 hours per day because they must leave time for rumination. Height and density are important assessments of quantity for animal requirements, because they determine how much feed an animal can access in a bite. Hence, if feed is very short, the animal may not be able to access enough feed in the available time, irrespective of whether the stocking rate is low. Table 3.6 shows the target amounts of dense green leafy pasture required to maintain different classes of animals. The table shows that about 1 cm of dense green pasture will adequately maintain dry sheep, but if ewes are lambing they will require pasture to be above 3 cm high.

Feed quality

Pasture quality is described by digestibility (%) or metabolisable energy (measured as MJ/kg DM) and is therefore an estimate of the energy value available to the animal. It can be estimated by looking at the pasture species present, stage of growth and knowing the fertiliser history of the paddock. The quality of the pasture present will affect both how much an animal can eat and how animals will perform. On very poor quality pastures, many classes of stock will be unable to eat enough, even if the feed is abundant.

Digestibility

Digestibility is a measure of how much of the feed is retained and used by the animal. If the digestibility of pasture is 75%, then for every kilogram of dry matter eaten, 750 g is retained by the animal and 250 g is excreted. Highly digestible feeds move through the digestive system quickly, whereas feeds with low digestibility will take many hours to pass through the rumen.

The speed that feed passes through the rumen will have an impact on how much an animal can eat. The quicker the feed moves through the rumen, the faster the animal will be able to replace it. For example, molasses is highly digestible and will only stay in the rumen for about half an hour compared with straw, which may take up to 55 hours to digest. This explains why animals that are eating dry summer feed or hay appear 'fatter' than when grazing highly digestible grass, because the rumen is full of a large amount of material for long periods of time.

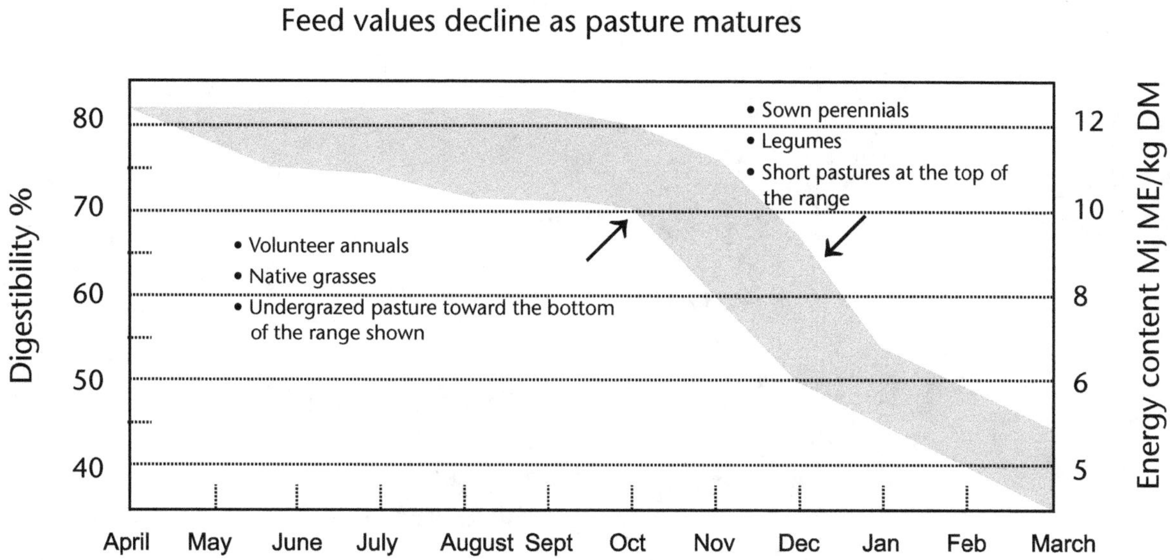

Figure 3.4. Change in energy and digestibility in pastures over the year in temperate Australia.

Therefore, the ability of the animal to maintain or increase weight on poorly digestible feed is low because of the small proportion of the feed that it can use and the very long time it takes to access energy from this feed.

As a rough rule, to be able to eat enough to maintain weight, a dry animal needs feed with a minimum digestibility of 55%. At 55% digestibility, nearly half the feed eaten is not used by the animal. This occurs from mid-summer on dry feed. Digestibility greater than 65% will generally meet the energy requirements of all classes of stock.

The energy value (ME) of a feed is calculated directly from the digestibility. Figure 3.4 shows how both the energy and the digestibility change with plant maturity over the year.

Stage of maturity – herbage is of highest quality during the vegetative stage from the autumn break until early spring. Once pastures start to become reproductive and send up seed heads, herbage quality declines, as shown in Figure 3.4.

Parts of the plant – in most plants, leaves are of higher digestibility than stems. The difference between the stem and leaf can be judged by how 'woody' the stem is; for example, there is a large difference (10–15%) in the digestibility of lucerne stem and leaf. In contrast, there is little or no difference in the digestibility of Persian clover leaves and stems. Figure 3.5 illustrates the range in digestibility of a lucerne plant.

Species – early in the growing season, there is little difference in the digestibility of different species. However, towards the end of the growing season, annual species decline in quality when they produce seed heads and die. Perennial grasses maintain quality longer and usually have some green material, particularly if rain occurs out of season.

Within the perennial grasses, tetraploid, hybrid, short-term or Italian ryegrasses are usually of the highest digestibility, followed by other improved grasses such as phalaris and perennial ryegrass, with unimproved species such as fog grass and sweet vernal of lowest quality.

Figure 3.5. Digestibility of different parts of a lucerne plant.

Clover – sheep and cattle are able to more efficiently use the energy provided by clover. In pastures with the same energy or digestibility, an increase in clover percentage will markedly increase animal production. For example, a pasture with 30% legume is expected to increase animal production by 20% compared with a similar quality pasture with no legume. Table 3.7 illustrates responses of lambs in a trial fed clover compared with grass.

Sheep are extremely good at selecting the most highly digestible parts of the pasture. This can make assessing pasture quality difficult, especially in summer, because what is seen is generally what the sheep have left behind because they will have selected a better diet than an average assessment of a pasture would indicate. This is particularly evident in a set stocked system where they able to select quite aggressively, and becomes less evident with the intensity of a grazing system.

Table 3.7. Performance of lambs fed grass or clover.

	Grass	Clover
Intake (g/day)	711	774
Live weight gain (g/day)	86	123
Wool growth (g/day)	9.4	12.6
Efficiency of conversion of ME for live weight gain (%)	33	51

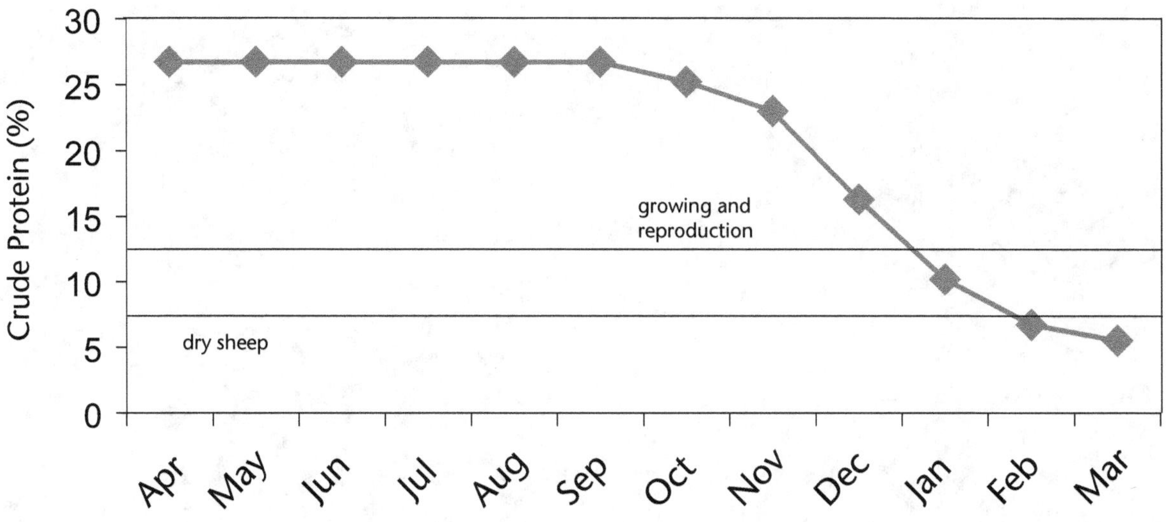

Figure 3.6. Change in crude protein in pastures over the year in temperate Australia.

Protein

Green feed is high in protein and a small amount of green pick will satisfy the minimum requirement of dry stock for 7 to 8% crude protein in their diet. Figure 3.6 illustrates the pattern of drop in protein levels of a pasture with the protein requirements for dry and growing animals.

Figure 3.6 illustrates that in a year with an early autumn break, protein requirements for dry stock are sufficient for most of the year and are insufficient only over the dry summer period for growing or reproducing animals. Irrigation, summer rainfall and summer-active plants such as lucerne that supply a green pick will fill the summer protein gap.

Table 3.8 provides pasture targets that would supply the energy and protein requirements for ewes over the various stages of pregnancy and lactation.

Note that on the poor-quality pasture (65% digestibility), animals with a high demand for energy and protein, such as ewes and lambs, cannot maintain weight, regardless of the quantity of

Table 3.8. Quantities (kgDM/ha) and qualities of pasture required for classes of sheep.

	High-quality pasture (kgDM/ha)	Moderate-quality pasture	Poor-quality pasture
	75% digestibility >30% legume <15% dead (rye/phalaris, etc.)	70% digestibility 15–30% legume 15–35% dead	65% digestibility 5% legume >50% dead (bent/onion/danthonia)
Dry Sheep	400	500	700
Pregnant ewes – Last month	800	1000	1500
Lactating ewes: **Wool sheep** **Prime lambs single** **Twin**	 1000 1000 1500	 1500 1500 2000	 not suitable not suitable not suitable

Figure 3.7. Example of 75% digestible pasture.

feed available. Animals on short green pastures need only half the quantity required on poorer quality pastures.

Assessing pasture quality

Figures 3.7, 3.8 and 3.9 show pasture varying in digestibility from 75% to 65%. They illustrate how pastures may look as they mature and as clover content decreases. Assessing pasture quality by eye can be learnt. Samples of pasture can be sent to accredited laboratories to be tested for feed value.

Fibre

Pastures are usually high in fibre content because fibre is the main component of cell walls of plants. As plants mature, the fibre content increases and becomes less digestible and reduces feed intake. The low fibre content of very lush, green pastures may increase the risk of milk fever in ewes because low fibre intake affects the way calcium is used in the body.

Crop stubbles

In many parts of southern Australia, crop stubbles are an important part of the feed supply for sheep over summer. The nutritive value of stubbles varies considerably and is primarily related to the amount of residual grain left after harvest, and also green weed component. Stubbles will vary considerably because they include grain, weeds, leaf, chaff and stem. Sheep will eat the most digestible feed first and leave the least digestible (the stem) until last, and weight loss will start to occur accordingly.

Figure 3.8. Example of 70% digestible pasture.

Figure 3.9. Example 65% digestible pasture.

Measurements made in trials suggest that sheep gain weight if there is more than 40 kg/ha of grain or 40 kg DM/ha of shot grain and/or weeds. These figures equate approximately to:

- **grain:** greater or equal to 15 grains/0.1 m^2 (the average intact grain heads contain approximately 16 grains)
- **shot grain/weeds:** greater or equal to 12 green shoots/0.1 m^2

(Note: 0.1 m^2 is a square area measuring 32 cm on each side.)

The true feed value of stubbles can vary widely, and careful monitoring of stock weight or body condition is always necessary. Weight changes have been reported from minus 176 g/head/day (after most of the grain is consumed) to gains of 65 g/head/day (when summer rains produced growth of green feed). There is a fine line between a useful crop stubble, where stock perform well, to one where an animal will lose weight. Lupin and legume stubbles provide a higher value grain and stubble, but lupin stubbles can be toxic to sheep when infected with a fungus, causing a disease called lupinosis.

Key messages – assessing the feed supply

- Pasture quantity, referred to as feed availability or feed on offer (FOO), is generally related to pasture height.
- Pasture quality for livestock is generally measured in terms of digestibility, which describes how much metabolisable energy a pasture can provide to the animal. Pasture quality usually declines throughout summer until the autumn break.
- In southern Australia, most pastures will supply adequate protein for 8 to 10 months of the year for dry animals, and longer if green feed is available.
- Assessing pasture height and quality provide a means of predicting animal performance.
- Pasture targets are available for different classes and requirements of stock.
- The quality of crop stubbles for animals varies enormously and will depend on residual grain and weeds, so animals need to be monitored.

Matching feed supply and animal requirements

Utilisation of pasture depends on stocking rate, stock type (sheep/cattle, ewes/wethers) and time of lambing or calving, sale of stock plus grazing management. The aim is not to fully use all pasture grown because this would lead to bare ground and poor pasture growth for much of the year. Estimated pasture utilisation by sheep flocks in southern Australia is only about 30 to 40% of feed grown. As shown in Figure 3.2, most of the feed is grown in the 6 months between May and October, with 70% or more of the total growth occurring in spring. Most strategies to increase pasture growth will also increase the spring flush and hence using spring feed is part of the challenge. Much of the feed built up over the spring is used to feed stock over summer.

Stocking rate

The stocking rate, as measured by the number of animals or dry sheep equivalents (DSE) per hectare, is one of the most critical factors influencing the profitability of a sheep enterprise. The stocking rate largely determines the animal production and the returns per unit area and often the return per

Table 3.9. Carrying capacity (DSE/ha) based on length of growing season (months), Olsen P (mg/kg) and paddock size.

Growing season (months)	5	6	7	8	9	10	11	12
DSE/ha – for paddocks less than 20 ha								
Olsen P 10 mg/kg	11	14	17	20	24	28	31	34
Olsen P 20 mg/kg	12	16	19	23	26	29	33	36
DSE/ha – for paddocks more than 20 ha								
Olsen P 10 mg/kg	8	11	15	18	21	25	28	32
Olsen P 20 mg/kg	10	13	16	20	23	27	30	33

unit of labour or on financial investment. Stocking rate affects individual animal performance by an amount that will vary from year to year and from one season of the year to another. During the height of the spring growth, a pasture will usually be greatly under-stocked, but the same stocking rate in mid-winter may impose severe demands upon the animal as well as the pasture.

Stocking rates vary from district to district, but also vary significantly within districts. This is predominantly because of differences in plant species, fertility, slope and attitude to risk. Trials in the 1950s were designed to estimate potential stocking rates for many districts; they ranged from 1 DSE (Victoria) to 1.3 DSE (Reg French, South Australia) per ha for every 25 mm rainfall above 250 mm. So, for an area that had an annual average rainfall of 650 mm, the potential stocking rate was estimated to be between 16 and 21 DSE/ha. A program run by the Grassland Society of Victoria (now Grasslands Society of Southern Australia), called the Grassland Productivity Program (GPP), set up trials over south-eastern Australia to help producers test and evaluate the most profitable stocking rates and management systems in their environment. The results across all sites showed the length of growing season, paddock size and soil phosphorus (see Chapter 2) improved the reliability of predicting carrying capacity (particularly for meat enterprises). Soil phosphorus level is important because it influences pasture growth potential and paddock size influences the ability to better use feed.

Table 3.9 shows how to estimate the **potential** stocking rate using growing season, soil phosphorus level (measured as Olsen P) and paddock size.

Table 3.9 provides some good benchmarks for **potential** carrying capacity over southern Australia. They are only likely to be achieved where there are no limitations to growth, such as fertility and soil structure problems or poorly productive pasture species. Most producers who participated in these trials could only reach potential stocking rates on parts of the farm not limited by slope and pasture species.

Low rainfall areas

In areas of low rainfall dominated by cropping systems, such as the Mallee and Wimmera, grazing systems tend to be based on annual pastures, complimented by crop stubbles for livestock over summer. In the Mallee, the growing season is estimated to be 6 months and crop stubbles provide the majority of the feed source to sheep. Studies have shown that ewes have maintained weight on stubbles from December to March, predominantly because of the contribution of summer-growing weeds. Therefore, cropping systems that have efficient harvesting processes (and hence leave less grain behind) and good summer weed control will have poorer chances of maintaining stock. A study in 2001–02 showed that farmers in the Mallee achieved stocking rates higher than traditional predictions (of 1 DSE/25 mm over 250 mm), but did not achieve the potential described in the Grassland Program estimates above, based on fertility and length of growing season.

The determination of the **optimum** stocking rate for an enterprise is therefore complex and must depend on several different criteria, which may conflict. These may include:

- the optimum potential production per unit area (or labour or investment)
- the maximum profits per unit area (or labour or investment)
- the stability and persistence of the pasture
- minimising stress on livestock
- the amount of supplementary feed required
- soil conservation
- the appearance of livestock.

The choice of an optimum must be a compromise between some of these factors. Most managers probably emphasise the profitability of the enterprise, the stability of the system (protection against soil erosion and droughts) and the appearance of their livestock. The compromise will depend a great deal on individual value judgements.

Stock type

The types of stock run have an impact on the capacity to use feed over the year. Ewes, as reproductive animals, have the advantage over dry stock in that their feed requirements change with the demands of pregnancy and lactation and this can be used to use and manage the feed supply. Wethers have similar requirements all the time, but can use spring feed to store fat reserves, which can be depleted over the summer months. It is unlikely that running a constant number of wethers throughout the year can use large amounts of spring feed without leading to overstocking in winter, and hence other means of increasing the stocking rate (such as agistment or cropping) may be required in such enterprises. Figure 3.10 shows an example of the monthly requirements of wethers (run at 10 DSE/ha) and also of ewes (also run at 10 DSE/ha or 7 ewes/ha) against a pasture supply curve.

Time of lambing

The time of lambing is critical in setting potential stocking rates. Generally, a late winter or spring lambing provides the best match between animal requirements and pasture production throughout the year. A rough guide for the best time of lambing is to aim to wean lambs when there is still 1 month of green feed remaining. This timing strikes a balance between matching peak ewe feed requirements with the time of peak pasture production, while still allowing time for ewes to recover condition between the end of lactation and the next joining, as well as avoiding weaning lambs onto low-protein dry pastures that will not meet their energy and protein requirements.

The best time of lambing is a balance between matching animal and pasture requirements throughout the year and optimising stock management and marketing. Prime lamb production

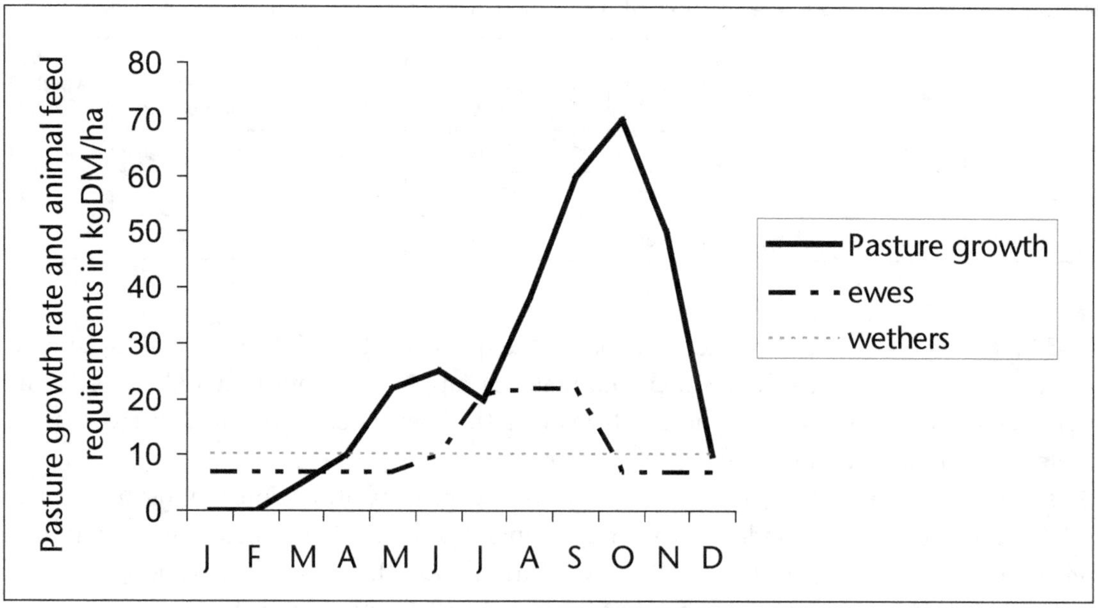

Figure 3.10. Energy requirements (as DSE/ha) of wethers and July lambing ewes at annual average stocking rate of 10 DSE/ha and pasture supply.

tends to favour an earlier lambing, in order to reach target sale weights for lambs at 4 to 5 months while still grazing green pasture. A wool flock, weaning 3 months after the start of lambing, can afford to lamb (and hence wean) later in the year compared with a prime lamb flock.

Figure 3.11 compares the feed requirements for lambing in April and July for crossbred ewes joined to Poll Dorset rams and running at six ewes/ha. The lines representing the DSE/ha for the two lambing times represent the feed requirements of the ewes over the year, which can be compared with the pasture supply curve. It is clear that the April lambing ewes, while not running out

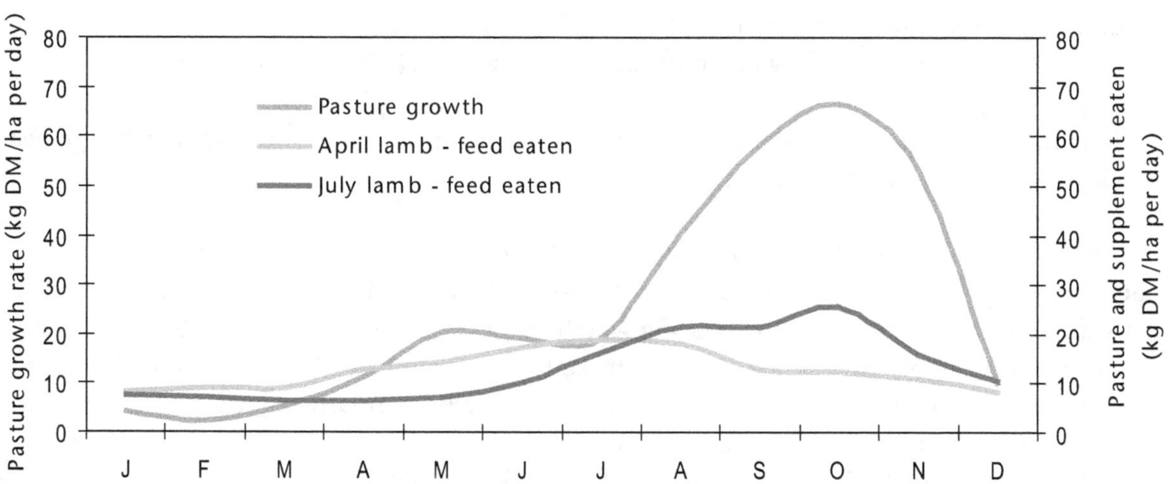

Figure 3.11. Feed requirements for ewes lambing in April and July (DSE/ha) and pasture availability.

Table 3.10. Estimated gross margins, feed costs and lamb returns for April and July lambing ewes at six ewes/ha.

		Average annual	Gross margin	Supplement costs	Weaning	Weight lamb sold	Live weight/ha sold
	Ewes/ha	DSE/ha	$/ha	$/ha	%	Live weight	Kg/ha
April lambing	6	13.3	219	$93	104	44kg	345
July lambing	6	13.6	362	$53	130	43kg	418

of feed in the 'average' year illustrated, do not use the spring flush, and the potential to increase the stocking rate is limited because feed demand matches supply in the autumn and winter. The July lambing ewes in this example use more of the spring flush and there is potential to increase stocking rate in the autumn and winter.

In setting a time of lambing, the important factors are pasture utilisation, the cost of supplementary feed, the stocking rate and lamb weaning weights. These decisions will also have an impact on other management decisions such as time of shearing and selling of stock. Although the values used for costs and income may not be current, Table 3.10 illustrates the relative gross margins estimated for the two lambing times. The July lambing group required less supplements and carried more lambs through to the sale weight resulting in better returns, more meat sold per hectare and a slightly higher stocking rate (because of the higher lambing rate).

Every district and enterprise will be different, but the principles are the same.

Grazing management

Grazing management can have an impact on both the growth and the utilisation of pastures. and also have beneficial impacts on weed control, environmental management and animal health, especially worm control. As discussed in Chapter 2, a change from a set stocked system to an intensive grazing system can allow considerable, sustainable increases in stocking rate. This is achieved primarily by:

- better utilisation of feed because stock are less able to select and hence bare out stock camps and leave large areas of rank feed
- increasing pasture growth, by maximising potential growth (with fertiliser) through maximising leaf area.

Rotational grazing requires extra labour and infrastructure and a greater level of management expertise compared with set stocking.

Rules of thumb for estimating optimum stocking rates

Target animal condition score and pasture composition at critical times of the year (particularly in early autumn) give some guide to assessing whether stocking rates are adequate. If animal condition or pasture supply is continually below these targets, then it is likely that the paddocks are overstocked. At the other end of the scale, if stock condition regularly significantly exceeds the minimum at the worst time of year and/or there is regularly a large bank of dry carry over feed at the autumn break, then it is likely that pastures are understocked.

Target – animal condition score

Dry stock should not fall below condition score 2–2½ at the poorest time of the year (late summer and/or winter) and should be above condition score 3½ at the end of spring.

Wethers should require no supplementary feeding in most (average) years.

Target pasture composition

Optimum stocking rates result in late autumn pastures (after the break) with the following characteristics:

<6–10%	bare ground
<10%	dead plant material
25–40%	sub clover
<30%	percentage of annual species in pasture
Negligible	winter grass (*Poa annua*)

Overstocking results in an increase of bare ground and of less-productive species such as winter grasses, capeweed and volunteer legumes such as trefoils.

Understocking results in species such as fog grass and large amounts of dry feed at the autumn break.

Feed budgets

Predicting rainfall and future weather patterns is extremely unreliable, but some simple feed budgeting can be very useful to predict whether current stocking rates and animal requirements are on track. This can identify possible opportunities when growth is good, and provide early warning signs to trigger decisions when pasture growth is poor.

Feed budgeting is simply making an estimate of how much feed will be available at some time in the future. It is therefore a function of knowing how much feed is currently available, adding on the expected pasture growth rate and deducting what the animals will eat, while allowing for some losses through trampling. Feed budgeting therefore requires estimates of:

- feed availability
- pasture intake
- pasture growth rates.

Estimating feed availability has been covered earlier in this chapter.

Estimating pasture intake

To predict animal performance, and to match animal requirements with pasture availability, some assessment of intake is required. With no limitations to the feed supply, an animal can eat a dry matter weight equivalent to 3 to 4% of its body weight per day. The higher intake (4%) is only possible for young or thin animals. As animals reach mature weight and increased fatness, possible intake reduces. Roughly, a 50 kg animal can eat about 1.5 kg of dry matter per day. A young growing lamb of 20 kg can eat about 0.8 kg dry matter per day.

Intake is further effected by feed availability (or height) and quality. When pastures are very short, intake is limited by the amount that can be harvested per mouthful and so maximum intake

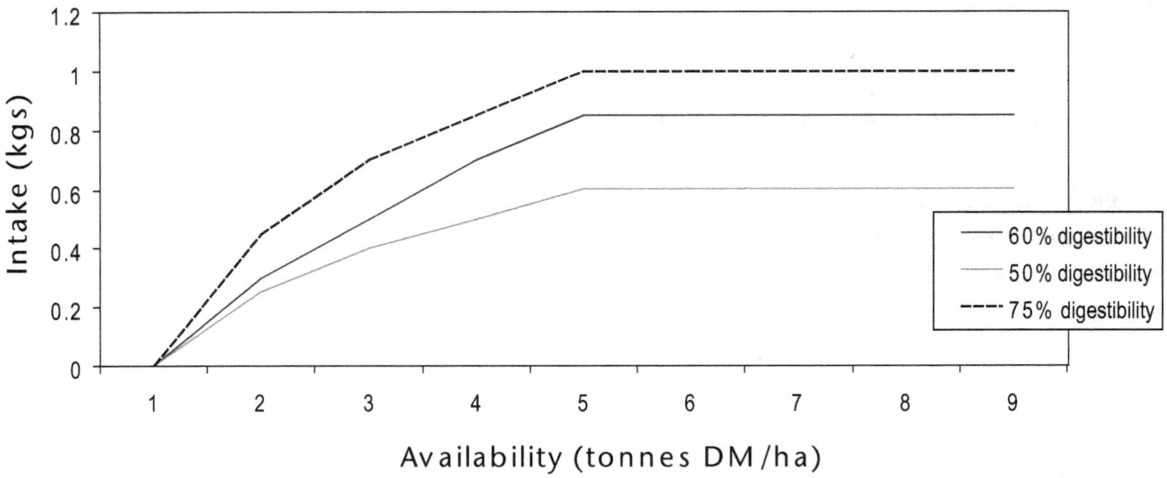

Figure 3.12. Intake affected by pasture availability and quality.

is not reached. When feed quality is low, as measured by digestibility, intake is affected by the time taken for low-quality feed to stay and move through the rumen. Figure 3.12 shows the effect on intake of both pasture height and digestibility.

Generally, pasture intake is not limited when there is more than 1500 kg DM/ha, or about 4.5 cm, of green feed in the paddock. Estimating intake is useful for predicting animal performance, as well as for feed budgeting.

A very simple guide is that, on average Victorian sheep pastures, intake is approximately 1 kilogram of dry matter per DSE and total pasture use is 1.2 kg DM/DSE (the extra 0.2 kg includes wastage due to trampling). This is an oversimplification because it will overestimate intake when pasture is very short or poor quality and will underestimate intake when pasture is in good supply. Nevertheless, this figure allows a quick and easy calculation of feed eaten throughout the year. Alternatively, a better estimate can be made using Figure 3.12. Monitoring feed availability will indicate whether this approximation is correct.

Estimating pasture growth rate

Table 3.11 provides a guide to pasture growth rates for southern Australia. Growth rates classed as 'very good' are more likely to be achieved in a high rainfall area with long growing season and productive pastures. Growth rates classed as 'poor' relate more to a native pasture with low fertility. These figures are indicative of 'average' years and hence are a guide only. Experience over time will give a manager a 'feel' for what is most likely on a particular property.

The following two scenarios provide examples of using feed budgets for determining:

- if there will be enough feed for ewes at lambing
- if there will be enough feed to get through summer.

Scenario 1 – Will there be enough feed for the ewes at lambing?

It is June and there is 500 kg DM/ha green feed in the paddock in which ewes will lamb. It is a 20 hectare paddock and is currently stocked with 100 ewes that will start lambing in July.

Table 3.11. Examples of pasture growth rates.

	Pasture Growth kg DM/ha/day			
	Poor	Average	Good	Very good
Jan	0	0	0	0
Feb	0	0	0	0
Mar	0	0	0	0
Apr	5	15	25	30
May	15	20	25	30
Jun	7	15	20	25
July	5	10	15	25
Aug	15	30	45	60
Sep	35	50	80	90
Oct	40	70	90	100
Nov	30	60	80	80
Dec	20	40	60	60

Pasture requirements – aiming for 1000 kg DM/ha in July and 1500 kg DM/ha in August to allow for some twin lambing ewes (see Table 3.8).

Will pasture targets be achieved if the paddock is continued to be stocked at this level and no other action is taken?

Table 3.12 shows the workings to estimate the end of month feed availability, allowing for pasture growth and feed eaten.

In this scenario, ewes require 1000 kg DM/ha by 1 July because they start lambing, but in this budget, we estimate that there will only be 680 kg DM/ha. This is because pasture growth is not much higher than animal intake.

The choices are:

- do nothing and suffer the penalties
- supplementary feed right through
- destock the lambing paddock.

Doing nothing will mean that the ewes will lose weight and may reduce milk production and hence lamb growth rates. Some weight loss may be acceptable depending on the condition of the ewes, but will incur penalties in poorer lamb performance. Very fat ewes or ewes with multiple lambs may also be at risk of pregnancy toxaemia.

Supplementary feeding would have to supply most of the ewe requirements to allow the pasture to reach target on time.

Destocking this paddock for a month would mean that nothing would be eaten, the grass would still grow at 15 kg DM/ha with the net result of 965 kg DM/ha for July at lambing, which is very close to the target.

A simplified 40-day budget of this scenario is outlined in Figure 3.13.

Table 3.12. Worked example of estimating monthly feed availability for Scenario 1.

		Start				Minus	Plus			Ends	
	A	B	C	D[a]	E (D*C)/20 ha	F = E x 1.2	G[b]	H = B – F + G	I = H x A	B + I	
Month		**Feed**	**Stock**	**DSE**	**DSE/ha**	**Eat**	**Pasture growth**	**Diff**	**Change**	**Equals**	**Target feed**
	Days	**kg DM/ha**	**No.**	**Per head**		**KgDM/ha/day**	**KgDM/ha/ day**	**Per day**	**Per month**	**Feed**	
June	30	**500**	100	1.5	7.5	**9**	**15**	6	180	**680**	1000
July	31	**680**		2	12.5	**15**	**10**	–5	–155	**524**	1500
August	31	**525**		2	12.5	**15**	**30**	15	465	**990**	

[a] refer to Table 3.1
[b] refer to Table 3.11.

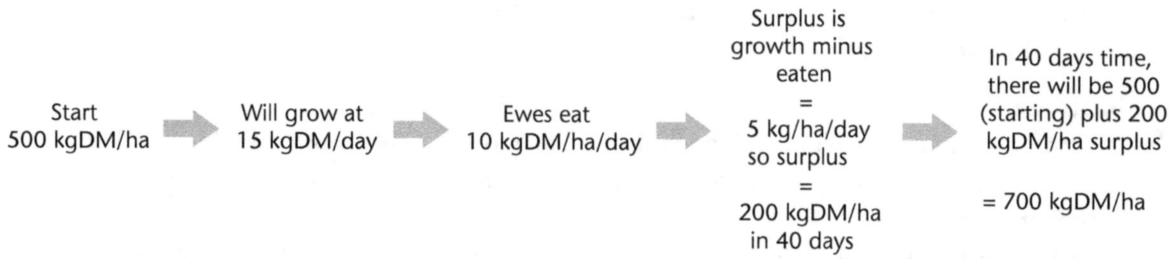

Figure 3.13. A 40-day feed budget for ewes before lambing.

Scenario 2 – Will there be enough feed to get through the summer?

Feed availability at the end of spring is 3500 kg DM/ha.

Stocking rate over summer is 8 DSE/ha.

Length of summer is 150 days.

The aim is to have 800–1000 kg DM/ha at the autumn break.

Not all pasture will be consumed because there will always be a level of wastage through trampling and decay. The amount of wastage will be higher over summer and highest where capeweed is dominant. It is difficult to forecast this wastage, but, as a generous estimate, a figure of 40% over summer is often used.

Length of summer: 150 days, paddock stocked at 8 DSE/ha

Start feed:	3500 kg DM/ha
Less wastage (40%)	–1400 kg DM/ha
Less eaten (8 DSE × 150 days) (@ 1 kg/DSE)	–1200 kg DM/ha
Left in autumn	**900 kg DM/ha**

Therefore in this scenario, 3500 kg DM/ha needs to be available at the end of spring to maintain animal intake and leave enough pasture cover. This type of budget provides a guide, as spring progresses, as to whether there is going to be excess feed (and opportunities for using this feed) or cause for concern requiring possible early grain purchases or stock sales. For example, if indications are that pasture availability will exceed 3500 kg DM/ha well before spring finishes, opportunities for using the extra feed could be considered before feed quality drops.

Although there are many estimates required in these feed budgets, they provide a means of monitoring progress and forecasting periods of low or excess feed, so decisions can be made early.

Key messages – matching feed supply and animal requirements

Stocking rate

- The stocking rate is one of the most critical factors influencing profitability.
- Optimising stocking rates will be a compromise between optimising pasture utilisation in spring, meeting environmental benchmarks and personal attitudes to risk.
- Stock type, time of lambing and grazing management will influence optimal stocking rates and the ability to match feed supply and animal requirements.
- Guides to potential stocking rates have been developed from on-farm trials, but achieving these assumes good fertility and productive species, and these rates are unlikely to be achievable over the whole farm.

- Pasture and animal indicators at particular times of the year are available to provide a guide on whether stocking rates are appropriate.

Estimating intake

- Animals can eat a dry matter weight equivalent to 3 to 4% of their body weight if feed quantity and quality is not limiting, with young sheep, thin or reproducing animals consuming the higher amount.
- Intake is unlikely to limited by pastures above about 1500 kg DM/ha (about 4.5 cm of moderately dense green pasture).
- Intake may not meet requirements if digestibility is less than 65% for growing, pregnant or lactating stock or less than 55% for mature, dry stock.

Feed budgets

- These provide a guide as to whether pasture targets for animal needs are likely to be met.
- They provide opportunities for addressing likely pasture over or undersupply (while prices are favourable).

Feeding supplements

When pastures or crops are unable to meet the nutritional needs of the animals, a supplement will be required. For many enterprises in southern Australia, some supplementary feeding is required annually. For example, spring born Merino weaners will usually need to be supplemented over most summers because of a lack of energy and protein required to support reasonable growth, unless summer growing pastures, such as lucerne, are available. Earlier lambing prime lamb flocks frequently need to supplement ewes as they get close to lambing if autumn rains have not arrived early enough to support the needs of pregnancy.

A cost-effective supplement should supply the nutrition not met by the available pasture, so it is important to know what is limiting. The most important criteria for evaluating feeds are the values of metabolisable energy/digestibility and protein.

Feeds are composed of water and dry matter, so feeds should be compared on a dry matter basis. Pasture will vary from 10% dry matter when short and green to 90% dry matter at the end of summer when dry. Grains and hay are usually about 90% dry matter, whereas silage may be 30 to 40% dry matter.

Energy value of feed

Energy is the most common limiting feed requirement in southern Australia. Grains tend to be the cheapest energy source fed as supplements for sheep, but all supplements considered should be compared on the cost of energy supplied in the dry matter.

Pasture tends to be described in terms of digestibility and supplements in terms of metabolisable energy, even though they are both a measure of the energy content of the feed. Table 3.13 shows some average values of energy (measured as ME MJ/kg dry matter) for a range of common feeds and pasture, as well as the amount required to feed one DSE per day.

Table 3.13 indicates that 0.6 kg wheat or 1.0 kg oaten hay would be required per day to maintain a 45 kg wether (one DSE).

Table 3.13. Energy values for some feeds and the amount required to maintain a DSE per day – as fed (which includes water content).

Feed	Metabolisable Energy MJ/kg fed	Amount (kg) required to maintain one DSE per day
Wheat, triticale, lupins, barley, maize, peas	12	0.6
Pasture immature, oats	11	0.7
Pasture hay, oaten hay	7	1
Pasture mature	6	1.2
Cereal straw	4	1.8[a]

[a] It is unlikely that an animal will be able to eat this much owing to the low digestibility and high fibre content.

Note that feed sources can vary in their energy values and these are a guide only. 'As fed' means that the water component is included. For the feeds in this table, 90% dry matter has been assumed. When feeds are measured as by FeedTest®, they are measured as per kilogram dry matter and so will be slightly higher.

How much an animal can physically eat in one day is affected by the digestibility and also by the fibre content of the feed. If the neutral detergent fibre (NDF) of a feed is known, this can be used to estimate the possible intake. This is useful for estimating intake from hay sources (grains are not measured for NDF). If the NDF percentage is known then:

$$\text{the maximum percentage of a sheep's live weight that can be consumed} = \frac{120}{\text{NDF\%}}$$

For example:

If hay has an NDF% of 70%.

$$\frac{120}{\text{NDF\%}} = 1.7\%$$

Therefore 1.7% of the animals' live weight is the most of this feed source the animal can eat.

A 20 kg animal can eat a maximum of 0.3 kg per head per day.

A 50 kg animal can eat a maximum of 0.85 kg per head per day.

If the energy value of the feed is also low (which is probable at high NDF%) then it may be that the animal cannot consume enough of the feed to maintain its weight.

For example:

The 50 kg sheep requires 7.6 MJ ME per day and, in the example above, can eat 0.85 kg hay.

If the hay has 8 MJ ME/kg dry matter then 0.85 kg of hay can only provide 6 MJ ME, which is not enough to maintain its weight.

Before making choices based on the cost of energy, the energy density and the fibre content of the supplement needs to be considered to ensure that animal requirements can be met,. The example below illustrates how to compare the cost of energy for two different feeds.

Example of comparing costs of different feeds

Oats at $250/tonne at 90% dry matter

1 kg oats costs 25 cents

1 kg oats has 10% water so if oats supplies 10 MJ ME per kg dry matter then

1 kg oats supplies 9 MJ of energy (10 × 0.9)

Cost per 1 MJ is 2.8 cents (25 cents/9 MJ ME).

Table 3.14. Cost of feeding 1000 wethers half a ration for 2 months.

	ME MJ/kg	Protein CP%	Cost $/Tonne	Cost c/MJ	Kg required/ head/week	Amount required for 1000 wethers for 2 months (T)	Total cost $
Wheat	13	13	200	1.5	1.9	15T	$3015
Pellets a	12	15	280	2.3	2	16.3T	$4573
Pellets b	10	15	260	2.6	2.5	19.6T	$5096

Wheat at $275/tonne and 90% dry matter
1 kg wheat costs 27.5 cents
1 kg wheat has 10% water so if
1 kg wheat dry matter supplies 13 MJ of energy
1 kg wheat supplies 12MJ ME (13 × 0.9)
Cost per 1 MJ is 2.3 cents (27.5 cents/12 MJ ME).

In this example, wheat is the cheapest feed per unit of energy, although it is more expensive per tonne.

To illustrate this at a flock level, Table 3.14 shows relative costs to feed 1000 wethers half a ration for 2 months.

This example shows that relatively small differences in costs of energy (1.5 compared with 2.3 c/MJ) can lead to significant savings when feeding large numbers of sheep for some time.

Prices and feed values will vary considerably, but as a guide, Table 3.15 compares a range of feeds for their energy costs. In this example, if wheat was $270/tonne delivered, then oats might be a comparative buy at about $230/tonne or less, and good quality lucerne hay at about $200/tonne or less.

Processing grains

Processing grain improves its digestibility for cattle by increasing the surface area of the grain and so improving the utilisation of cereal starch by rumen micro-organisms. However, there is no advantage of processing grain for sheep because they chew grains very effectively. Processing grains is more likely to be disadvantageous to sheep through increasing the availability of starch and hence increasing the likelihood of acidosis.

Table 3.15. Comparing the cost of energy in different feeds over a range of feed prices.

Feed	Energy value (MJ/kg as fed)[a]	Cost per unit of energy (cents/MJ) Feed cost ($/tonne):							
		90	120	150	180	210	240	270	300
Wheat, triticale, lupins, peas	***12***	0.8	1.0	1.3	1.5	1.8	2.0	2.3	2.5
Oats	***10***	0.9	1.2	1.5	1.8	2.1	2.4	2.7	3.0
Lucerne, early clover hay	***8.5***	1.1	1.4	1.8	2.1	2.5	2.8	3.2	3.5
Oaten, mid-season pasture hay	***7***	1.3	1.7	2.1	2.6	3.0	3,4	3.9	4.3
Late grass hay, cereal straw	***6***	1.5	2.0	2.5	3.0	3.5	4.0	4.5	5.0

[a]approx. 90% dry matter, except hay/straw at 85% DM

Protein

Protein may be limiting if there is no or little green feed in the pasture. Protein supplements can sometimes be useful as a supplement when there is ample dry standing feed, but, in this scenario, energy is also likely to be low. Provision of a protein source can, however, increase the amount an animal can eat in a day and therefore increase the energy intake. For example, lupins have a very high protein content (as shown in Table 3.16) and a supplement of lupins on stubbles or dry feed has shown an increased utilisation of dry feed and reduces the weight loss of animals.

Table 3.16 provides averages and ranges of energy, protein and fibre for a number of feed supplements. The table shows that **most** grains and supplements provide enough protein for dry stock. The variation in protein in most feeds show that some, fed on their own, would not maintain a dry animal and others may allow for good weaner growth without the need to provide a higher cost high-protein supplement.

Table 3.16 highlights the range in energy and protein of different feeds. Therefore, it is very important to test feeds before purchasing or formulating a ration, because the ranges in nutritional values can vary enormously. For example, buying good-quality hay versus a poor hay could mean

Table 3.16. Indicative average values and ranges for energy, protein and fibre for a range of feeds.

	Average ME MJ/kg DM (range)	Average crude protein (%) (range)	Neutral detergent fibre (NDF) (%) (range)	Comments
Oats	10.4 *(6–15)*	8.8 *(5–15)*	29 *(22–35)*	Low starch, high fibre
Barley	12.3 *(11–14)*	9 *(4–15)*	20 *(14–36)*	High starch, moderate fibre
Wheat	13 *(11–14)*	13 *(7–23)*	12 *(8–24)*	High starch, low fibre
Triticale	13 *(12–13)*	11 *(7–19)*	15 *(11–23)*	High starch, low fibre
Lupins	13 *(11–15)*	32 *(21–43)*	23 *(9–29)*	High fibre, low starch
Canola meal	12 *(10–16)*	37 *(27–42)*	30 *(27–35)*	May have high by-pass protein. High in phosphorus – may aggravate low calcium. If cold pressed, the oil levels will be very high and will reduce intake
Sheep pellets				Vary enormously. Some have additives such as calcium and phosphorus and buffers.
Hay – legume	9 *(6–11)*	17.6 *(6–26)*	47 *(33–72)*	
Hay – grass	8 *(5–11)*	8 *(1–18)*	67 *(43–83)*	
Hay – cereal	8 *(6–10)*	10.2 *(1–18)*	63 *(45–84)*	
Silage – grass, legume	10.4 *(6–12)*	15.3 *(4–27)*	51 *(29–81)*	Very variable dry matter content *(17–86%)*

Table 3.17. Impact of increasing protein levels for a maintenance diet. (Source: GrazFeed)

Amount required per day to maintain weight	Ration	Energy in diet ME/kg DM	Protein CP%	Protein loss g/day	Protein used g/day
590 g/day	Wheat	13.8	12%	4.9	60
590 g/day	Wheat	13.8	14%	16.2	60
590 g/day	Lupins	13.8	32%	110	60

Note: with energy constant and higher protein, the same amount of the ration must be fed and there is no advantage in extra protein

halving the amount required to meet an animal production target. For a more comprehensive comparison of some of the less common feed types, refer to Tables 4.8 and 4.9.

In comparing feeds on cost per unit of energy, as outlined in previous examples, consideration of protein or other requirements should be included. If the cheapest source of energy is deficient in protein, either choose the next cheapest energy source that is adequate in protein or consider purchasing a high-protein supplement to mix with the cheap energy option. For example, it may only require 10 to 20% of a supplement such as peas or lupins to bring a low-protein grain up to requirements.

The supply of protein in supplements is often overvalued because an over supply of protein will often not be used. Tables 3.17 and 3.18 illustrate that when protein levels are above requirements, protein is excreted. In the first example, wethers are required to maintain weight only and so an increase in protein (12–32%) provides no benefit and is excreted, as illustrated by the amount of protein lost.

Protein is required for growth and Table 3.18 illustrates the benefit of increasing protein levels for a 30 kg weaned lamb, growing at 200 g/day. Where wheat is provided at 12% or 14% protein, there is a benefit in using the higher percentage in that slightly less grain is required to sustain growth (880 g/day versus 900 g/day).

Generally, when protein exceeds requirements, it is lost through urine and faeces. In the prime lamb example, there is no advantage in supplying feed with protein levels above about 15%. There is some advantage in lupins over wheat because, in this example, more of the protein bypasses

Table 3.18. Impact of increasing protein levels for a lamb production diet. (Source: GrazFeed)

Amount supplement required per day to maintain weight	Ration	Energy in diet ME/ kg DM	Protein CP%	Protein loss g/day	Protein used g/day
900 g/day	Wheat	13.8	12%		118
880 g/day	Wheat	13.8	14%		118
890 g/day	Wheat @ 90% Lupins @ 10%	13.8	16%	9	116
900 g/day	Wheat@ 80% Lupins @ 20%	13.8	18%	24	116
800 g/day	Lupins 100%	13.8	32%	138	124

Note: 30 kg weaned prime lamb – required to grow at 200 g/day (high weight gain)

rumen fermentation and is directly, and more efficiently, digested in the sheep's intestine (see below). This may be outweighed by the greater cost of the lupin-based ration.

Bypass protein

Most plant protein eaten is used by the rumen micro-organisms and converted to microbial protein, which can then be used by the animal. When the protein in the diet exceeds the ability of the microbes to use it, then ammonia is excreted in the urine. This means that ruminants are relatively inefficient users of high-protein feeds. However, some proteins avoid degradation in the rumen and may be available for absorption further down the gut. These proteins are called by-pass or protected protein. Cottonseed meal and fish meal can be used to provide a source of bypass protein. Trials in New England, New South Wales have shown good production responses (from animals requiring to grow or produce lambs) to cottonseed meal supplements fed on dry pastures, before the feed quality dropped below maintenance, when an energy supplement was required. Supplements that are too high in bypass protein would be disadvantageous to the animal because the rumen micro-organisms would not have enough protein for their survival.

Urea

Urea is not a protein but is a source of nitrogen that can be used by rumen micro-organisms to produce protein. Rumen micro-organisms use the ammonia from urea to build their own protein requirements and this protein becomes available to the sheep later in the digestive process. Urea can be difficult to feed, it can be toxic and is less suitable for growing stock that could perform better by being fed true protein supplements. One per cent of urea by weight of the ration will increase the diet crude protein by 2.8%, but levels above 2% of urea are likely to reduce feed intake.

Feeding grain supplements

As explained in the first part of this chapter, sheep are ruminants that rely on the microbial population in the rumen to be able to digest and use much of their feed. Therefore care must be taken when introducing new feed to sheep because the rumen micro-organisms need time to adapt to new energy sources in particular. Pasture is high in cellulose and fibre, but many of the cereal grains are high in starch and low in fibre. Sheep are quite good at adapting to a new diet if it is introduced slowly. Introducing some feeds too quickly can cause a build up of lactic acid from fermentation of grain carbohydrate in the rumen. This condition is called acidosis, which is the most common cause of death in feedlots or where sheep are fed high-grain diets.

Good quality hay and silage (high in fibre) and oats and lupins (low in starch) can be useful when a supplement needs to be introduced quickly (for example, to newly shorn sheep experiencing a cold snap).

Introducing sheep to grain

Feeding should be started at the rate of up to 50 grams per head per day for adult sheep (25 grams for weaners), and increased slowly to the required ration over 3 or 4 weeks (see Table 3.19). Oats and lupins have a higher fibre content than the other grains and the full ration may be introduced over a shorter period of 2 to 3 weeks with little risk of causing digestive upsets. Table 3.19 outlines a daily feed program for introducing sheep to a grain ration.

Table 3.19. A program for introducing sheep to grain.

Feeding days		Amount of grain per feed	
		gram per head	kg per 100 sheep
1, 2	feed daily	50	5
3, 4	feed daily	100	10
5, 6	feed daily	150	15
7, 8	feed daily	200	20
9, 10	feed daily	250	25
11, 12, 13, 14	feed daily	300	30
15, 17	feed on alternate days	600	60
19, 21	feed on alternate days	850	85
23, 26, etc.	feed every third day	1300[a]	130[a]

[a]gradually adjust to suit final ration

If the mob to be fed has not previously experienced grain feeding, introduce some previously grain fed sheep to the mob: the inexperienced sheep will follow the lead of the sheep that are familiar with grain. Sheep that have not previously experienced grain feeding are best introduced to grain in small paddocks.

Table 3.19 is only a guide to introducing sheep to grain. Early in the introductory program it is unwise to progress to higher levels of grain feeding until all of the sheep have taken to the ration or shy feeders have been drafted off.

It is especially important to avoid sudden changes in the ration. Sheep that have become accustomed to one type of grain cannot immediately adjust to another. Deaths and a high incidence of tender wool can result from a sudden switch of feed. Even the same grain type obtained from a different source or new batches of pellets from the same company have caused losses. It is important to determine at the beginning of a grain feeding program how long a batch of feed will last. This will allow time for planning a gradual changeover from one feed batch or type to another. (See 'Practical tips for supplementary feeding' at the end of this chapter.)

How much grain to feed

It is relatively easy to calculate a full supplementary ration for a particular class and weight of sheep, but it is very difficult to estimate what proportion of this ration to feed if stock still have access to pasture or stubble.

The simple answer is to start with approximately one-third of a full ration and monitor the sheep for weight gain or loss. The ration can then be altered accordingly. The starting grain ration can depend on what fodder is being fed, the condition of the sheep, the aims of feeding and how much paddock feed is available. Pasture targets supplied in Table 3.8 indicate when pastures will not supply the needs of some classes of sheep, and hence when responses to supplementation should occur.

The CSIRO has developed a computer feeding model (GrazFeed™) which estimates responses of animals to pasture and also to supplements. It is a useful tool for estimating and formulating effective supplements and predicting animal production. It requires an ability to adequately assess both the pasture quality and quantity and the animal's nutritional status.

Fibre

Fibre is unlikely to be limiting for sheep on pasture. Short lush pastures are the lowest in fibre, but is unlikely to be cost effective to provide a supplement. Responses to a fibre supplement given to cattle grazing lush green pastures have been reported, especially for high-producing dairy cows, but such responses have not been recorded for sheep. There may be cases where fibre has a role to play as a supplement where mineral absorption is poor, such as calcium for pre-lambing ewes, but there is no reported work to show that this has been the case. Supplementing animals with a feed that is of poorer quality than the feed they are grazing may have the undesirable effect of reducing the overall quality of the diet being consumed (if animals substitute the supplement for the pasture). Substitution may be desirable if there is a need to reduce pasture intake for other reasons.

Fibre may be limiting when feeding a high proportion of the diet as cereal grain. In this case, straw and hay provide good sources of fibre. Past research has shown sheep survive well on total wheat diets for maintenance, but, in practice, it has been observed that both performance and survival can be improved by including hay or straw at around 10% of the diet when feeding a high proportion of the diet as cereal grain.

Treating straw

Cereal straw as a crop by-product is seldom of good enough quality to maintain even dry animals. Cereal straw has low nitrogen content (and hence low protein value) and low digestibility. It is also deficient in sulphur, phosphorus and calcium. It is relatively high in energy, but most of this is indigestible (as it is bonded to lignin) and so there is interest in treating straw to increase the availability of this energy. Straw can be treated chemically with urea to increase digestibility by breaking down some of these indigestible components. When urea is dissolved in water and exposed to bacterial enzymes the ammonia combines with the water to form an alkaline solution. The alkali acts on the fibrous part of the straw to increase its digestibility. The urea also provides extra nitrogen, which can boost the low nitrogen content of the straw. Inclusion of molasses with urea has given better results than urea alone, but both diets have not stopped weight loss.

Several experiments have investigated the potential use of urea as a method of increasing the digestibility of straw for sheep, but compared with results from similar trials with cattle, they have been disappointing. As with the crop stubble, the addition of lupins to a straw diet has reduced weight loss more than treated straw (see Figure 3.14).

Feed blocks or licks

There are many commercial feed blocks available for a range of claimed uses and benefits, including utilisation of dry feed and correcting a range of nutritional deficiencies. Although they may save labour (compared with weekly feeding out of grain or dosing animals individually), the decision on whether to use licks to provide supplementation should go through the same scrutiny as for the choice of any other supplement. Are blocks adequately rectifying a deficiency? Is there a deficiency present? Is a lick an effective means of supplying the nutrient and how does it compare in

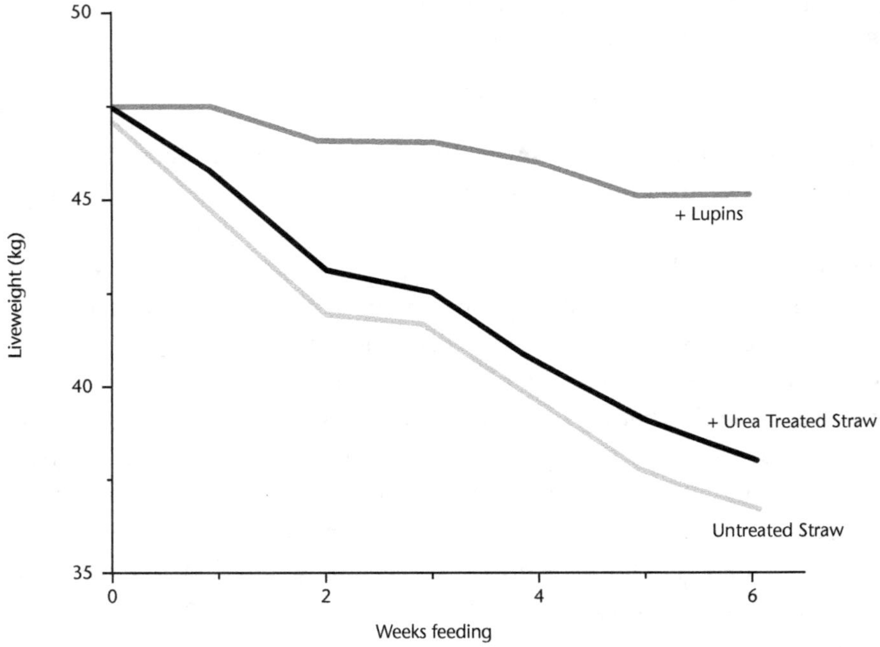

Figure 3.14. Effect of feeding lupins or urea treatments to sheep eating chaffed straw.

cost to an alternative source? Often provision of a feed block or lick is a 'peace of mind' action. If animals do have a deficiency in a trace element or mineral, then it is unlikely that all stock will get their requirements from a lick. It is extremely difficult to supply adequate and safe supplementation with this voluntary feed system.

As for most supplements, animals need to be trained to use licks. Research has measured ranges in intake of individual animals from a lick, within a mob, from 0 to 400 g/head/day. To rectify a trace element deficiency, for example, where the requirement will be similar across most animals in a paddock, some sheep will be deficient and others will receive a toxic dose.

Key messages – feeding supplements

- Supplementary feeds vary considerably in their levels of protein and energy and should always be tested.
- The value of supplements should be determined by the concentration of energy and protein in the dry matter.
- Knowing what is limiting will ensure that the appropriate supplement is fed and not wasted.
- Sheep need to be introduced carefully to grains, particularly those high in starch, such as cereal grains.
- Getting the ration right will require monitoring of live weight or condition.

Assessment of body condition of sheep

Feed requirements of livestock vary with live weight, level of production, physiological status (dry, pregnant or lactating) climatic conditions and body condition. The body condition of a mob of sheep and production targets will determine whether animals need to gain or maintain weight or if they have scope to lose condition. It is important to gauge the condition of different classes of stock to ensure that the feed supply adequately matches the animals' requirements. Body condition of sheep can be assessed by monitoring live weight and/or fat reserves.

Live weight

Monitoring live weight change is a reliable method for assessing the response to pasture or supplements. The change in live weight, measured on two or more occasions, is usually more useful than a single weight.

When weighing, it is necessary to make allowance for the estimated weight of wool, wetness of the fleece after rain and loss of gut fill due to prolonged periods of yarding. The easiest way to allow for these effects is to weigh stock after a similar time off feed and not to handle wet stock.

Measuring live weight is primarily useful for:

- **Assessing lambs and weaners** – young lambs need to grow and Merino lambs in particular have few body fat stores, so measures of weight and weight change are the most useful gauge of their nutritional status. Target weights can be set for survival, production and markets (for example, turn off weights for prime lambs).
- **Maiden ewes** – reaching critical (minimum) joining weights for young ewes maximises their chances of getting in lamb. Generally, this is approximately 75 to 80% of their mature weight (for example, 38 kg would be the minimum joining weight for ewes that have a mature weight of 50 kg).

Monitoring animals for feeding

When supplementary feeding at a high rate (for example, during drought), weighing a number of stock regularly will give the most accurate guide to weight gain or loss. This can mean significant savings in terms of reduced feeding if weights are higher than needed, or reduced production losses if weights are lower than needed.

For adult sheep above their critical weight or condition targets, it may be feasible to allow some controlled weight loss. This should be monitored by weighing regularly. The **maximum** weight loss that can be sustained without risk to animal health is about 2% of live weight per week, or approximately one kg per week. Young animals weighing less than approximately 25 kg cannot afford to lose any weight and must be fed for at least slow weight gain.

Animal health treatments

Knowledge of accurate live weight is important for administering the correct dosage for some animal health treatments such as drenches. Drenching to correct live weight is particularly critical to avoid under drenching (because it will assist in the development of drench resistance) and also over drenching (because this is costly and high doses of some chemicals can be toxic).

Assessing fat reserves

Live weight on its own does not give a full picture of the body reserves available as stored fat. Animals of different frame sizes (even within a mob) will have quite different reserves of fat at the same weight. Live weight can be misleading for assessing the condition of sheep that are in late pregnancy, because there will be considerable weight associated with the lamb and placenta. The ewe may be losing body reserves to ensure the survival and growth of the lamb.

The amount of fat stored provides a buffer for weight loss in tough times. Therefore knowing how much fat reserve an animal has will assist in knowing whether an animal can lose weight, or has a requirement to increase or maintain weight to achieve production targets. Assessing the fat reserves of sheep is a quick, cheap and easy procedure.

Fat is laid down and stored in different places on the sheep, such as over the long ribs, the short ribs and the tail, but also internally over the kidneys and as intramuscular fat. Fat is often deposited over the tail in an irregular fashion, so it is an unreliable place to assess overall fatness. The short and long ribs are the best place to assess for fat depth as an indication of fat reserves.

There are two systems used in the sheep industry to assess fat reserves – condition scoring and fat scoring.

Condition score

The 'condition score' system was developed as a method of assessing the 'condition' of an animal for monitoring nutritional status. It entails feeling the fat cover on the short ribs and backbone of the animal (see site A in Figure 3.15). A score is given on a scale of 1 to 5 according to the degree of fatness. In practice, it is possible, and useful, to assess sheep to the nearest half score.

Fat score

The fat score system assesses the fat depth 10 cm down from the backbone over the twelfth (second last) long rib (see site B in Figure 3.15). This is called the GR site and is where sheep are assessed for fat content after slaughter, so is a very useful measure for market specifications.

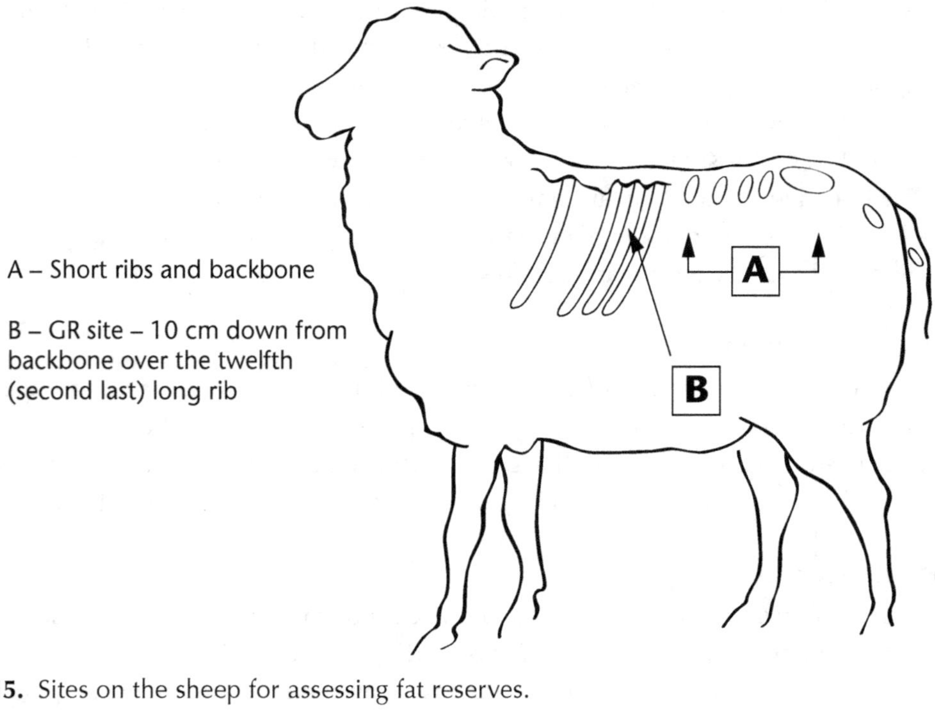

Figure 3.15. Sites on the sheep for assessing fat reserves.

Table 3.20. What to feel when fat and condition scoring.

Score	Condition	Fat
	Short ribs (A) and backbone	**Long ribs (B)**
1	End of short ribs are very obvious. Finger tips fit between and under short ribs. Backbone is prominent and sharp. Eye muscle tends to feel hollow and very small	0–5 mm fat over the GR Individual ribs easily felt. Impossible to feel tissue over ribs.
2	End of ribs rounded but possible to feel between them. They are covered in flesh but it is easy to press under and between them. Backbone is prominent but smooth. Eye muscle has reasonable depth but is still concave.	6–10 mm fat over the GR Individual ribs are easily felt but there is some tissue cover.
3	End of ribs can be felt and are well rounded. It is possible to feel the rounded ends but not between them. Backbone can be felt but smooth and rounded. Eye muscle is smooth and rounded.	11–15 mm fat over GR Individual ribs can still be felt but there is considerable tissue present.
4	It is only possible to feel or sense one or two short ribs and only possible to press under them with difficulty. Backbone is a smooth slightly raised ridge above full eye muscles.	16–20 mm fat over GR It is just possible, with pressure, to feel the ribs and fluid movement of the tissue.
5	It is virtually impossible to feel under the short ribs and no rib ends can be felt. The spine, if felt at all, will only be felt with reasonable pressure.	21–25 mm fat over GR Ribs are barely felt and tissue movement is very fluid

Note: GR site 10 cm down from the backbone over the 12th long rib

Fat scoring also uses a scale from 1 to 5. Each score is equivalent to an additional 5 mm of fat (so a sheep with a fat score of 5 might have 25 mm of fat at this site compared with an animal of fat score 1, which may have 5 mm fat).

The sites are illustrated in Figure 3.15. Ideally a combination of the two systems should be used, especially at low condition or body reserves. A change of one score in both systems is related to a body weight change of about 6 to 9 kg, depending on the frame size of the sheep.

Table 3.20 describes what to feel for using both methods.

Both fat and condition scoring can be easily learned and used: they do not require equipment, and allow for differences in frame size and stages of pregnancy of animals. Condition scoring is more sensitive when sheep are at the lower end of the nutritional status (less than condition score 2.5). As a consequence, this is the preferred method to assess the nutritional status of stock. Fat scoring is necessary to describe fat depth when marketing animals.

The disadvantage is that the systems are subjective. It is possible to calibrate an operator or oneself for fat scoring using the feedback on fat depth from sheep after they are slaughtered.

Minor differences between breeds may lead to some difficulties, particularly if there is a difference in the distribution of subcutaneous and internal body fat.

Table 3.21 outlines a guide to recommended target condition scores for classes of sheep at different times of the year. Crossbreds should be at the higher end of the scale.

Table 3.21. Target condition score for different classes of stock at different times of the year.

Animal	Target condition score
Wethers/dry sheep	
Leanest time of year (end summer or winter)	Low 2
End spring	3–3+
Ewes	
Joining	2.7–3
Point of lambing	2.7–3
Lactation	Minimum 2.5
End of spring	3–4
Rams	
Pre joining	3–4

Monitoring changes in weight or condition

A quicker way to monitor changes in weight or condition of a mob, without doing the whole mob, is to tag or brand 25 to 30 sheep from each mob and weigh them regularly. Individual tagging and weighing the same sheep each time will give a more accurate estimate of weight gain and loss. However, if it is easier to randomly draft sheep each time, 10% of the mob, or a maximum of 80 sheep, would give a good estimate. This will give confidence in identifying body weight changes of 2 kg or more.

A practised eye may feel confident in assessing sheep in a paddock, with no 'hands on' assessment, but this 'eyeball' assessment can be misleading, especially for late pregnant ewes or sheep with reasonable wool length. A heavy pregnant ewe can appear quite 'fat' but often a 'hands on' assessment may reveal that the fat cover is very low.

Key messages – assessment of body condition of sheep

- Monitoring animal condition using fat reserves or weight change is crucial to ensure that their nutritional requirements are being met.
- Monitoring fat reserves is particularly useful for adult sheep and pregnant ewes.
- Weighing is crucial for young animals, to monitor suitability of rations and for administering correct doses for animal health treatments.

Ewe nutrition

The ewe's energy requirement varies widely over the year, from the equivalent of one to three or four DSE, depending on the number of lambs she rears. Protein requirements also double during lactation. The nutrition and the condition of the ewe over the reproductive cycle are the main influences on the number of lambs conceived, born, weaned and their subsequent production. Meeting ewe feed demands, and balancing these with the feed supply, is the key to optimising production and stocking rates. The time of lambing in a flock has an impact on potential stocking rate, the management of reproduction and the likely needs for supplementation. The key parts of the reproductive cycle are joining, pregnancy, lactation and weaning.

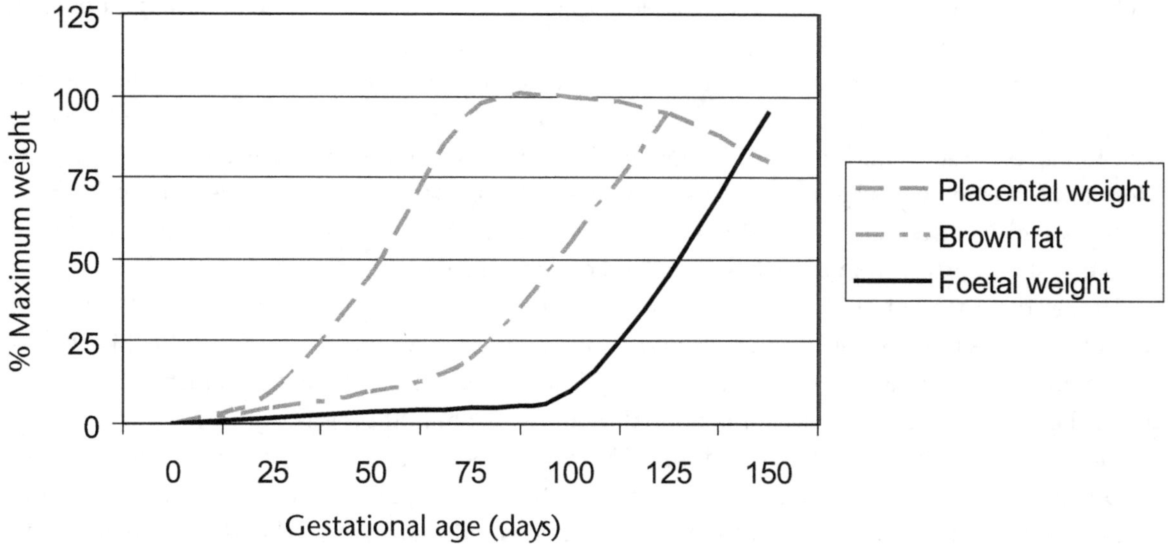

Figure 3.16. Changes in the ewe from joining to pregnancy.

Figure 3.16 illustrates the changes that occur during pregnancy and lactation to illustrate the crucial times for ewe management.

Management of the ewe through any of these periods has an impact on subsequent condition (for example, weaning lambs late will reduce the time for the ewe to recover body condition before joining). Other factors influencing reproductive performance are discussed in Chapter 5.

The target condition scores throughout the reproductive cycle are provided in Table 3.21 and pasture targets to meet feed requirements are outlined in Table 3.8.

Joining

Live weight at joining has a significant impact on the ovulation rate, and hence the potential number of lambs born. Poor condition at joining can lengthen the time it takes the ewe to conceive and hold conception. Ewes in very poor condition may not conceive. At joining, maiden ewes need to be at least 75 to 80% of their mature body weight. For mature ewes at joining, a target fat or condition score is 2 plus to 3 plus. As ewes approach condition score 3, more ewes are likely to get in lamb quicker.

Chapter 5 outlines the likely response to increasing weight or condition to increase ovulation rate, before or at joining.

Pregnancy

Ewes' requirements for feed will increase by 30% or more in the last trimester of pregnancy because this is when the foetus grows substantially and milk production is initiated. Eighty per cent of total foetal growth occurs in the last 6 weeks of pregnancy. However, most of the placental growth occurs in mid pregnancy, and placental growth also affects potential foetal growth. Under-nutrition of the ewe results in low birth weight lambs and premature lambs that have low survival rates. A reduction in ewe energy intake of up to 25% below requirement in late pregnancy may not cause significant problems. However, a 40% deficit in ewe energy requirement can reduce lamb birth weight by 20 to 30%. This may be acceptable in single lambs, but may jeopardise survival of twin lambs.

Where maternal energy reserves are used because of inadequate feed, milk production and mothering behaviour may suffer more than lamb weight. Colostrum and milk production are initiated in late pregnancy and these affect not only the lamb's growth rate, but also mothering ability. More twin-bearing ewes have been shown to desert at least one of their lambs when nutrition has been poor in late pregnancy. Good nutrition is crucial in the last month of pregnancy to ensure good colostrum production in both Merino and crossbred ewes.

Higher birth weight lambs (4 kg or more) have a much higher chance of survival. Lamb birth weight is affected by ewe nutrition over mid and late pregnancy, with most of the effect in the last month. Because twin lambs are the most at risk from low birth weight, ewes identified as bearing twins should be given preferential nutritional treatment from mid pregnancy.

Lifetimewool research has reported that if Merino ewes lose significant condition (for example, drop from condition score 3 to 2) early in the first half of pregnancy and then **maintain** weight, lamb birth weight can be significantly penalised (that is reduced by 0.3 kg) and the lambs will cut less wool of stronger fibre diameter later in life. If the same amount of weight was lost in late pregnancy, there would be similar effects, but the birth weight could be less again. If ewes lose condition in early pregnancy and **regain** condition in the second half, there are no long-term effects on fleece weight and fibre diameter, and birth weights may be slightly improved.

The effect of nutrition on birth weight of crossbred lambs is likely to be similar, but low birth weight is less of a problem for crossbreds than for Merinos. If ewes are fat (greater then condition score 3.5), they should be able to lose some weight in the first half of pregnancy with little repercussion, providing nutrition is good for the last half of pregnancy and for lactation.

Lactation

The ewe's energy requirement to maintain weight and produce milk will be 2.5 DSEs for a Merino ewe feeding a single lamb to as high as 4.5 DSEs for a crossbred ewe feeding twins (see Table 3.1). This increased energy demand is matched with an increase in need for protein. In most cases, ewes cannot eat enough feed to supply these nutritional needs and hence weight loss during lactation is likely.

Where pasture growth is unlikely to provide the requirements for optimum production at all times of the year, the options are to metabolise fat, suffer production penalties or use supplementary feed.

Supplementary feeding for late pregnant and lactating ewes

This increased energy demand often comes at a time when pastures contain insufficient energy to satisfy the ewes' requirements. Furthermore, protein may be a limiting factor if pastures are dry.

Ewes grazing dry pasture can only physically consume enough of it to provide 4 to 5 MJ of ME per day. High-energy supplements, such as grain are needed until pasture growth is sufficient to support high energy demands. If pastures are dry, protein will be limiting. If there is some green pick in the pastures, up to 50% of the total energy requirements, and most of the protein necessary for late pregnancy, may be available. Therefore the energy deficit can be made up by supplementing with the cheapest source of high-energy feed available.

Immature green pasture has a high energy content (about 12 MJ/kg DM). It usually takes at least 5 to 8 weeks after the autumn break before there is enough green pasture to meet all the energy requirements of the lactating ewe.

The best method of determining when to provide a supplement is to condition score a random sample of ewes as they approach late pregnancy. A pregnant ewe will put on 6 to 7 kg of uterine weight in the last 6 weeks of pregnancy, so that to maintain condition she must gain at least 7 kg in this period.

The decision on what to feed should be based on the following:

- what is deficient – energy and/or protein
- the cost per unit of energy (MJ) of the feeds
- the availability of the feed.

Hay is not a suitable supplement if there is no higher quality feed in the diet. Hay has a low energy density and ewes are unable to eat enough of it to meet their requirements.

Lucerne hay alone is also an inadequate supplement for late pregnant ewes if there is no green pick available. It is, however, a very useful supplement when used with grain.

If no autumn break occurs, then late pregnant ewes may be getting less than 25% of their total energy needs from the pasture.

Weaning Merino lambs

The optimum time for weaning is 12 to 14 weeks after the start of lambing. The reasons for this are:

- by 12 weeks, a Merino ewe provides only 5 to 10% of the lamb's nutritional requirements as milk
- the rumen in lambs reared on pasture reaches adult proportion and function by 8–10 weeks of age
- it is inefficient for ewes to graze pasture to produce milk to feed a lamb, when the lamb can do well on pasture alone (the ewe also needs time to recover condition following pregnancy and lactation to ensure successful joining)
- ewes are more efficient grazers than lambs, so the quality and quantity of the lamb's intake is affected by competition from the ewes
- ewes contaminate the paddock with worm eggs, which infect lambs from 7–10 weeks of age, depending on the season
- live weight and greasy fleece weight at 28 weeks are higher in lambs weaned at 12 weeks old compared with lambs weaned at 20 and 28 weeks (see Table 3.22).

Table 3.22. Effect of age of weaning on body weight and fleece weight.

	Weaning at		
	12 weeks	20 weeks	28 weeks
Live weight at 28 weeks (kg)	31	28	27
Lamb greasy fleece weight (kg)	1.6	1.4	1.4
Ewe greasy fleece weight (kg)	4.1	4.0	4.0

Imprint feeding

Sheep need to be trained to recognise grain or supplements. Once trained, sheep will adapt to supplements more quickly because they will remember the experience for several years. Lambs should be trained to eat grain or other supplements while they are still with their mothers. Although feeding may seem unnecessary if lambing occurs in spring on abundant good quality feed, the benefits are worth the small cost involved. Lambs fed with their mothers for four to six feeds over a 2-week period will consistently come up to a feed trail in later life. The amount fed can be small (for example, 50 grams per head) but the trail must be long enough to allow all ewes and lambs to feed.

Key messages – ewe nutrition

- The condition or weight of ewes at joining will significantly affect the number of lambs conceived.
- The nutrition and condition of ewes over pregnancy and lactation will have an impact on lamb birth weight (and survival), growth rate and weaning weight and lifetime wool production.
- Ewes should not fall below condition score 2.5 over pregnancy and condition score 3 (or higher for crossbreds) will improve survival of twin lambs.
- Lambs should be weaned at a time to enable ewes to recover condition before the following joining.
- Lambs should be trained to eat supplements while on their mothers.

Ram nutrition

In most flocks, rams are only used for a short period each year. The recommended joining length is 5 to 6 weeks for autumn joining because this allows for two ewe oestrus cycles. Sperm production takes about 8 weeks, so ram management needs to concentrate on health and nutrition for at least 8 weeks before joining. Poor feeding or disease (such as worms, footrot, foot abscess or flystrike) that occurs within 8 weeks of the start of joining can adversely affect sperm production. Ram condition needs to allow for rapid weight loss during joining. For some breeds, over fatness may be a problem, so reasonable targets are condition score 3.5 to 4.

If ewes are being supplemented before and over the joining period, rams need to be adapted to the same diet before being put in with the ewes.

Key messages – ram nutrition

- The nutrition and health of rams needs to be good and maintained from at least 8 weeks before the start of joining.
- The target condition score pre-joining should be 3.5 to 4.

Weaner nutrition and management

Pre-weaning growth rates of prime lambs are critical to the time required to reach finishing or selling weights (for a store lamb operation). Therefore meeting ewe targets (condition score and feed) for lactation are the key to managing weaning weights for a prime lamb enterprise. Finishing lambs after weaning is covered in 'Finishing prime lambs'.

Table 3.23. Target weights for weaners when pasture dries off.

	Mature weight (kg)				
	45	50	55	60	70
Weight at drying off pasture of pasture (kg)	20	22.5	25	27	31.5

Merino weaners are the most difficult class of sheep to manage. Generally, the aim with Merino weaners is to maximise survival and reach target joining weights. If keeping first and second cross lambs after weaning, the aim is to reach economic lamb market specifications and, for ewes retained for breeding, to reach the target joining weight.

Good weaner management is needed for:

- good maiden ewe lambing percentages (weaner ewes must be well grown at their first joining)
- maximised production from weaners for least cost
- minimised weaner deaths for least cost.

Important considerations are:

- age at weaning
- weight at weaning
- supplementation, how much and with what
- parasite control.

Age at weaning for Merinos

The optimum time for weaning Merino lambs is 12 to 14 weeks after the start of lambing. This is both for the benefit of the ewe and the lamb. The poorer the season, the greater is the need to wean early and feed differentially.

Weight at weaning for Merinos

By 12 to 14 weeks of age, most Merino lambs will weigh between 15 and 22 kg. Lambs at these weights can be weaned successfully.

More critical than the weight at weaning is the weight at the start of summer or when the pasture feed dries off. When pasture dries off, weaners should average 45% of the adult body weight. Target weights for weaners of different mature sizes are given in Table 3.23. Weaners below these target weights have little or no body reserves to carry them through periods of low feed quality and quantity.

Wethers will have a mature body weight 10 to 15% higher than ewes of the same strain.

Weaners below these target weights when pasture feed dries off are at risk. Given there is always a wide range in weights within a mob of weaners, there will be a fair proportion of the mob at risk, even if the average of the mob is at the target weight. Figure 3.17 shows a typical distribution of weaner weights. To improve survival rates of Merino weaners and provide adequate feed, weaners that are below the target weight should be drafted off and preferentially fed.

Production penalties for weaners

Lifetime reproduction may be restricted if sheep have poor nutrition in their first 6 months of life (including the time on their mother).

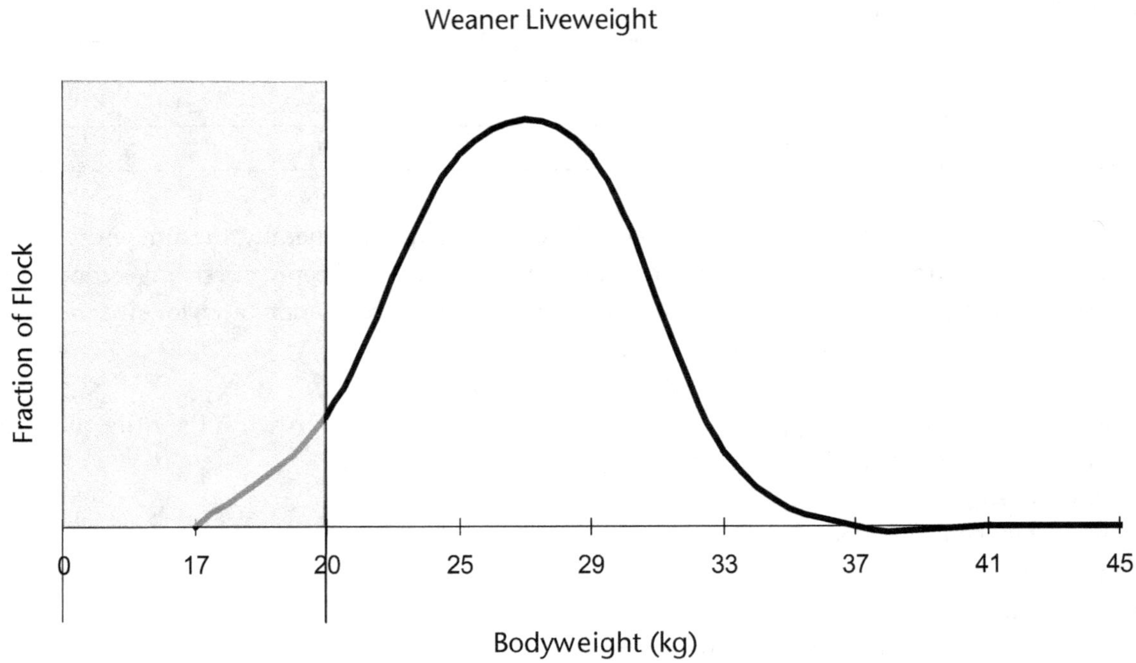

Figure 3.17. Example of distribution of weight at weaning illustrating the proportion at risk when the average target weight is 22.5 kg.

Severe nutritional restrictions in the 6 months after weaning can lead to 25% lower mature body weight. These low mature body weight sheep will eat less and are less likely to achieve target weights for sale or joining.

If weaners have restricted feed intake only in their second 6 months of life, they can compensate with no effects on lifetime performance.

There is no point in feeding Merino weaners for maximum growth unless there is a worthwhile market target. This is because sheep can catch up, or make compensatory growth, as feed becomes available.

Compensatory growth of weaners

Weaners usually make rapid growth through winter and spring when feed quality and quantity improves. This growth can compensate for periods of low growth or maintenance of body weight over summer. Figure 3.18 illustrates the weight changes of a mob of weaners that were supplementary fed over summer compared with a mob that was not supplemented. The live weights of the two mobs were similar at 18 months of age.

Feeding weaners for survival

Energy is the major dietary requirement for maintenance. Supplements for weaners should have an energy concentration of 8 MJME/kg DM or more. Weaners cannot eat enough of the low energy concentration supplements for maintenance.

On dry pasture, there may be a requirement for protein, although in most cases, provided energy is adequate, weaners can derive sufficient protein for maintenance from the microbial products of

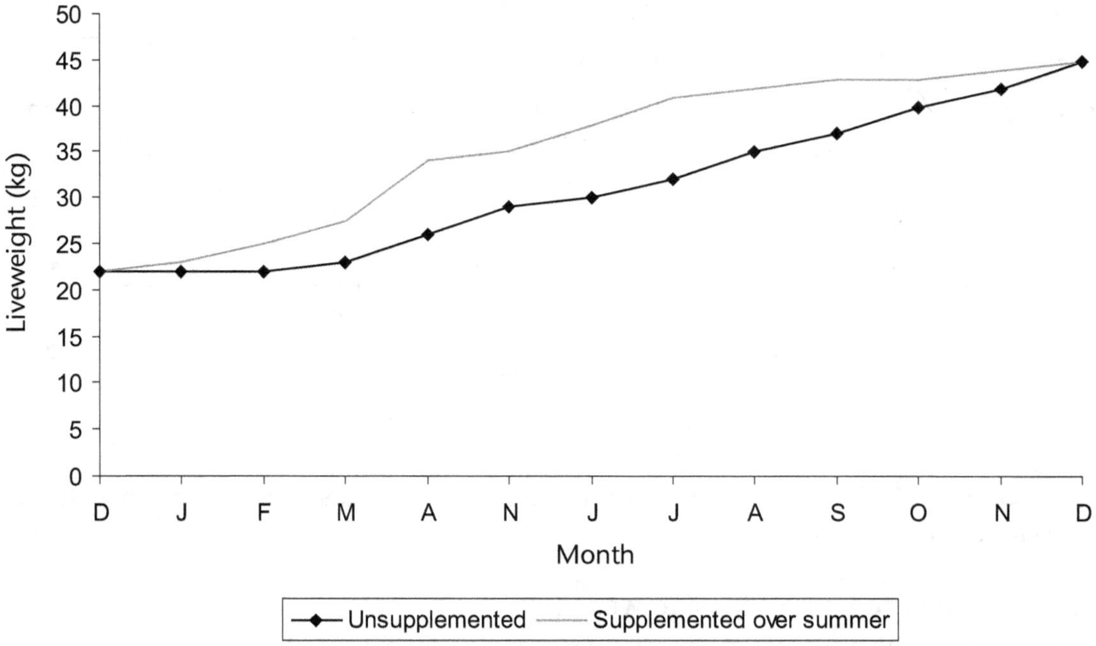

Figure 3.18. Comparison of post weaning growth rate of supplemented and unsupplemented weaners.

rumen fermentation. If weaners are below 20 kg, adding protein may be worthwhile. Table 3.24 shows the energy and protein requirements of weaners at different weights.

Target weights for weaners for survival and adequate weights for first joining are given in Table 3.25.

Supplementing weaners

If no pasture is available, full maintenance rations need to be fed. Table 3.26 shows the amount of good hay, oats or lupins needed to fully maintain weaners of various ages and weights.

If some pasture is available, it is difficult to know how much to feed because the intake of nutrients from pasture cannot be determined with certainty. Feeding should start with 30 to 50% of a full ration on a trial-and-error basis. The amount can be adjusted by monitoring live weights. Shy

Table 3.24. Protein and energy (as DSE) requirements for growing lambs.

Stock	Live weight (kg)	Required weight gain (g/day)	Dry Sheep Equivalent (DSE)	Approximate protein requirement CP (%)
Weaner	20	100	0.8	16
	20	200	1.2	18–20
	25	0	0.8	9–11
	25	100	1.0	12–14
	35	0	0.9	8–10
Finishing lamb	35	>200	2.6	15–18

Table 3.25. Summary target weights for survival and reproduction.

Adult weight (kg)	Birth weight (kg)		Weaning/ pasture drying off (kg)		Autumn break (kg)	Late winter (kg)	Joining (kg)
45	4	Grow at 200 to 300 g/ day	20	Grow at about 1 kg per month	22.5	27	34–36
50	4.5		22.5		25	30	37–40
55	5		25		27	33	41–44
60	5		27		30	36	45–48
70	5.5		31.5		35	42	52–56
% adult weight			45%		50%	60%	75–80%

feeders can be drafted off for separate treatment or can be minimised by feeding three times the daily requirements at 3-day intervals.

Monitoring the progress of Merino weaners

The best way to monitor weaners is by weighing a sample of the flock. Minimum acceptable weights for small and medium framed Merinos are:

- weaners less than 20 kg should be hand fed if necessary to achieve weight gains of about 1 kg per month (weaners weighing less than 20 kg may have less than 1 kg of total body fat and are not able to withstand prolonged weight loss; their requirements for maintenance must be fully met by current intake)
- weaners 21–23 kg can be maintained at that weight for 1 to 2 months
- weaners heavier than 24 kg can safely lose weight at about 0.5 kg per month until they fall below 24 kg.

These recommendations represent a **minimum** to prevent long-term effects on productivity and many sheep farmers would be dissatisfied by the appearance of such sheep.

Weaners that are on a poor level of nutrition will be more prone to disease. Managing gastrointestinal worms is extremely important because high worm burdens will not only threaten survival of weaners but also reduce growth rate and feed conversion efficiency. Research has shown that supplementing pasture fed Merino weaners with protein has resulted in similar weight gains as an

Table 3.26. Daily rations for maintenance of weaners (kg/head/day) with no pasture available.

Feed	ME Content (MJ/kg DM)	Age and weight of weaners: 5 months 22.5 kg (kg/hd/day)	8 months 27.5 kg (kg/hd/day)
Good hay[a]	8.6	0.6	0.7
Oats[a]	10.0	0.5	0.6
Lupins	13.0	0.4	0.4

[a]providing protein is adequate

effective drench. Although strategic drenching and grazing management to reduce worm burdens are likely to be more efficient means of maintaining growth in weaners compared with protein supplementation, this illustrates the interaction between nutrition and worm infection. Good worm control is an essential component of ensuring good weaner growth and survival.

Key messages – weaner nutrition and management

- Weights at weaning and when pastures dry off in summer are critical to weaner survival.
- Merino weaners below 20 kg body weight are at most risk and need to be fed to gain weight.
- Monitoring weight is probably the most important management tool for weaners.
- Ensure good parasite control.

Finishing prime lambs

Finishing lambs means producing lambs to meet market specifications, which are based on measures of fat and carcase weight.

Traditionally, most prime lambs in Australia are finished on pasture or fodder crops because they are the most cost effective feed options. Lambs may be finished on their mothers as 'sucker' lambs or weaned and fed high-quality feed to sustain high growth rates. Late in the season, lambs may need to be provided with specialist pastures or crops such as brassicas, lucerne or clover, which have high energy and protein levels at a time when most perennial and annual pastures are diminishing in quality.

Silage

Where grain and lamb prices allow, there may be opportunities to finish lambs on grain and other preserved fodders. High-quality silage can sustain good growth rates for lambs, but trials have shown that increasing the percentage of grain in the diet increases the efficiency with which animals convert feed to body weight. Table 3.27 shows an example of intake, growth rates and feed conversion efficiency for second cross lambs (Poll Dorset over Border Leicester/Merino ewe) offered clover silage and barley. Feed conversion efficiency is a measure of how much feed is required to achieve a kilogram of weight gain.

Table 3.27. Percentage of barley added to silage increases live weight, intake and feed conversion.

	Percentage of barley in the diet				
	0	25	50	75	75[a]
Initial live weight (kg)	31	31	31	31	31
Final live weight (kg)	40	40	40	40	40
Live weight gain (g/day)	108	163	197	208	243
Intake (kg/day)	1.09	1.16	1.18	1.14	1.24
Feed conversion ratio (FCR) (kg feed/kg wt gain)	10.2	7.3	6.1	6	5.2
Days to reach 40 kg live weight	83	55	46	43	37

[a]lupin and barley mix to increase protein

Table 3.28. Crude protein requirements of balanced rations for lambs.

Energy in the ration (as dry matter) (MJ/kgDM)	Crude protein (%) requirements for 20–40 kg lambs		
	20 kg lamb	30 kg lamb	40 kg lamb
13	19.3	16.1	13.8
12	17.5	14.7	12.9
11	15.7	13.3	11.9
10	14.0	11.9	10.8
9	12.4	10.6	9.6

Intake and growth rate from silage will variously enormously depending on the quality of the silage produced. Poor quality oaten silage has recorded live weight gains of 3 g/day compared with 142 g/day on lucerne silage. The addition of grain to both of these resulted in improved lamb growth rates. Short chop length (for example, 10–30 mm compared with 750 mm) can improve growth rates in lambs.

Protein energy balance

The most commonly available grains such as cereals and pulses have good levels of available energy, but crude protein can be highly variable. For example, lupins are a rich source of crude protein (32% CP), but, on the other hand, the crude protein of oats can be as low as 5%.

The lamb's protein requirement varies as a result of the energy content of the ration and lamb live weight. Small lambs with a high energy intake need up to 20% CP in their diet, whereas large lambs on a lower energy content ration will need less protein. Where rations are being mixed on property, a feed analysis is critical in producing the right balance of energy and protein. Table 3.28 provides a guide to the protein requirements in feeds varying in energy density (as MJ/kgDM).

Fibre

As with all ruminants, lambs need roughage (fibre) to ensure the efficient functioning of the digestive system. With poor-quality roughage such as straw or grass hay, 10% by weight of the ration is sufficient, but if the roughage is of higher quality (more available energy) then it can make up 30% of the ration.

The provision of high-quality roughage is important when introducing lambs to a grain-based diet to ensure they all start eating. In an ad lib feeding situation, lambs will select around 30% of roughage and 70% of the grain, by weight of ration.

Minerals and vitamins

Lambs need a range of minerals to maintain good health: the most important mineral in a grain finishing system is calcium. Calcium deficiency can be corrected by adding 1.5–2% by weight (of ration) of finely ground limestone (calcium carbonate). Selenium and cobalt supplementation may also be necessary if lambs come from a deficient area.

Vitamin deficiencies are unlikely to be an issue if lambs are fed on grain for short periods. The only likely exceptions are vitamins A and E if lambs have not had access to green feed in previous months.

Animal health

Grain poisoning (acidosis) is the primary health issue and is managed by careful introduction of grains, care with changing the source type and batch of grain and the inclusion of roughage. A buffer, such as sodium bicarbonate or sodium bentonite, can be added to high-cereal-based diets. Sodium bicarbonate is an alkali and so buffers against acid production. Bentonite is a clay that has a high ion-exchange capacity and swells in the rumen to slow digestion. Commercial buffers in pellet form are also available as grain additives.

Other diseases common to sheep that are confined in small areas and fed concentrates include bladder stones (urinary calculi), coccidiosis, pink eye, salmonellosis, internal parasites and pulpy kidney. Where possible, preventative measures such as vaccination and drenching should be undertaken to prevent these diseases.

Budget

Finishing lambs on supplements or specifically grown fodder crops needs to be carefully budgeted. A more profitable option may be to sell lambs as stores because other finishing systems may be more efficient (for example, proximity to grain, feedlot or a later pasture finishing area).

The main budget considerations are the value of the lamb at the beginning (sale price), the cost of the feed to reach the desired weight and the value of the lamb at the end. Establishing a price contract to sell the finished lambs is critical to ensuring a profit. Consideration must also be given to any incidental costs such as likely deaths and shy feeders, disease treatments, shearing, labour and equipment. Feed costs are by far the greatest cost (excluding lamb purchases) and commonly fluctuate with sheep prices. The quantity of feed required is calculated using an estimated feed intake, feed conversion ratio (FCR) and desired weight increase.

Growth rates and feed intake

The growth response to feeding is affected by breed, genotype and sex. Generally, Merino lambs grow more slowly than crossbred lambs. Merinos are capable of daily growth rates of 110–200 g/head/day with a FCR of 7:1, whereas crossbreds can achieve up to 350 g/head/day and a FCR of at least 6:1.

As a guide, lambs can eat a maximum daily dry matter weight equivalent to 4% of their body weight. This assumes that the ration is highly digestible. For example, a 30 kg lamb could eat about 1.3 kg of grain-based supplement (90% dry matter) per head per day. Assuming a FCR of 6:1, this lamb would consume about 100 kg of feed to gain 15 kg live weight.

Genetic selection

There are large genetic variations within breeds in animals' ability to grow fast and lay down muscle rather than fat. Many trials have shown the benefits of sourcing lambs that have been bred using sires that are genetically superior for growth, muscling and leanness.

Lambs from parents selected for high growth have the potential to reach target weight faster, and therefore at less cost. Lambs that have also been selected to lay down more muscle and less fat will do this more efficiently, because the energy required to lay down muscle is less than that required for fat deposition.

Therefore the lambs that are most suited to grain finishing are first or second cross that are over 35 kg, with a fat score of 2 and with the genetic potential for growth. It is also advantageous for lambs to have been previously trained to eat concentrates.

Nutrition and meat quality

Meat quality covers a range of characteristics including colour, flavour, tenderness and general appeal – all of which have an impact on consumers' choice of meat. All practices in the supply chain can affect these attributes, including nutrition in the finishing phase.

Nutrition should be good in the last 2 weeks of finishing to maximise muscle glycogen. Low muscle glycogen at slaughter can cause dark cutting meat, which occurs when the pH of the meat does not fall sufficiently after slaughter.

Crossbred lambs need to grow at 100 g/day. Merinos need higher growth rates of 150 g/day, because they are more prone to deplete muscle glycogen stores between leaving the farm and slaughter.

Although roughage in the diet is important for the reasons outlined earlier in this chapter, the temptation to feed high levels of fibre to reduce scouring (leading to soiling at slaughter) will reduce the dressing percentage and actually increase effluent output from travelling livestock. Lambs finished over the summer, or during a dry period, may be deficient in vitamin E, which affects the production as well as the shelf life of meat.

Lambs that have had a nutritional setback from early weaning and/or drought conditions can exhibit compensatory growth with no real disadvantage. Generally, there will be no impact on the carcase (although they may deposit slightly more fat over the GR site than otherwise) and total fat or muscle will be no different when compared with lambs that have been continuously fed well.

Key messages – finishing lambs

- Finishing lambs is producing lambs to meet market specifications, which are based on measures of fat and carcase weight.
- Most market specifications require ideal lambs to be fat score 2–3.
- Finishing lambs off their mothers requires high-quality feed supplied by pasture or specialist pastures such as lucerne or brassicas for out-of-season supply.
- There may be opportunities to finish lambs on grain or supplements, but careful budgeting and feed planning is required.
- Approximately 6 kilograms of high-quality feed is required to produce an extra kilogram of live weight gain on a crossbred lamb, but there is large variation between lambs.
- Lambs from parents that have been selected for high growth and muscle will be the most cost effective to finish.
- Good nutrition in the last 2 weeks pre-slaughter is recommended for good eating quality.

Table 3.29. Increase in wool production of wethers fed high and low nutrition.

	Low nutrition	High nutrition
Greasy fleece weight	8.3 kg	10.4 kg
Fibre diameter	19.7 µ	22.9 µ
Live weight	55 kg	63 kg

Nutrition and wool production

As a general rule, it is not economical to feed a supplement to grow more wool, because of reduction in wool value, due to the concurrent increase in fibre diameter, and the cost of hand feeding cannot be recovered by additional wool produced.

Fleece weight and fibre diameter are positively correlated and so both increase with increasing nutrition. Table 3.29 shows this effect of nutrition on fleece weight, fibre diameter and live weight. In this example, the same sheep (same genetics) were fed either in a low or high nutrition environment. High nutrition caused an increase of three microns in fibre diameter.

Poor staple strength of wool can result severe discounts in some years, particularly in finer wools. The measure of strength that wool is discounted at changes with both the fibre diameter and with time (for example, it has increased from 32 N/kTex to 38 N/kTex). The point at which the staple breaks (base, middle or tip) can also have an impact on the price discount. Although there is some potential to improve staple strength through genetic selection, the main contributing factor is nutrition affecting fibre diameter along the staple. Figure 3.19 shows an example of the change in fibre diameter along a staple profile (for a medium wool sheep) with changing pasture production over a year.

Staples are more likely to break where there are sharp changes in fibre diameter along the staple. Trials have aimed to improve staple strength by manipulating nutrition. Improving nutrition during summer and autumn when feed is limiting may do this, but the associated increase in fibre diameter, along with the cost of additional feed, negates any benefit. The other option is to reduce

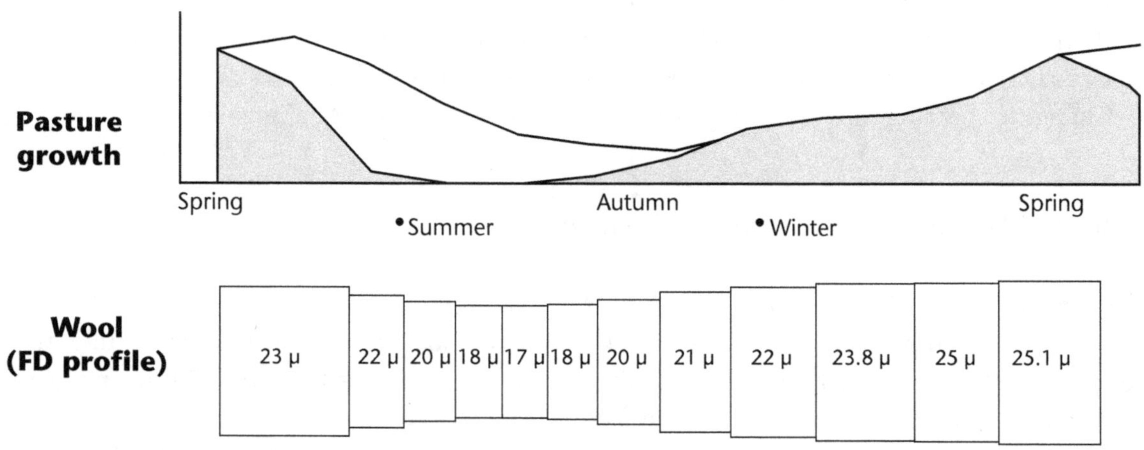

Figure 3.19. Example of fibre diameter (FD) changes along a staple with changing pasture growth and quality.

fibre diameter through lower nutrition in the spring time: this has the added benefit of reducing fibre diameter, but the wool cut per head is reduced.

Sheep that are shedded to produce superfine wool are provided with a low and even feed intake to keep wool growth (and hence fibre diameter) constant through the year. However, controlling the quality and quantity of feed intake of sheep grazing pasture is much more difficult. On-farm trials in Western Australia have been quite successful in doing this, but similar ones in Victoria were not. The difficulty was compounded by the need to control feed over the whole 12 months and manage appropriate weaner growth rates for survival and future reproductive targets. Most of the trials have been done with hoggets because they are more prone to lower staple strength due to the extra demands of body growth, and hence there is a need to balance feed restriction with requirements for body growth. There are more important factors to manage on most sheep farms.

Ewes are less likely to produce tender wool if late pregnancy and lambing occur on green pastures. This is because of the associated increased nutritional demands due to lamb growth in the last 6 weeks of pregnancy and then into lactation. Hence, autumn-lambing ewes are more prone to tender wool than winter- or spring-lambing ewes because of the associated demand on the ewe at a time when, in most years, nutrition is poor.

Another alternative to try and manage staple strength has been to shear in autumn, based on the assumption that poor feed availability in summer/autumn leads to a decrease in fibre diameter in autumn. If the large increase or decrease in fibre diameter occurs close to shearing, then the point of break will be at the tip or very base of the staple and therefore not such a problem for processing because the majority of the staple length would be left intact. However, seasonal and year-by-year variation has shown that there is not always one point of break at the same time every year and so trying to change shearing time will not always be effective. Shearing in March, when live weights of young sheep are likely to be low, may also decrease their chances of survival.

Sulphur, as sulphur amino acids, is important for wool production but, in trials, sulphur supplements have not increased wool production. Ensuring that soil sulphur levels are adequate for pasture growth should be the priority.

Key messages– nutrition and wool production

- Feeding for wool production or to increase staple strength is usually not cost effective.
- Ewes that lamb onto green feed are less prone to poor staple strength.
- Changing shearing times may improve staple strength, but is not guaranteed.

Practical tips for supplementary feeding

Ground feeding

Laying a trail of grain on the ground is the simplest and most practical method of feeding grain, despite there being some wastage. A long thin trail (though not too sparse) has the advantage that it allows all animals to have access to the feed. Feed space for all animals can often be limited when using troughs, and hence make it harder for shy feeders. The grain trail reduces the effect of the more-dominant sheep consuming more than their share and allows the shyer sheep easier access to grain.

Figure 3.20. Simple trough made from corrugated iron tech screwed to wooden posts.

The apparent wastage may not as great as it may appear. A scatter of grains left on the ground may look significant, but actually represent a very small proportion of the amount fed out. Boggy, wet ground can be a problem, particularly with pellets. If feeding in wet, boggy conditions rations need to be increased to allow for wastage (which can be as high as 50%). Grains, and especially pellets, should be fed on higher ground that dries quickly.

Because of the possibility of weed seed contamination of bought in grain, it is recommended to feed in the same area. If weed seeds germinate, they are easily located and can be quickly eliminated.

Troughs

When stock are confined in containment feeding, as in drought conditions or in feedlotting, feeding in troughs is recommended to reduce wastage and for animal health reasons. Trough space required is 10–15 cm per lamb and 15–20 cm per adult (double this if troughs are single sided).

Troughs can be made quite simply from robust materials like corrugated iron or rubber belting.

Feed-out trailers

There are a variety of feed trailers available. Some feeders have built-in weighing systems, electrical actuator opening feed-out shoots and augers for filling self feeders. Some feeders are sprung and some are not. When buying a feed trailer, consider the amount and type of use and number of animals being fed. For example, an unsprung feed trailer may bounce to bits in rough country. Give some thought to the capacity required to feed either all mobs or all mobs in one section of the farm with one fill. Save time by avoiding, where possible, having to come back to the silo to re-fill.

Some spinner super spreaders can be converted to feeders relatively easily by removing the spinner to allow grain to flow from the outlet. Some can be restored for use as spreaders and used interchangeably as super spreaders and grain feeders, though this is usually a nuisance.

Do not leave the spinner on and spin out the grain. There is little point in spreading the grain all over the paddock, increasing wastage and making the sheep work harder to get it.

Some feeders, which use a slide mechanism for opening the hopper and releasing the grain, require that the slide is pulled fully open before being released and 'catching' on the pawl at the correct setting for the flow rate required. This means that there is usually an initial large dump of grain at the start of the trail. The 'dump' effect can be reduced by ensuring the feed out vehicle is moving before opening (or closing!) the slide so that the dump is drawn out into a trail.

Filling the feeder

If feeding large numbers of animals, a considerable amount of time can be spent waiting for the bin to fill. If buying an auger to fill the feed trailer, it is important to consider the time it takes to fill the trailer and bins. A larger auger or larger motor on the auger may be a worthwhile investment to reduce the time spent feeding animals.

Calibrating a feed trailer

It is important to know how much feed is being fed and this requires calibrating the feeder. Feeders will need recalibrating with different grains, and sometimes for different batches, because they may vary considerably in flow speed. This is because of the different bulk densities of the grains such that, for the same volume of grain, some will weigh significantly more than others (for example, lupins are heavier than oats). The simplest way to calibrate is to weigh and time some feed coming out of the trailer. Place a container (wool pack or tarp) under the feed out chute and open the chute for 60 seconds. Then weigh the grain in the container, divide the weight of grain by 60 this gives us the flow per second; for example, 60 kg/60 seconds = 1 kg per second. This can be done for a full, part-full and near-empty hopper and the rate averaged, because the flow rate will change with the amount in the bin.

Self or lick feeders

There are many models of self feeders available (including home-made models). A 'lick' feeder means that animals need to lick the grain in order to get it, which restricts their intake somewhat. Full-flow, as implied, means that animals are on full ad lib feeding. Animals help themselves and so frequency or timing of feeding and, to an extent, the amount of feed is not controlled. Hence, they are most used for production feeding. Many of the lick feeders can be wound down to reduce intake. Lick feeders may give more control over the feed consumed by the animals than a self feeder. When using self or lick feeders:

- ensure sheep are well adjusted to a high-grain ration (at least 75% of the full ration), before filling self feeders

- provide sufficient feeders to allow all animals reasonable access to the feed
- check that the feeders do not become blocked by the animal's saliva mixing with the dust in grain or pellets.

Getting sheep onto supplementary feed

Whether using a lick feeder or a self feeder, animals need to be well adapted to the feed before a full ration is fed. Imprint feeding is by far the most effective of all the methods in encouraging adaptation of weaners to grain or concentrate feed, but this does not preclude the use of other 'tricks' to increase the rate of adaptation.

The following methods will assist adaptation even if weaners have already been fed as lambs on their mothers:

- Lay the grain trail on, over and around a line of hay (rolled out from a roll or a line of broken-up 'biscuits' from small bales). Weaners seem to come reasonably readily to the hay and, in doing so, get accustomed to coming across and eating the grain from the trail.
- Lay the grain trail (or grain and hay) close to a water point so that weaners cannot fail to come across it in the course of coming to drink.
- Include a few adult sheep in the mob that have had repeated, and preferably recent, experience with grain feeding. By example, these animals help to bring the less-experienced weaners to the trail and encourage them to feed.

Increasing rations or changing feeds

Generally, the introduction to grain rations should start with daily feeding until it is observed that most animals are regularly eating from the trail or troughs and the entire ration fed out is being eaten. Frequency of feeding can then be reduced to every second day and then to twice-weekly, but the key to successfully managing the transition to less frequent feeds (with a greater quantity being feed out at each feed) is to not to make sudden changes, but rather change the level of feed gradually.

Do not change from feeding say 250 g/head/day of barley on Monday to 500 g/head on Tuesday (intending to feed 500 g again on Thursday). Similarly, do not just miss a day, then change straight to 500 g/head/day on Wednesday (because the danger of acidosis, at least at a sub-clinical level, is considerable). Rather, feed 300 g/head on Tuesday, 400 g/head on Thursday and then again on Saturday and then 500 g/head on Sunday and every second day from then on.

The small reduction in the calculated quantity per head per day over the short period of change will be more than compensated for by the lack of any rumen upset that could be caused by changing too rapidly.

Similarly, if changing grain sources or pellet batches, either shandy the mixes (which may not be possible) or go back to daily feeding at the lower rate and gradually increase the ration.

Adding limestone to grain diets

When needing to supply calcium to grain diets, ground agricultural limestone – calcium carbonate ($CaCO_3$) – is the cheapest option.

Do not use 'builders' lime (calcium hydroxide – $Ca(OH)_2$) – also known as 'brickies lime', 'slaked lime', 'hydrated lime' or 'LIMIL') or 'quicklime' (calcium oxide – CaO - also known as 'burnt lime' or 'unslaked lime'). Both are extremely corrosive and will cause ulceration and burns to the mouth and throat if ingested by animals, as well as damage to the eyes and skin, not to mention possible similar injuries to the operator.

Limestone should not be added to the silo when filled because it is very corrosive. The simplest method is to add it to the feed-out wagon at 1.5–2% of the weight of the feed as grain is being augured in. For example, if the feed wagon holds one tonne of grain, add 20 kg of limestone, using buckets, as the grain is delivered into the hopper. An occasional 'stir' or 'swirl' with a square-mouth shovel will assist mixing; further mixing will occur as the feed vehicle is driven down the paddock. The limestone does not need to be perfectly mixed with the grain: do not be concerned with the apparent lack of the white powder on much of the grain and some limestone on the ground in and around the grain-trail. Experience and research shows that if added to the grain at 1.5 to 2% of the weight of the feed, sheep will take in sufficient calcium for it to be effective.

Many producers prefer to place lime in a trough with salt to encourage animals to eat it. Start with a 50/50 mix of salt and lime then reduce the salt content as the sheep become accustomed to the mix. The salt in the mix should be used to regulate consumption. If the sheep are eating too much of the mix, decrease the salt; if consumption needs to be increased, add more salt.

The most valuable tools for managing sheep nutrition are scales and a drafting gate. Monitoring of weight and/or condition is vital to ensure that the ration being fed is appropriate to meet the target weight for the group. Shy feeders, light weaners or poorer condition ewes in a mob can be drafted off and fed separately with more and/or better feed. Both these practices have the potential to reduce feed costs, and improve survival and production. Practical limitations, such as the availability of paddocks, ease and convenience of feeding and feed availability, can limit the number of groups that can be sensibly managed.

CHAPTER 4

Drought management

Severe feed shortages caused by fire, flood and drought force some critical decisions regarding managing stock and protecting pasture, soil and water resources.

In a drought, there is usually some feed that can be scavenged by sheep, but they will need a ration to supplement available grazing if they are kept on farm. Protection of the natural resources such as soil, water and pastures are also important considerations in retention and feeding of livestock. In addition, managing financial reserves during drought is essential to ensure long-term viability.

Drought management options

A number of options are open to a sheep producer in a drought. All financial and physical resources available should be itemised so that the effects of the various options can be budgeted and planned for.

Doing nothing is not acceptable farming practice. Sheep cannot be allowed to starve. Appropriate quantity and quality of feed must be provided; otherwise sheep should be sold.

Agistment

Agistment may be an option, especially early in the drought, but the cost (including transport), ease of management from a distance, feed quality and animal health need to be considered.

Selling stock

Decisions to sell stock need to be made before sheep are unsaleable. Shearing and joining dates will have an impact on critical selling times. Taxation may also have an effect on selling policy.

Usually at least some stock are sold to reduce feeding costs, and so classes or mobs of sheep should be prioritised for sale. If the season deteriorates, trigger dates can be set to sell the next mob. Priorities should include cast-for-age and cull sheep, with the aim of retaining as many breeders as possible to allow numbers to be built up quickly after the drought. Wethers therefore may be high on the priority list, but, if some are retained, then it should be the most productive ones (that is one-and-a-half- to three-and-a-half-year olds).

Deciding whether to keep and feed, pay agistment (if available), to delay joining or not joining at all (to reduce nutritional requirements) or to sell sheep will depend on a number of factors. Common practice in southern Australia is to implement a combination of strategies, which include selling some stock and feeding the most productive or core breeding ewes. Generally the decision will be based on:

- an educated guess on how long the drought will last based on historical rainfall records
- the cost to feed stock
- water resources
- financial and physical resources
- consideration of the value of stock sold at the beginning of drought and the cost to buy back after the drought
- the value of the genetics and health status of stock
- personal commitment – feeding through droughts can be very monotonous and seemingly unrewarding, but selling core stock in the middle of, or late in, a drought can be extremely costly.

Comparing options

Tables 4.1 to 4.4 show how to compare the costs of feeding versus selling and buying (on a dollars per ewe basis). They are examples only, because costs and values vary constantly, but they illustrate a process that can be used to help with decision making.

Tables 4.1 and 4.2 provide an example cost comparison of feeding or selling a spring lambing Merino ewe. The assumption is that most of the ration is grain, with an average per tonne cost of $350, with 0.5kg of hay per ewe per week.

Table 4.1. Example of comparing the costs to keep and feed a Merino ewe versus selling and buying back.

Feeding – per ewe		Feed Price	$350/tonne
	Rate feeding (kg/ewe/week)	Monthly[a] Feed cost/ewe	Cumulative Feed cost/ewe
Oct	0.0	$0.00	$0.00
Nov	2.5	$3.75	$3.75
Dec	4.0	$6.20	$9.95
Jan	4.0	$6.20	$16.15
Feb	4.0	$5.60	$21.75
Mar	4.0	$6.20	$27.95
Apr	4.0	$6.00	$33.95
May	4.0	$6.20	$40.15
Jun	3.5	$5.25	**$45.40**

Table 4.2. Expected cost of feeding.

Feeding – per ewe		Feed price	$350/tonne		
	Rate feeding	Monthly[a]	Cumulative worst case		Cumulative expected feed cost
	(kg/ewe/ week)	feed cost/ewe (A)	feed cost/ewe	Probability of effective rain (B)	cost/ewe A–(A × B)
Oct	0.0	$0.00	$0.00		$0.00
Nov	2.5	$3.75	$3.75	0%	$3.75
Dec	4.0	$6.20	$9.95	0%	$9.95
Jan	4.0	$6.20	$16.15	0%	$16.15
Feb	4.0	$5.60	$21.75	2%	$21.64
Mar	4.0	$6.20	$27.95	22%	$26.47
Apr	4.0	$6.00	$33.95	50%	$29.47
May	4.0	$6.20	$40.15	80%	$30.71
Jun	3.5	$5.25	**$45.40**	95%	**$30.98**

[a] the monthly grain cost per ewe calculated on the number of days in each month

Table 4.1 shows the monthly cumulative cost of feeding grain to ewes in drought, starting in October. The assumption is that there will be some paddock feed, but that stock are confined and fully fed from December.

In this scenario, it will cost $45.40 to feed a ewe for 8 months. This calculation describes the worst case scenario: the full cost of feeding until the end of June. In reality, it is likely that some rains will arrive before this time and the expected feed cost will be lower. Table 4.2 takes account of the probability of an autumn break. The monthly feed cost minus the product of the monthly feed cost and the probability of effective rain in that month will give the expected feed cost. The probability of the autumns rains occurring are calculated from historical rainfall records. In the example below, long-term rainfall records (80 years) were used for a farm near Benalla in north-east Victoria.

Using this feed cost, Table 4.3 then compares the gross margin of feeding ewes versus selling and buying back after the drought. All costs and income need to be calculated on an individual basis. The selling price is usually known when the decisions are being made, but the buy back price needs to be made based on some experience and view of the likely shortage of stock in the future.

With the costs associated with keeping and feeding, less the wool income, the net cost to feed a ewe through is $6 (ignoring capital costs). When considering fleece value post drought, conservative budgets should be used because wool has excessive dust and is tender. Alternatively, selling for $20 and buying back for $60 would be at a cost of $40/ewe. For a spring lambing flock, the assumption would be that ewes were purchased in lamb. If not, this will have a longer term impact on future income and flock rebuilding. It is also important to consider the worst-case-scenario costs, particularly in terms of cash flow. By retaining sheep, can the farm support the additional debt of feeding, especially if the drought continues? Although the expected cost of feeding (assuming

Table 4.3. Gross margin: feeding versus selling a spring lambing Merino ewe.

	Keep and feed	Sell now
Sale price	$0	$20
Costs		
Feeding *(from Table 4.2)*	$30.98	$0
Shearing	$5	$0
Crutching	$1	$0
Animal health	$2	$0
Sundry	$1	$0
Income (fleece)	$34	$0
Buy back price	$0	$60
Gross margin/DSE	**– $6**	**–$40**

some rain) is more likely, the scenario of feeding through to winter must be considered. Producers should make the calculations based on the probability of effective rainfall in their own region.

Tables 4.4 and 4.5 provide an example cost comparison of feeding or selling a May-lambing crossbred ewe. The assumption is that most of the ration is grain, with an average per tonne cost of $350, with 0.5 kg of hay per ewe per week.

Table 4.4 shows the monthly cumulative cost of feeding grain to ewes in drought, starting in October. The assumption is that there will be some paddock feed, but that stock are confined and fully fed from December. The timing of the autumn break will be unknown. The expected feed cost is based on the probability of effective rain used in Tables 4.2 and 4.4.

Table 4.4. Example of comparing the costs to keep and feed a May-lambing crossbred ewe versus selling and buying back.

Feeding – per ewe		Feed price	$350/tonne		
	Rate feeding	Monthly	Cumulative		Expected feed cost
	(kg/ewe/ week)	feed cost/ewe (A)	feed cost/ewe	Probability of effective rain (B)	cost/ewe A-(A x B)
Oct	0.0	$0.00	$0.00		$0.00
Nov	2.5	$3.75	$3.75	0%	$3.75
Dec	4.5	$6.98	$10.73	0%	$10.73
Jan	5.0	$7.75	$18.48	0%	$18.48
Feb	5.0	$7.00	$25.48	2%	$25.34
Mar	5.0	$7.75	$33.23	22%	$31.38
Apr	6.0	$9.00	$42.23	50%	$35.88
May	9.0	$13.95	$56.18	80%	$38.67
Jun	5.0	$6.75	**$62.93**	95%	**$39.01**

Table 4.5. Gross margin: feeding versus selling a May lambing crossbred ewe.

	Keep and feed	Sell now
Sale price	$0	$35
Costs		
Feeding *(from Table 4.4)*	$39	$0
Shearing	$5	$0
Crutching	$1	$0
Animal health	$2	$0
Sundry	$1.00	$0
Income (fleece)	$11	$0
Buy back price	$0	$100
Gross margin/DSE	–$37	–$65

Because of the larger size of the ewe and the extra requirements needed for lambing, the worst-case scenario cost to feed a ewe for 8 months in this example is $62.93 and the expected cost is $39.01. Again, both these figures should be considered when making decisions about what and how many stock to keep.

The difference in this scenario is that there will also be a lamb to sell later in the year for ewes that are kept and fed, providing income. It is unlikely that ewes could be purchased to lamb before the following year and so the sell-now scenario will have an impact on cash flow and future income.

These are examples only and the calculations must be done to represent the relevant prices and stock requirements. The costs and stresses associated with feeding, as well as cash flow and available water, must be considered in the calculations. However, doing these calculations, and making a decision, provides ease of mind and a real knowledge of ongoing and total costs involved.

If financial reserves are low, feeding and hoping for an early break is not a good option as feeding costs, as well as the cost of selling sheep for little return, may be incurred before the drought breaks.

Time of shearing and weaning

The time of shearing may need to be altered to reduce shearing losses or to hasten the selling times of cull stock. Recently shorn, drought-affected sheep are at risk of dying in cold, windy weather. Extra feed may be required before and after shearing to guard against possible losses if the weather following shearing is unfavourable.

The time of weaning may be altered to simplify the management of ewes and lambs. Lambs in drought conditions should be at least 6 weeks old, at least 9 kg in body weight at weaning and they must receive high-quality rations.

Financial recovery

Decisions on whether to sell or keep stock should also consider the future ability to recover financially and to create income. Strategies usually include selling some stock and feeding the most productive. One strategy could also be to sell some stock, and to breed numbers back up rather than buy back in. Although this decision will require less cash outlay, it may take longer to get back to a productive stocking rate and good cash flow. For this reason, it is essential to budget the impact of decisions out at least 3 to 5 years, considering peak debt and the recovery period. Every drought will pose a different set of criteria on which to base decisions, but recovery and making profit in more normal or better years is crucial to long-term viability.

Drought feeding

If stock are kept and fed through a drought, then the decisions will involve what to feed, how much and how. Chapter 3 outlines the nutritional requirements of sheep and how to monitor animal condition. In drought, it is common to let stock drop in condition and to maintain them at a lower weight, compared with a season where pasture is supplying the majority of feed. Therefore the recommended rations for drought feeding may be lower than maintenance feeding in a normal year.

Stock can be confined and fed full rations in small areas (called containment areas or drought-lots) to protect pasture and soil resources. In this situation, stock will require full rations and so feed costs are likely to be higher. This system has a number of advantages and has been used successfully by many producers in past droughts (see the section on stock containment later in this chapter).

When to feed

The aim is to avoid excessive weight loss. It is cheaper and easier to maintain stock at a reasonable weight, than to feed for weight gain later on. Ideal condition scores or weight will vary with different market and production scenarios, and the comfort level of the individual producer. One 'rule of thumb' is to feed a half ration – normally of grain– when half the sheep in the mob have fallen to condition score 2 or below. If condition still falls, lift the feeding rate to a full ration. Stop feeding after the drought when the pasture availability increases above the critical minimum levels outlined in Chapter 3.

For weaners, feeding should be aimed at reaching or maintaining critical live weights for survival and future reproduction as outlined in Table 3.25.

How much to feed

Table 4.6 outlines full drought rations for some classes of stock, assuming there is no paddock feed. Note that these rations will vary, and adjustments to the rations are listed below.

If the sheep are fed in paddocks where some pasture residue is available, the actual amount to be fed may be a proportion of the total ration. If fed in a containment area, a full ration will be required.

After deciding on the proportion of the ration to be fed, adjustments will still need to be made according to the changing body weight and condition score of the sheep, weather conditions and pasture availability.

Adjustments to rations

The rations are a guide only and must be adjusted according to the requirements and responses of sheep. Some general adjustments include:

- Larger framed sheep – increase the ration by 10% for each 5 kg of extra body weight. For example, 3.5 kg of wheat per week should maintain a medium-framed sheep in condition score

2 at about 40 kg live weight. A further 0.4 kg per week would be required to keep a sheep at about 45 kg.

- Rams – increase the ration by 10%, plus a further 10 to 20% for 4 to 6 weeks before mating.
- Cold weather – increase ration by about 20%, or even more under severe conditions.
- Sheep exposed off-shears to cold conditions (especially if cold, wet and windy) may need their ration increased by the equivalent of two dry sheep rations and should be given whatever shelter can be provided. Hay is the safest feed for sudden boosts in the ration. Hay can be gradually replaced by grain over 2 to 3 weeks if the increase has to be sustained.
- Muddy ground – increase rations by 0.5 kg/head to allow for wastage when trailing grain on muddy ground.
- Paddock feed – even a comparatively 'bare' paddock can provide some of the requirements of a dry sheep. Past droughts have shown that sheep continue to get some nutrition from very bare paddocks, but the care and stability of these paddocks must be considered.
- Continue to monitor the condition and/or body weights of the sheep.

How to feed

Grain supplements can be trailed onto the ground. Troughs may be used to reduce the wastage if long-term feeding is likely. Up to 10% of grain may be wasted if fed on the ground and this should be allowed for when working out the ration.

Adult dry sheep should not be fed more than three times a week, so that shy feeders get an opportunity to eat some grain. Table 4.7 shows the percentage of sheep gaining weight and the percentage death rate in sheep fed the same weekly grain ration either on a once per week or on a daily basis.

Young sheep do better if fed more often (daily) as long as shy feeders are removed. Hay and grain can be alternated.

Introducing sheep to drought rations

Feeding should be started while sheep are still in good condition and while there is still some paddock feed. Grain should be fed out in thin trails so that it can be eaten along with pasture residue. As outlined in Chapter 3, sheep have to be introduced to cereal grain rations gradually (see the section on introducing sheep to grain rations and Table 3.19). Grain poisoning or acidosis is the most common cause of deaths in sheep fed high-grain rations in drought.

Choosing a drought feed

The most limiting nutrient in drought is energy. Generally the feed that provides the cheapest source of energy is the best to buy. However, some classes of stock (weaners, late pregnant ewes) may have special requirements.

A guide to the energy, protein and fibre values of various traditional feeds is given in Table 3.16.

Protein, minerals and other supplements

In most cases, there is no advantage in paying high prices for extra protein for drought feeding. Most grains can supply most drought needs for protein: 7% crude protein (CP) for dry stock and 12% CP for growth or reproduction.

Supplementation of protein, minerals and vitamins are likely to be needed in the following instances:

Table 4.6. Total drought rations for sheep – weekly energy requirements for maintenance and minimum dietary protein concentrations for different classes of sheep, assuming that no paddock feed is available and the weather is mild.

Class of stock	Energy requirement MJ/week	Minimum crude protein % DM		Feed	Ration kg per head per week	Remarks[a]
1. Adult dry sheep, ewes in early stages of pregnancy in store condition						
• 40 kg live weight – medium frame	42	6		Wheat	3.5	
			or	Oats	4	
			or	Hay (good)	5	
			or	Hay (poor)	7	
• large-framed, or crossbred ewe at 60 kg live weight	57	6		Wheat	4.75	
			or	Oats	5.7	
			or	Hay (good)	7	
			or	Hay (poor)	10	
2. Pregnant ewes, last 4–6 weeks before lambing • medium framed	62	8		Wheat	5	Some hay (or dry paddock feed) is desirable but, if in short supply save until after lambing and increase grain ration by 0.5 kg as a substitute.
				Oats	5.6	
			plus	Hay (good to average)	1	
			or	Hay (good) alone	7	
• large-framed, or crossbred – 60 kg live weight	84	8		Wheat	5.5	
				Oats	6	
			plus	Hay (good to average)	1.5	
			or	Hay (good) alone	10	
3. Ewes with lambs at foot[a] • medium framed	84	10		Wheat	7	Rates apply to mobs with normal lambing patterns from start of lambing. If lambing is concentrated, increase rations by 1 kg grain plus 1 kg hay for first 3–4 weeks following the lambing peak, for full milk production. Wheat alone is not a satisfactory feed for lactating ewes.
			plus	Oats	9	
			or	Hay (good)	1.5	
				Hay (average)	2	
			or	Hay (good) alone	10	
• large-framed	120	10		Wheat	8.5	
			plus	Oats	10	
			or	Hay (good)	2	
				Hay (average)	2.8	
			or	Hay (good) alone	14	

Class of stock	Energy requirement MJ/week	Minimum crude protein % DM		Feed	Ration kg per head per week	Remarks[a]
4. Lambs • early weaned lambs up to 15 kg live weight gaining 1 to 2 kg per hd/wk	35	12	*plus*	Mixed cereal grain (3 parts) and lupins (1 part) Hay (good) @ 10%	Feed to appetite (3.5)	Combine the mixed grain feed with hay and feed the combined ration.
• Weaned lambs greater than 15 kg live weight	35	10	 *or* *plus* *or*	Wheat Oats Hay (good) plus grazing (about 1/3 ration) Hay(good) alone	2 2.3 3 4.5	Give access to the best grazing at all times. If no useful grazing is available, give extra 1.5 kg good hay (or 2 kg average hay). If hay is very scarce, reduce to 0.3 kg and increase grain by 0.8 kg (per week).

Notes

See 'Adjustments to rations' below for allowances needed for different weight animals or breeds, cold weather.

[a]Energy requirements for lactating ewes assume that ewes maintain body condition. If lambs are kept on the ewes longer than 6 to 8 weeks, requirements will increase because the lamb requirements increase.

Barley, maize and sorghum – as for wheat. Sheep pellets and millet – as for oats. Lupins and peas – as for wheat, but their high protein contents can be used to best advantage by adding small quantities to enrich poor-quality feeds, especially for weaners or for pregnant or lactating ewes.

Pregnant or lactating ewes are unable to eat sufficient poor-quality hay (as a sole or major part of their ration) to maintain body condition. Likewise, it is difficult to induce weaned lambs to eat any poor-quality hay. Even dry sheep may need a small quantity of grain (0.5–1 kg) to supplement a ration of poor-quality hay.

Table 4.7. Effects of daily or weekly feeding on weight gain and death rate in adult sheep.

	Sheep fed daily	Sheep fed once per week
Deaths (%)	30	12
Sheep gaining weight (%)	42	14

Death rates can be high when feeding daily unless the shy feeders are removed

- the diet of late pregnant and lactating ewes and weaners fed low-protein cereal grain can be improved by adding a higher protein feed such as lupins or peas
- poor-quality roughage fed can be improved by adding a higher protein feed such as lupins or peas
- calcium (administered as limestone) and sodium (as salt) are required for sheep on cereal grains
- vitamin A and vitamin E can be beneficial to sheep confined to feedlots and weaned lambs up to 1-year old if they have not eaten green feed for 3 to 4 months.

When making feed choices, compare feeds on the cost per megajoule of energy (an example of this calculation is worked though in 'Feeding supplements, Energy value of feed' in Chapter 3, and also in this chapter in 'Water content of alternative feeds'). If protein is limiting, then compare the cost of available feeds that provide adequate protein or the cost (including storage and labour) of adding high protein sources to a cheap energy feed.

Pellets that provide appropriate energy, protein and minerals (and possibly buffers) should still be considered in the energy–cost comparison. Some additional cost may be acceptable because of the ease of management and/or peace of mind derived from providing a balanced ration.

Alternative feeds

When grain is expensive or in limited supply, many other feed sources can be used as part of a main diet to reduce the cost. There are a range of alternative nutrient sources that may be available and cost effective.

When considering other feed sources, the principles are the same as for comparing traditional grain or roughage sources. Take into account the water content because this will affect the cost of the nutrient and cartage. Always work out the cost per megajoule of energy and compare with other readily available energy sources.

Other considerations include:

- **continuity of supply** – especially if time is taken for stock to adjust to new feed sources
- **storage life and requirements** – some by-products (like bakery and fruit pulp) may not store for very long
- **equipment needed to feed out** – some products may require extra or special machinery to feed out
- **palatability to stock** – the new feed should be introduced carefully and accompanied by some of the standard food source. Legume or canola silage made too wet, may have an unpleasant smell and often may only be eaten the day after it is offered

- **restricted animal material** – there is a strict ban on the feeding of animal materials to ruminants, including sheep. It is an offence to feed any animal material to a ruminant, with the exception of tallow, gelatine or milk and milk products. 'Animal material' means any material, tissue or blood taken from an animal and includes meal obtained from rendering tissues and blood from animals. In addition, compounded feeds and meals containing animal materials, and/or the documentation, invoices, tags and labels accompanying these products, must contain a statement that the product contains animal material and must not be fed to ruminants, or a statement to similar effect.
- **chemical residues or other banned compounds** – maximum residue limits (MRLs) are set for a range of chemicals in food commodities and animal feeds. MRLs must not be exceeded in food products such as milk or meat. Withholding periods (WHP) for any chemical use are designed to ensure that the food products will not have residues that exceed Australian MRLs, provided the agricultural (fungicides, herbicides and insecticides) and veterinary chemicals are used in accordance with their label directions. A few traps of which to be aware are:
  - failed crops may have been harvested before the WHP expired and so remain contaminated
  - some chemicals may have different WHP for harvesting, grazing and/or fodder conservation
  - some crops, not normally used for stock feeding, may have chemicals applied which do not carry specific label warnings related to consumption by livestock.

  Chemical residues may not be a problem in most alternative feeds, but are serious enough to be of concern in a minority of others. Therefore, feedstuffs should not be fed to stock without knowledge of chemical use or certification that it is appropriate for feeding.
- **animal health problems associated with feeding** – rations containing significant levels of some alternative feeds may need extra minerals; for example, limestone and salt are added to rice straw to provide calcium and sodium. Many alternative feedstuffs are high in fat or oil, which can reduce rumen function and inhibit feed intake. Fat or oil levels should be at or below 5% of the total ration.
- **contaminants such as metals, plastic or dead animals** – many food waste products, such as bread and lollies, may contain plastic bags, plastic ties, and so on. These may be labour consuming to remove and may be a future problem if not removed. Dead mice in cereal silage or stored grain present a risk of botulism, which can be fatal to stock.

Each drought brings a new list of possible alternative feed sources. Some examples of feeds that have been used successfully are listed below. Tables 4.8 and 4.9 outline some of the risks and feeding recommendations for a range of alternative or uncommonly fed products.

Generally, it is wise to ensure non-traditional supplements are not the sole ration and are not fed to hungry stock.

Energy by-products

Energy by-products include:

- **fruit by-products** – include apple pomace (by-product of apples used for cider or vinegar), citrus pulp, grape and tomato pomace. Pesticides are the main concern for these products, as well as the nutritive value because it will vary widely.

- **bakery waste** – ensure no meat products are included
- **fats** – the level of fat in ruminant diets should not exceed 5%
- **onions** – can be fed, but can cause anaemia in sheep if fed at more than 50% of the total ration.

Protein by-products

These include brewers grains and many crops that are grown for oil such as copra (or coconut meal), safflower, soybean, linseed and cottonseed meal. All these can be good protein sources, but may be expensive energy sources.

Roughage by-products

Many types of straw are available from crops and are useful as a roughage (fibre) source. Sawdust has been used with wheat diets to improve sheep survival rates. Waste paper has also been used as a source of roughage, but the risk of toxic contaminants such as lead and cadmium can be high.

Water content of alternative feeds

Many alternative feeds have a high level of water, which needs to be taken into account. An example comparing barley, a traditional feed, with onions, a more novel feed, is provided in the following example.

Which is the better buy – onions or barley?

Onions at $50/tonne delivered and are 11% dry matter and have 13 MJME/kg dry matter.
Barley at $250/tonne delivered and is 90% dry matter and 13 MJME/kg dry matter.
Which is the better buy?
One tonne of onions at 11% dry matter provides 110 kg dry matter. At $50/tonne, 1 kg of dry matter costs 45 cents ($50/110kg)
1 kg provides 13 MJ so 1 MJ costs = 3.5 c/MJ (45 cents/13)
One tonne barley at 90% dry matter provides 900 kg dry matter
At $300/tonne, 1 kg of dry matter costs 33 cents/kg ($300/900 kg)
1 kg provides 13 MJ so 1 MJ costs 2.6 c (33 cents/13)
Barley is the better buy in this example
Note that onions should not be fed as a sole ration.

Stock containment

Overgrazing due to season, fire or flood can reduce ground cover and expose areas of soil. As the ground cover starts to drop below about 70%, bare patches can become quite large and start to join up to create opportunities for soil movement, by wind or water. The slope of the land and stability of the soil will affect the critical level of ground cover. Hillsides are most prone to overgrazing, which can lead to bare areas and domination with plants such as capeweed. Most of the plant nutrients are in the topsoil, so any loss of topsoil can be very costly.

Many farmers remove stock from pastures during drought and use droughtlots or stock-containment areas. Containment areas are large yards in which stock are lot fed for survival or maintenance. Containment areas enable pastures to be destocked before they become too bare. Lot feeding for production, such as for finishing prime lambs, is a separate issue and is not covered in this chapter.

Feeding in stock-containment areas should be considered:

- to protect vegetative cover on pastures or failed crops
- to protect areas vulnerable to erosion
- to facilitate stock feeding, watering, monitoring and handling
- where stock are losing weight on full drought rations in paddocks
- where weeds in bought feed are of concern.

The short-term cost is likely to be higher because a full ration is required for the entire time stock are contained compared with part rations that can be fed in paddocks. However, the long-term saving in pastures and soils, though difficult to quantify, should outweigh the additional feed and capital structure costs.

Site selection

The site of the containment area is important and it should be set up as a permanent structure, like sheep yards, for future emergencies (drought/fire/flood) or for possible lot feeding opportunities. The site should have:

- a moderate slope and a well-drained, stable soil such as a clay or clay loam
- ready access to the house and feed storage
- no important remnant vegetation
- shade, shelter and good drainage
- access to good-quality water and clean facilities
- minimal problems with noise and smell.

The stock containment area should be setback 500 metres from watercourses and water storages. A nutrient filter can be established on the downslope side of the site.

Size and structure

Areas between two-and-a-half and five square metres per sheep have been used successfully.

A good size mob is about 500 sheep. If considering containing more than one group, good sub-divisional fencing will be required, as well as boundary fencing. Provision of a separate yard for feeding makes it easy to fill feed troughs and use additives before the stock get assess to the feed. A separate feeding yard can also provide a means of monitoring the vigour of the stock as they run into the feeding area and make it easy to draft off shy feeders or poor doers for preferential treatment. Other considerations include vehicle access, ease of filling feed troughs, water and cleaning.

Monitoring the body weight and condition score is extremely important. An easy means of handling sheep for weighing or condition scoring should be included in the planning.

Protect trees with guards or they will be ringbarked.

Water supply

A good reliable water supply is essential. Troughs need to be checked daily and cleaned regularly. Generally stock will be fed diets very low in water content and must be supplied with water at all times.

Table 4.8. Feed quality values and feeding risks and management of less common forage supplements.

Fodder	Quality values (range in brackets)				Risks	Management
	DM%	MJ ME/Kg DM	CP%	NDF%		
Canola silage	46.9 (24.8–75.7)	10.5 (6.1–12.4)	18.1 (6.6–26.0)	38.0 (25.6–58.2)	Potential nitrate and sulphur (S) toxicity. S-methyl cysteine sulphoxide (SMCO) toxicity causing haemolytic anaemia. Occasional 'Brassica-type' problems. Oil levels high if pods forming.	Introduce slowly to well-fed animals; avoid feeding to hungry or empty sheep. Limit intake to one-quarter to one-third of diet to avoid toxicity related problems. Watch stock for 'brassica-type' signs of respiratory distress, photosensitisation, blindness and sudden excitability, digestive problems, pulpy kidney and bloat. Total S intake to be less than 0.4% of diet DM. If silage is very dark in colour and foul smelling, it was probably too wet at ensiling.
Canola hay	84.8 (61.3–93.5)	10.4 (3.6–13.0)	16.8 (5.5–27.2)	37.9 (25.4–68.6)	Fire risk and mouldy hay. Hay has the potentially to be stalky, unpalatable and capable of puncturing rumens. Oil levels high if pods forming	Stems may have been too wet at baling. Visually inspect hay for mould. Inspect hay for high levels of leaf loss. Rumen damage is much less of a problem if hay is roller/super conditioned prior to baling to crush stalk, or if chopped short at feed out.
Millet silage	41.2	9.7 (8.5–11.6)	14.4 (5.9–26.6)	58.2 (44.0–65.0)		No sorghum prussic acid type problems, but nitrate poisoning can still occur in stressed plants.
Millet hay	79.9	8.5 (5.5–10.6)	8.7 (2.5–23.3)	66.2 (48.3–80.1)		No sorghum prussic acid type problems, but nitrate poisoning can still occur in stressed plants.
Pea hay	84.9	9.7 (5.1–11.6)	14.9 (4.5–21.6)	42.9 (29.1–70.8)	Mould in pods	Quality will vary according to leaf loss and mould presence.

Fodder	Quality values (range in brackets)				Risks	Management
	DM%	MJ ME/Kg DM	CP%	NDF%		
Sorghum silage	NA	8.1 (3.2–10.5)	9.8 (2.1–18.2)	62.3 (52.7–79.1)	Potential prussic acid (hydrogen cyanide or HCN) poisoning in stressed plants. Sorghum plants are low in sulphur content. Plants have high nitrate levels when stressed.	Drought, frost, high nitrogen and low phosphorous levels increase risk of prussic acid poisoning. Greater risk in fresh regrowth. Try not to cut/graze under about 0.75 – 1 m height. Crops mower-conditioned and chopped into a stack will have reduced levels of prussic acid; less so with baled silage. Similar reasons also can cause nitrate poisoning, nitrate levels being highest in bottom third of the stem. Ensiling will reduce N levels substantially, but be wary. Introduce partly gut-filled animals slowly to sorghum forage or silage slowly. Add sulphur salt blocks when ration is largely sorghum only.
Sorghum hay	Not available	9.0 (6.8–10.4)	9.9 (1.7–18.7)	62.2 (45.3–75.8)	Potential prussic acid (hydrogen cyanide HCN) poisoning in stressed plants. Sorghum plants are low in sulphur content. High nitrate levels when plants are stressed	See comments as per sorghum silage for causes of prussic acid and nitrate poisoning. Haymaking will not markedly reduce risk of prussic acid or nitrate poisoning. Super-conditioning may slightly reduce the risk of prussic acid poisoning. Introduce partly gut-filled animals slowly to sorghum forage or silage. Avoid making hay with stressed plants. Add sulphur salt blocks when ration is predominately sorghum.
Straw – rice	85.2 (52.2–93.5)	6.7 (5.3–8.9)	4.0 (1.9–5.0)	63.4 (53.4–68.5)	Contains silica and oxalate.	High levels of silica can predispose animals to urinary caculi. Add limestone and salt to provide calcium and sodium.

Note: The nutrient levels and ranges are a guide only and can vary considerably with property and season

Table 4.9. Feed quality values and feeding risks and management of alternative feed supplements.

By-product energy/ protein concentrates	Quality values (range in brackets)				Risks	Management
	DM%	MJME/ kgDM	CP%	NDF%		
Bakery waste bread	61	14	16	Low	Risk of acidosis (grain poisoning). Often high in fat and low in fibre.	Introduce slowly to avoid acidosis. Limit intake to 10–20% of diet depending on the product. The baking process may predispose animals to decrease their milk fat content. Other wastes, such as doughnuts, cakes and biscuits, may differ slightly in nutrient value so obtain a feed analysis.
Brewers grain	31.7 (13.9–93.0)	10.5 (7.7–11.9)	21.9 (16.9–35.2)	[7]55.1 (41.6–61.6)	Fat 8.0% (3.7–12.2)	May be expensive to transport due to low DM%. May be difficult to store and feed. Fat levels can be high, but not normally a problem. Brewing process makes protein less soluble than most protein supplements so useful with silage, which has high soluble CP levels.
Carrot pulp	10.0 (8.0–15.5)	12.7 (8.8–14.2)	9.8 (6.5–15.3)	27.6 (26.1–29.1)	Potential risk of acidosis.	May be expensive to transport due to low DM%.
Citrus pulp	14.3 (10.6–17.3)	12.7 (9.6–14.5)	8.6 (6.0–11.9)	25.1 (17.9–34.1)	High calcium: phosphorus ratio.	Introduce slowly. Composition varies depending on fruit type and whether peels and seeds included. High energy feed, but can be low in CP, NDF and DM content. Commonly mixed with silage, hay and grain as a fully balanced ration. Ideally fed in troughs or on a feed-pad. Can be difficult to store and feed unless set-up for it. Distinctive smell and taste. Can be expensive to transport. High in calcium and low in phosphorus, so should be counter-balanced by other feeds low in calcium and high in phosphorus.
Citrus pulp silage	15.6 (15.1–16.5)	***11.9*** (10.5–13.1)	9.5 (8.9–9.8)	Not available		As per citrus pulp comments above. Can be unstable on opening so feed out quickly. Use inhibitor additive at ensiling. Can be ensiled with other products
Chocolate by-products	91.3 (53.4–98.1)	15.6 (14.1–17.9)	7.0 (0.1–11.0)	Not available	Fat 18.7% (7.0–29.8) Note wide range in DM, ME and CP and potential of very low CP and high fat content	Potential risks of acidosis unless feed is introduced slowly and feeding rate is increased slowly. Must be fed in conjunction with a fibre source and some additional protein.
Choc malt mix	94	14	24	35	May be difficult to manage, store and feed out. Potential risk of acidosis.	High-quality blends of sugar, starch, protein and fibre. Introduce and increase feeding rates slowly. Very palatable, but may be difficult to manage, store and feed out, particularly under warm conditions.

By-product energy/ protein concentrates	Quality values (range in brackets)				Risks	Management
	DM%	MJME/ kgDM	CP%	NDF%		
Copra meal	90	12	20	52	9% oil	A highly nutritious and palatable co-product from copra processing.
Grape marc: raw	[7]55.1 (19.6–93.9)	6.5 (2.3–12.1)	12.2 (5.4–18.5)	47.6 (20.3–60.6)	8.8% oil (6.0–10.5)	Note huge range in DM, ME, CP and NDF. This data probably also contains some pressed grape marc data. High tannins in grape marc tend to bind much of the protein, so allow for lower CP levels in rations. May contain high copper levels depending on pesticide used. May be difficult to store and feed.
Grape marc: pressed	50	10	13	33		Excess alcohol (ethanol) removed by distillation compared with raw grape marc from wine industry. Other comments as per raw grape marc.
Grape marc silage: raw	35.8 (28.1–46.4)	8.1 (4.3–11.1)	17.9 (11.7–23.3)	Not available		See comments as per raw grape marc.
Palm kernel extract (meal)	90 (88.0–94.5)	11 (10.5–11.5)	16 (14.5–19.5)	68	8–10% oil. High Phosphorus: calcium ratio and sodium levels are low. Low effective fibre (NDF ~30) due to small particle size. PKE can have high levels of copper and iron.	Ruminants are good at dealing with high levels of iron, but sheep particularly accumulate high levels of copper in the liver, which can cause toxicity. Indications of levels of copper that may cause toxicity are 12–36 ppm (for sheep), 100–300 ppm (for cattle) and 30–100 ppm (for goats). Do not to feed sheep more than 50% PKE in the diet. Calcium and sodium may need to be supplemented. Introduce slowly as may be unpalatable until they adapt. Store in silos with steeply sloped bottom and wide throat to avoid blockages. Must be kept dry to avoid heating and mould growth.
Potato by-products	23.1 (10.9–62.3)	13.3 (10.8–14.8)	11.2 (6.7–25.8)	[7]15.3 (2.9–37.4)	Approx. 2% oil. Potential risk of acidosis.	Note wide range in moisture, ME and CP. Very high in starch and therefore highly digestible. Potential risks of acidosis unless levels introduced and increased slowly. Must be fed in conjunction with a fibre source and some additional protein.
Potato slivers	20	14	6	11	Potential risk of acidosis.	See comment as per potato by-products.
Tomato pulp	27.3 (16.6–30.2)	7.7 (4.1–9.3)	19.4 (5.0–22.6)	58.8 (1 sample)		Note wide range in moisture, ME and CP.

Note: The nutrient levels and ranges are a guide only and can vary considerably with property and season

Average daily water consumption is about 4 litres/head, but intake can change dramatically with the weather. On very hot days, intake can be increased to as much as 9 litres/head/day.

Trough space is less important than flow rate. The water trough space allowance does not need to be more than required in a paddock, because sheep will take turns to drink at the trough. A good rule of thumb is that the flow rate should supply enough water for the whole mob for 24 hours in 2 to 3 hours.

The maximum desirable salt levels for stock water are given Table 3.4. Bore water should also be tested for toxic minerals such as magnesium.

Feeding management

Sheep can be fed on the ground in large areas, but it is recommended to feed in troughs in containment areas to reduce contamination and wastage. Allow 15 to 20 metres of double sided trough for 100 sheep. The larger allowance is needed for large and/or full woolled sheep.

Feed troughs can be bought or made cheaply from materials like roof capping, folded roofing iron and suspended cloth.

Full drought rations for different classes of stock are provided in Table 4.6. To keep stock in better than condition score 2, rations should be increased by 20 to 30%. Because the stock will not have access to any pasture, roughage (fibre) should be included at 10 to 30% of the weight of the ration. Roughage can reduce the problems of grain poisoning and can be safely increased quickly to supply additional energy (if of sufficient quality) in cold conditions and for shorn sheep.

Sheep should be started on grain in the paddock before introducing them to the containment area. If this is not possible, most of the diet in the first 2 weeks in containment should be hay, and the grain ration increased gradually. Grain can be started at 50 g grain/head/day and the rest of the ration made up with hay, building up to the desired grain ration over 2 weeks. The best hay should be fed first and the hay fed before grain.

It may take a while to get the ration right and, because the cost of feed is especially high during a drought, monitoring body weight or condition score should be done regularly. Over and under feeding is costly. There will always be a number of sheep that do not take to containment feeding; these sheep should be identified and removed to pasture or smaller yards or sold.

Livestock management

Adult dry sheep are easiest to manage in containment and it is preferable to avoid confining lactating ewes and their lambs. Adult sheep, weaners and hoggets should be confined separately.

Dust in wool is often reported as a major concern with sheep fed in containment areas, with wool yields as low as 40% of greasy fleece weight reported. However, in drought situations, similar wool yields may also occur in paddock fed sheep, so the real cost may not be as great as thought. Wool at this low yield will usually be penalised on style, so the discounts are more likely to be higher for fine wool. Strategies for reducing dust may include establishing wind breaks on the windward sides, using shade cloth, or spreading and compacting sawdust or similar on the ground. Compaction of the site using high stocking densities does not always reduce the dust because the soil type tends to have a greater impact.

Releasing sheep from containment areas

Releasing sheep onto short green pastures can cause digestive problems because of the sudden change in feed type. Release sheep when they have a full stomach and continue to feed hay or supplement for a few days. Ewes that are lambing may need the full ration for a few weeks.

Farmers' experiences with feeding in containment areas

Farmers who have used containment areas for drought management generally believe they are worthwhile and likely to be part of future drought management strategy. Managing a stock-containment area involves transition from a broad acre to an intensive system, where the manager is in total control. All feed and water is supplied by the manager, and constant vigilance and attention to detail is required. Containment allows for better control over weight loss and gain and provides the means to come out of a drought with valuable land assets and stock numbers intact.

Observation and reports from surveys of farmers who have used stock containment areas or drought-lots as part of their drought management strategy indicate the following:

- 85% of deaths in containment areas are due to acidosis (grain poisoning) and shy feeders (poor doers)
- reducing the death rate from acidosis includes introducing sheep to grain carefully and slowly, provision of roughage (fibre) and taking care with changes in batches or type of feed
- shy feeders must be removed from containment areas frequently, especially as the mob size increases
- mortality can be reduced and production improved by having stock densities at about 2000 sheep per hectare (one sheep per 5 square metres) and mob sizes fewer than 1000
- one survey reported no advantage of feeding in troughs compared with feeding on the ground with mob sizes fewer than 2000 head (at low mob sizes there were fewer deaths with ground feeding, possibly due to lower risk of smothering and better access to feed)
- roughage (fibre) reduced death rates and the number of poor doers. This response was better with straw than with hay.

Animal health

Sheep must be vaccinated against enterotoxaemia (pulpy kidney) if the ration is high in grain. They should be drenched onto the area and worm tested regularly.

Pregnancy toxaemia can be a problem in containment areas. Pregnancy toxaemia can be reduced by recognising the higher demands of ewes in late pregnancy. A full late pregnancy ration should be reached by 6 weeks pre-lambing and at least 20% of the ration should be hay.

Accidents, pulpy kidney, shy feeders, suffocation and flystrike are other causes of death in containment areas. Disease can spread very quickly if sheep are confined in small areas. Diseases such as flystrike and pink eye need to be identified and dealt with quickly.

Stock need to be monitored daily and sick animals removed. Avoiding stresses – including boggy ground, overcrowding, dust and irregular feeding – will help reduce diseases such as salmonellosis and pneumonia. Regular cleaning of feed and water troughs is also recommended.

Sacrifice paddocks

Sacrifice paddocks have also been used successfully for containment. These tend to be poor-quality small paddocks that will be re-sown after the drought. The advantages of sacrifice paddocks are that, as part of a pasture renovation program, re-sowing can take advantage of the built up nutrients from large numbers of stock. Sacrifice paddocks are characterised by much lower stock densities, which may have the disadvantage of less soil compaction and more dust, but this also very dependent on soil type.

Storing fodder for drought

One of the biggest costs and risks associated with droughts is the high price of supplements. The typical trend in drought is for feed grain prices to increase from late winter, peak by the middle of spring and remain reasonably static thereafter. Roughage prices tend to continue to increase as the drought continues.

Every drought is different in terms of the cost of feed, with several important drivers determining price including domestic supply and demand, international supply and demand, import parity and other agricultural commodity prices and profitability. Crops for ethanol production and uncertainty about climate change further complicate the supply and price of grain for stock feed. The cost of feeding stock in drought is always significant, so strategies to offset this cost should be closely examined.

The physical strategies to reduce the cost of feeding during drought are the storage of grain, hay or silage. The alternatives are to use forward contracts, futures and options to manage price risk and off-farm investment built up in good years to fund drought feeding.

When considering storing feed for the long term, the suitability of the stored feed to supply stock with an adequate ration should be taken into account. The major issues with storage of feed relate to the cost of feed, cost of storage, quality of the feed and deterioration over time.

Cost of feed

When planning to store feed for the long term, it is important that the cost of the stored feed is low when it goes into storage. This is usually possible with grain, but accumulating grain may not be economically viable every year.

The cost of harvesting hay or silage is highly variable and needs to be calculated on a farm-by-farm basis. The cost will depend on whether the farm's own equipment can be used or contractors need to be brought in. If only considering harvesting costs, hay or silage appears cheap but all associated costs should be considered, including the replacement of nutrients. One tonne of dry matter of mixed pasture hay removes about 3 kg of P, 15 kg K and 3 kg S and the equivalent of 25–40 kg of lime. Therefore if 5 tonnes of hay is cut from a paddock, the equivalent of 300 kg/ha of a product such as super potash 1 & 1 as well as 150 kg/ha lime would be needed to restore nutrients to the paddock. Additional costs may be incurred if pasture manipulation (such as herbicide to control capeweed and insecticide to control red legged earth mite) is adopted to improve the yield and fodder quality. If a custom crop is sown, such as short-term rye grass, the cost of sowing the paddock should also be added. The value of any grazing should be removed from the capital cost of sowing the paddock. An additional cost to consider is the opportunity cost of lost grazing. The opportunity cost is highly variable, depending on season and land use. In very good seasons, the cost may be minimal (if taking on agistment is not an option), but in situations where paddocks are shut for hay and stock are penalised, the cost could be significant.

The total cost of making hay should therefore include:

- harvesting and carting to storage
- nutrient replacement
- pasture manipulation
- sowing specialised crop
- opportunity cost of fewer stock.

Knowing the true cost of making home-grown fodder provides good information for making cost comparisons and decisions on buying, selling and storing fodder.

Cost of storage

The cost of conventional storage is generally higher for grain, intermediate for hay and lower for silage. The variation between farms is large, depending on the existing and new infrastructure. The capital cost of a new silo is significant. For example, a 40 tonne silo that costs \$6000 equates to a capital cost of storage of \$150/tonne. A 150 tonne hay shed costing \$10 000 has a capital cost of \$67/tonne and a 300 tonne silage pit costing \$2000 to make and line with plastic costs only \$6.70 per tonne. The capital cost and working life of new infrastructure should be considered if investing in the long-term storage of grain or fodder.

Quality of stored fodder or grain

The quality of stored feed should be considered. The metabolisable energy of stored fodder or grain is the most important factor. Fibre is also important, and is often very expensive in drought.

Deterioration of fodder in long-term storage is a critical issue. Long-term storage of hay must ensure that is not spoilt by water. Round bales exposed to weather will suffer unacceptable levels of water damage. Damage caused by rodents can be significant, especially in cereal hay and lucerne. If stored correctly, hay quality should only deteriorate by 3 to 5% in the first few years.

If silage is stored correctly, the feed quality may drop by 5 to 10% initially, but should remain stable thereafter. Inoculants, depending on the type of silage, can help to maintain quality. The storage technique is critical to long-term success. Common issues include pit site selection to prevent water damage, proper lining with plastic, including internal partitioning for square bale silage for short-term feeding, and covering with at least 1 metre of soil for long-term protection. Wrapped silage is not a realistic option for long-term storage.

Stored grain can be spoilt by insects, but this can be prevented. Some producers have used storage bunkers as a cheaper alternative, though careful siting and planning is required.

Planning the amount of storage

The amount of storage required depends on the number of stock retained and the duration of the drought. For example, feeding for 8 months may require about 100 kg of grain and 15 kg dry matter of roughage per DSE, so requiring 100 tonne of grain and 15 tonne of hay for 1000 DSE.

Table 4.10. Example of the value of long-term storage of hay, grain and silage at two buy-in prices and allowance for capital depreciation and opportunity cost.

	$/T wet Buy in	1	2	3	4	5	6	7	8	9	10	% increase
grain	*$140*	$160	$168	$176	$184	$193	$203	$213	$223	$234	$245	75%
	$180	$202	$212	$222	$233	$244	$256	$269	$282	$296	$310	72%
hay	*$80*	$89	$93	$98	$103	$108	$113	$119	$124	$131	$137	71%
	$100	$110	$115	$121	$127	$133	$140	$147	$154	$162	$170	69%
silage	*$80*	$85	$89	$93	$98	$103	$108	$113	$119	$125	$131	64%
	$100	$106	$111	$116	$122	$128	$135	$142	$149	$156	$164	64%

If planning to feed roughage only, more energy is required to maintain an animal (because more energy is required to digest lower quality feeds). It may require about 10% more energy if feeding hay of moderate quality compared with feeding a more digestible grain.

Opportunity cost

Given there are significant costs of either purchasing or growing grain, hay or silage, and significant infrastructure costs involved with proper storage, the opportunity cost with long-term storage, should be included in a costing. There may also be taxation implications of storage. Often the investment in storage will occur in good seasons when profits are made, so the after tax cost may be less, depending on individual circumstances.

Three scenarios are outlined in Table 4.10:

- wheat with 13 MJ/kg DM stored in 40 tonne silos costing $6000 (life 40 years)
- hay with 9 MJ/kg DM stored in 150 tonne shed costing $10 000 (life 50 years)
- silage with 10 MJ/kg DM stored in 300 tonne bunker costing $2000 (life 20 years).

The costs of the three supplements ($/tonne wet) over a 10 year period, using two comparative buy-in prices for each feed are outlined in Table 4.10. Grain storage is depreciated at $4/year, hay at $1.34 and silage at 34 c/year to reflect the differential capital costs. All feeds include an annual opportunity cost of 5%.

In this example, when a conservative opportunity cost is used, grain and fodder increase in 'value' or 'cost' by 60–70% in 10 years. If a higher opportunity cost is used, such as 8%, then the original cost of feed doubles in 8 to 10 years.

Table 4.10 illustrates that for large quantity, long-term grain storage to be worthwhile, it would have to cost more than $245/tonne or $310/tonne in the drought if bought originally at $140 or $180/tonne, respectively. This does not take into account a potential loss of quality.

Grain, hay and silage vary in feed quality, dry matter content and deterioration of quality over time. Hence they need to be compared on a true cost of energy.

If grain drops 1% (in energy content) by year 3, it drops from 13 MJ/kgDM to 12.8 MJ/kgDM

If hay drops 5% by year 3, it drops from 9 MJ/kg DM to 8.6 MJ/kg DM.

If silage drops 7% by year 3, it drops from 10 MJ/kg DM to 9.3 MJ/kg DM.

Table 4.11. Estimated cost per unit of energy (c/ MJ) adjusted for dry matter and deterioration over time (all deterioration is assumed to occur in the first 3 years).

	Buy-in price ($/tonne wet)	Year 1 energy MJ/kgDM	Year 1 cost c/MJ	Year 10 energy MJ/kgDM	Year 10 c/MJ
Wheat	*$140*	13	1.4c	12.8	2.2c
	$180	13	1.8c	12.8	2.7c
Hay	*$80*	9	1.0c	8.6	1.8c
	$100	9	1.3c	8.6	2.3c
Silage	*$80*	10	1.6c	9.3	2.8c
	$100	10	2c	9.3	3.5c

Table 4.11 shows the estimated change in cost of energy (using the values estimated above). The buy-in costs and increased cost in year 10 can be used as a guide as to the likely future cost that fodder would need to exceed to make long-term storage worthwhile.

In this example, silage is a higher cost feed owing to greater deterioration in quality and its higher water content. The ability to buy in cheap grain, or make cheap silage or hay, will determine which options would be more profitable.

Frequency of drought

In regions where droughts are more frequent, long-term storage is more attractive than in regions where serious droughts occur perhaps every 20 years. The capacity to rebuild reserves, however, is also likely to be less where drought occurs more often.

The storage of substantial fodder reserves is more attractive if grain prices remain high owing to changes in world markets. This is especially so when the frequency of drought is higher. Other price risk and management strategies to manage feeding costs should also be considered.

Key messages – drought management

- The best way to manage a drought will be different for everyone.
- Planning for droughts will include assessment of finances, cash flow, water, feed and labour resources, present and future value of stock, personal attitude and commitment.
- If feeding stock, feeds should be costed on an energy basis and forward budget carefully.
- Alternative feed sources are worth considering, but there are other issues to take into account.
- Feeding stock in containment areas has been used successfully to preserve valuable pastures and soil.
- The cost of stored fodder or grain increases by 70% in 10 years or over 100% if higher interests costs are charged.
- Storing fodder for drought is more attractive if world grain prices remain high.

CHAPTER 5

Reproductive management

In order to understand how reproduction in sheep flocks can be managed and improved, this chapter covers:

- the reproductive cycle of the ewe
- the factors affecting reproduction
- some strategies to increase lambing percentage.

Many factors influence reproductive performance, including nutrition, live weight, breed, season and disease. An investigation into the cause of poor reproductive performance must consider all these factors.

Reproductive performance

Flock reproductive performance is measured by lamb marking percentage, which can be improved in many flocks. The extent to which lamb marking percentage can be improved must be balanced with overall profitability. In wool enterprises, raising the reproductive performance above the industry average may not be profitable unless the returns from stock trading are significant or the cost of improving reproductive performance is small. The marginal return from increasing fertility will be greater in flocks when all progeny are sold, particularly prime lamb flocks where wool value is low. When meat, surplus store stock or individual animal product value is high (as with studs) the financial return from increasing reproduction performance will be greater.

Other benefits of increasing reproductive performance include assisting the rate of genetic gain by increasing the selection potential and the opportunity to run a younger flock.

Lamb marking percentage

Average lamb marking percentage is shown in Table 5.1. The potential for improvement is shown by the low averages and the wide range of lamb marking percentages.

Table 5.1. Average lamb marking percentages.

Breed	Average	Range
Merinos	78%a	56–100%b
Prime lamb production	99%a	60–143%a

[a] same figures for Victorian wool industry 2003 & 2004/5 South West Monitor Farm Project (SWFMP)
[b] taken from 2004/5 SWFMP

The oestrous cycle

The oestrous cycle is a reproductive cycle. The oestrous cycle comprises a regular series of times at which the ewe is sexually active and receptive, called oestrus or heat, separated by times in which there is no sexual receptivity. The length of the cycle from heat to heat is about 17 days, with a range of 14 to 19 days. Sheep are seasonal breeders and are stimulated to cycle by decreasing day length.

During heat, ewes become sexually attractive to rams and are ready to mate. The actual heat (or oestrus) lasts about 24–36 hours in adult ewes. Maiden ewes have a shorter (10–15 hours) weaker heat. Some ewes (most commonly maidens) exhibit a 'silent heat' in which they may ovulate, but they do not necessarily show overt signs of heat.

The release of ova or eggs from the ovary or 'ovulation' occurs during oestrus. If the ewe is mated, fertilisation can occur. The fertilised ovum develops into a foetus. If a ewe does not become pregnant, she will have another oestrus or heat in about 17 days.

Ewes that become pregnant have elevated levels of progesterone, which prevents the release of eggs or ova generally until after lactation. Gestation (the time from fertilisation of the ovum to birth) in sheep is 150 days or about 5 months.

A ewe that is not exhibiting cycling activity is described as being in 'anoestrus'.

Young sheep reach puberty at 7 to 8 months.

Reproductive wastage

In most cases, nutrition is the major factor influencing the reproductive performance of sheep, but sometimes there are major 'reproductive failures', which are discussed at the end of this chapter.

Fewer than 69% of all eggs produced by Merino ewes become weaned lambs; therefore reproductive wastage in Merinos is high. Table 5.2 shows the results of a South Australian study where reproductive wastage was measured at each reproductive stage. Maiden ewes tend to have higher rates of reproductive wastage. Reproductive wastage is generally less in crossbreds.

Wastage in reproductive opportunities occurs from joining through to weaning. The loss of 'potential' lambs from conception to late pregnancy is generally uneconomic and difficult to prevent. The most significant losses are from mid pregnancy to marking, with most losses at birth and during the 3 days after birth (neonatal losses). There is a high economic and biological cost associated with birth and neonatal losses as during pregnancy: the ewe has diverted energy to the foetus, which will have affected ewe wool production, for no gain in animals weaned.

This study found some evidence that, for autumn-joining flocks, a number of days with temperature higher than 35°C may lead to reduced conception rates. This may be an important issue in pastoral areas when ewes are mated during the hot months.

Table 5.2. Proportion of reproductive losses at each reproductive stage for Merinos

Reproductive stage	Proportion of losses
Ewes failing to mate	8.4%
Ewes mated, not pregnant	13.2%
Ewes with more than one ovulation, but only one foetus (partial failure multiple ovulations – PFMO)	20.4%
Ewe death	2.3%
Lamb loss mid-pregnancy to marking – single	18.3%
Lamb loss mid-pregnancy to marking – twin	34.0%
Lamb loss marking to weaning – single	1.3%
Lamb loss marking to weaning – twin	2.0%

Factors affecting reproductive performance

Conception is the onset of pregnancy. The conception rate is the percentage of matings that result in conception. The pregnancy rate is the percentage of foetuses to ewes mated (accounting for multiples including twins). There are many important factors that affect ewe conception and pregnancy rates. The most important factors are:

- breed
- genetics
- time of year
- nutrition and condition score.

Breed

Merinos have lower potential ovulation rates compared with British breeds and their crosses. Merino ewes usually produce fewer than 1.4 eggs per ovulation. First cross ewes can produce up to 1.9 eggs per ovulation.

There is significant variation between strains of Merino. South Australian Merinos have high ovulation rates out of breeding season when compared with Peppins and Saxons. Booroola Merinos (Peppin strain) are the exception and have very high ovulation rates, but this is compensated for by higher lamb mortality.

Genetics

The heritability of fertility is low, so improvement in fertility based on genetic selection is slow. To achieve genetic gains in fertility in a commercial flock, rams must be purchased from a stud that selects for fertility. There is little value in selecting ewes for fertility unless the ram source does the same.

From 1959 to the 1980s, research flocks were established to identify just how much gain in reproduction could be made by genetic selection. The main flock was at the Trangie Research Centre. Rams were selected because they were born and reared as twins, and ewes were selected on the basis of their ability to rear lambs. Any ewe failing to rear a lamb at any age was automatically culled from the flock. The breeding objective included strategies to maintain other characters (such as fleece weight and fibre diameter).

Over the time of this experiment, the lambs weaned per ewe joined increased by an average of 1.8%. After 10 years of selection, the lamb survival to weaning of both singles and twins was increased by 7%.

Inbreeding

Inbreeding influences the breeding potential of a strain with up to a 10% decrease in lambing percentage in some strains.

Booroola gene

Some breeds and strains of sheep have incorporated the Booroola gene to increase fecundity (the ability to produce offspring frequently and in large numbers). The positive benefit of this is that ewes with the Booroola gene have more multiple births; the downside is higher neonatal lamb losses because the lamb birth weight is lower. This can be managed by ensuring ewes with multiple births are provided with excellent shelter and good pasture. Good nutritional management is critical to ensure that the lamb survival from these ewes is adequate.

Culling dry ewes from the flock

Some breeders choose to cull dry ewes from the flock every year in an attempt to increase flock fertility. In wool enterprises, dry ewes still produce wool and provide an economic return for the farming business. The value of culling dry ewes is more relevant in flocks selling more young sheep, where the relative value of stock trading is more important.

When selecting ewes for a particular trait, returns are influenced by two main factors:

- the superiority of the selected group
- the superiority of the next generation.

Superiority of the selected group – ewes that rear a lamb as maidens rear 10% more lambs in their lifetime. If 30% of maidens are culled each year for failing to rear a lamb, then the 70% left will rear 10% more lambs, increasing lambing percentage from 75 to 82.5% when all the ewes in the flock have been subjected to this type of selection.

Superiority of the next generation – the ability to rear lambs is poorly inherited (as with most other traits affecting reproduction in most species). **The heritability of ability to rear lambs is around 0.05** (compared with a heritability of about 0.6 for fibre diameter).

As with selection for any trait, it is ram selection that has a major influence in genetic improvement in the flock. The improvement resulting from ewe selection will be negated unless rams are selected on the same traits.

Culling dry maidens does not necessarily lead to improved reproductive performance. Dry maidens often have better subsequent performance than those that reared a lamb. They are often dry because of low body weight at joining.

Wet/drying maidens at marking

Attempt can be made to identify dry and wet ewes at lamb marking, but this is a reasonably inaccurate practice. For example, other lambs may suckle a ewe that has lost her lamb suggesting she is a 'wet' ewe even though she has already lost her lamb.

Dry ewes have no udder development and no birth stains.

Lambed and lost ewes have a large dirty doughy udder, with distinct demarcation between the two halves of the udder. The udder secretion is variable but usually watery.

Lambed and reared ewes have clean teats, and the udder is warm and resilient and does not have a distinct demarcation between halves of the udder. The udder contains milk.

Beware of wet/drying adult ewes if clover disease is present.

Wet/drying is most accurate when the lambing period is 6 weeks or less and is done strictly at the end of lambing.

Time of year

Sheep are seasonal breeders and are stimulated to cycle by decreasing day length. Cycling activity increases after the longest day (summer solstice) as the day length reduces. The maximum number of ewes cycling and the maximum twin ovulations occur from March to June. Merinos are the least obvious seasonal breeders because they can cycle all year, followed by the short wool British breeds (such as Poll Dorsets), which are moderately seasonal. Long wool British breeds such as Border Leicesters or Romneys are strongly seasonal and will only cycle between February and June.

Ovulation rate increases with shortening day length as illustrated in Table 5.3. Ewes joined in the autumn – taking advantage of their natural breeding season (associated with decreasing day length and resulting in increased ovulation rate) – have a higher pregnancy rate, including a higher percentage of twins.

Time of joining

The time of joining (hence lambing) is one of the most important management decisions that can optimise farm profitability. Generally, to maximise farm profits, it is important to optimise stock

Table 5.3. Pregnancy rates of Merino and first cross (Border Leicester × Merino) ewes in different months.

Ewes	Lamb date	Mating date (day of year)	Pregnancy rate (%)	Singles (%)	Twins (%)
Merino	April	1 Nov (300)	80	70	5
	May	1Dec (330)	90	80	5
	Jun	1 Jan (1)	110	90	10
	Jul	1 Feb (32)	120	80	20
	Aug	1 Mar (60)	130	70	30
	Sep	1 Apr (90)	130	70	30
	Oct	1 May (120)	120	80	20
Crossbred	April	1 Nov (300)	105	65	20
	May	1Dec (330)	120	60	30
	Jun	1 Jan (1)	135	55	40
	Jul	1 Feb (32)	148	48	50
	Aug	1 Mar (60)	156	40	58
	Sep	1 Apr (90)	156	40	58
	Oct	1 May (120)	145	45	50

numbers while minimising the amount of supplementary feed. Matching ewe feed demand with pasture supply is discussed in detail in Chapter 3. As a guide, for Merino producers in southern regions, this will coincide with lambing 3 to 4 months (or joining 8 to 9 months) before the end of the growing season. For meat production, this will coincide with lambing 4 to 6 months (or joining 9 to 11 months) before the end of the growing season. The specific time will depend on target markets and balancing lamb market weight and prices for different market sectors.

Length of joining

If ewes are at target live weight or condition at joining (condition score 3), then they should conceive in the first two cycles (2 × 17 days = 34 days), so a 5 to 6 week joining period (35–42 days) gives all ewes the opportunity to come on heat twice.

Short joining periods:

- allow earlier weaning and therefore more opportunity for the ewe to get back to target condition for next joining
- result in older average lamb age and greater weight at weaning
- result in heavier crossbred lambs at sale
- result in heavier Merino weaners in summer with less ill-thrift and improved survival
- result in less tail-end lambs, which are costly because they have a high risk of dying over summer and autumn if feed quality is poor.

If joining occurs before January (out of the natural breeding season), the joining period may need to extend to 6 to 7 weeks because all ewes may not be cycling when rams are introduced to ewes.

Teasers can stimulate ewe cycling activity, allowing ewes to be mated outside the natural breeding season so the joining period can remain short (see the section on teasing and the ram effect later in this chapter).

In pastoral regions with large paddock and mob size, joining periods may need to be extended and higher ram numbers used to ensure adequate ewe–ram contact.

Nutrition and condition score

Apart from genetic merit, nutrition is the most important factor determining conception rates of ewes. Conception rate is directly related to the condition score or live weight at joining. There have been many trials that demonstrate the close relationship between live weight and conception rate. Higher levels of ewe nutrition will generally lead to higher pregnancy rates and more multiple births. Poor live weight at joining can lengthen the time it takes the ewe to conceive and hold conception. Ewes in very poor condition may not conceive. As ewes approach condition Score 3, more ewes are likely to get in lamb quicker. For good conception rates, the live weight of maiden Merino ewes must be equivalent to 75 to 80% of mature body weight.

Increasing live weight can have a direct effect on ovulation rate. For each additional kg of ewe live weight at joining, the number of lambs born increases by 1.5 to 2.5 per 100 ewes joined. Autumn lambing crossbred ewes and small Merinos are likely to be at the lower end of the response range and spring lambing crossbred ewes and highly responsive ewes are likely to be at the higher end. A rule of thumb is that one extra condition score (or eight to 10 kg live weight) will contribute up to 20 additional lambs per 100 ewes.

Table 5.4. The ration of barley needed to maintain ewes at a weight of 40 or 47 kg in a stock containment area (no paddock feed available), with the expected increase in lambing percentage and estimated cost per additional lamb born.

Weight of ewes	Condition score of ewes	Ration of barley needed to maintain target weight	Extra number of lambs born per 100 ewes joined	Ration cost for 6 months per 100 ewes with a feed cost of $300 per tonne	Ration cost per lamb weaned at 80% survival of lambs born
Kg	CS	Kg/head/ week		$	
40	2.0	3.5		$2520	
47	2.8	4.0	10[a]	$2880	$42
			14[b]	$2880	$31.50
			18[c]	$2880	$25.20

[a]**10 extra lambs per 100 ewes joined** (for an extra condition score) equivalent to one-and-a-half extra lambs per 100 ewes joined per kg live weight (this response might be expected in some Merino flocks and autumn lambing crossbred flocks)

[b]**14 extra lambs per 100 ewes joined** is equivalent to two extra lambs per 100 ewes joined per kg live weight (this response expected in some Merino flocks and winter lambing crossbred flocks)

[c]**18 lambs per extra 100 ewes joined** is equivalent to two-and-a-half extra lambs per 100 ewes joined per kg live weight (this response expected in spring lambing crossbred ewe flocks and highly responsive Merino flocks)

Cost–benefit of feeding ewes to improve pregnancy rate

For each extra kilogram in ewe live weight, approximately 70 grams of barley or wheat per day is needed for maintenance. Table 5.4 illustrates how to estimate the cost of maintaining ewes at a higher live weight to improve the lambing percentage.

In this example, feeding for 6 months with the cost of grain at $300 per tonne, the extra cost per lamb weaned, at 80% survival of lambs born, is between $25.20 and $42.

The outcome will vary depending on:

- the cost of grain
- additional cost of any roughage added to the diet
- the response in conception rate, because this varies considerably between farms (recent research has shown a very high response to heavier ewe live weights, with an average response of 20 extra lambs conceived per 100 Merino ewes maintained at one condition score heavier – with a range from five to 35 extra lambs),
- length of maintenance feeding (6 months used in this example)
- lamb survival.

There may be additional benefits of having heavier ewes, such as fewer deaths and more wool. The cost will be higher if the money is borrowed.

The most important cost-free strategy to improve ewe condition score or live weight at joining is to wean lambs early (12 to 14 weeks after the start of lambing for Merinos and about 1 month longer for first-cross ewes). This ensures that ewes put on weight in late spring while pasture quality is still good, in preparation for next joining.

In addition, ewes less than condition score 3 can be drafted off and provided with high-quality feed to ensure they improve condition for joining. The lighter portion of the ewe flock is likely to give the best response to feeding compared with ewes in good body condition.

To reduce overall reproductive wastage, join ewes in condition score 3 and maintain them in condition score 3 until lambing. Fat ewes (condition score greater than 4) tend to have more lambing difficulties associated with reduced lamb survival. Thin ewes (condition score less than 2.5) have small lambs that are more likely to die of starvation, exposure, hypothermia, which is compounded by poor ewe mothering ability.

Flushing

Flushing is short-term, high-level feeding before joining to improve ovulation rate. There is a common belief that a rising plane of nutrition at joining will lift conception rate. The main effect may not be the change in live weight before joining, but the effect of the increased live weight at joining. Generally, it is more cost effective to maintain ewe live weight rather than attempt to increase live weight before joining.

Even so, short-term flushing is often attempted, especially when ewes are grazing on poor-quality dry pasture. The most common strategies employed are to provide good-quality green feed or a high protein supplement, such as lupins, for a period of up to 2 weeks.

Conception rates following joining on a green pick are often elevated; some believe this is due to a short-term boost in protein levels; others believe it is associated with increased live weight at joining. Research shows highly variable response to supplements and green feed and research findings can be found to support either school of thought.

Teasing and the ram effect

The introduction of rams or teasers into a flock of non-cycling ewes can induce ewes to cycle. The effect is known as the 'ram effect'. A teaser is often a vasectomised ram. A wether treated with testosterone preparations to prevent pizzle rot may also cause a ram effect.

Teasers are most relevant for autumn-lambing flocks and for out-of-season lambing in some British breeds that are highly seasonal breeders. For flocks where joining occurs in autumn, the use of teasers is not so relevant.

Because ewes are seasonal breeders the ram effect is seen until January in Merinos and until February in Corriedales and crossbreds and other British breeds.

British breeds (other than Dorsets) respond to the ram effect by ovulating with a silent heat. Only from late December onwards – immediately before the natural breeding season – are silent heats followed by a normal cycle.

The ram effect works in the following way. After teasers or rams are introduced at a rate of 1% into a mob of non-cycling, isolated ewes, a large proportion of ewes respond by ovulating 2 to 3 days later. They do not show signs of heat (silent heat) and hence do not mate. Other ewes may experience silent heat 7 to 9 days after the introduction of the rams or teasers. These silent heats are followed about 17 days later by a true heat. In practice, this means that most of the previously non-cycling ewes will be on heat 19 to 26 days after teasers or rams have been introduced.

For teasing to work, ewes must be isolated from the sight, sound or smell of rams, billy goats or teasers for 2 months.

A veterinarian can advise on how to access teasers for a sheep flock.

Introducing the teasers to the ewes 2 weeks before the rams are normally put out will ensure a compact lambing from the corresponding lambing time after rams are put out.

If entire rams are used to induce cycling in ewes that have been isolated from rams, the ram effect will occur. The result at lambing is a trickle of lambs in the first 2 weeks (because some ewes were already cycling and these ewes mate and become pregnant in the first 2 weeks) followed by a rush of lambs, which were conceived in the true heat period which followed 19-26 days after the introduction of the rams.

Management of ewes during pregnancy

The key factor is to ensure adequate nutrition of ewes during pregnancy. Lamb birth weight and survival will be greater if ewes are maintained in adequate condition.

Early pregnancy

In early pregnancy, the ewes' feed requirement is similar to a dry sheep. Stress, disease and poor nutrition can lead to embryo loss at this time. Good general husbandry practices should be adopted. The critical factor is to ensure that ewes do not lose weight rapidly.

Mid to late pregnancy

A mid-pregnancy condition score of 3 is the optimum condition score in an average year to ensure that the placenta develops appropriately to ensure the developing lamb has access to adequate nutrition for growth and survival both pre-and post-birth.

Economic modelling shows that ewes pregnant with a single lamb can afford to lose some condition in mid pregnancy (condition score 2.7) as long as they are at condition score 3 at lambing. This is only practical on properties where ewes lamb late enough to ensure pasture availability is good in late pregnancy; otherwise the preferred strategy is to ensure ewes are maintained in condition score 3. Ewes carrying twin lambs should be maintained in condition score 3 or more.

Ewes carrying a single lamb should graze pasture with at least 800 kg DM/ha from mid pregnancy and at least 1000 kg DM/ha at the point of lambing. Twin-lambing ewes should have at least 200–500 kg DM/ha more (1200–1500 kgDM/ha) at lambing if they can be identified (by scanning) and separated from the rest of the flock.

Ewes' feed requirements increase by 50% or more in the last 6 weeks of pregnancy because this is the time that the unborn lamb grows substantially and milk production is initiated. Eighty per cent of total foetal growth occurs in the last 6 weeks of pregnancy. Under-nutrition of the ewe results in low birth weight lambs that have low survival rates. A reduction in energy intake of up to 25% below requirement in late pregnancy is unlikely to cause problems. However a 40% deficit in requirement can reduce lamb birth weight by 20 to 30%. This may be acceptable in single lambs, but may jeopardise survival of twin lambs. Where maternal reserves are used because of inadequate feed, milk production and mothering behaviour may suffer more than lamb weight. An added benefit of ensuring that pregnant ewes meet condition score targets is to optimise lifetime wool performance of progeny. Research has shown that a 10 kg difference in ewe weight during mid pregnancy resulted in progeny that produced 190 g extra clean fleece weight that was 0.3 micron finer.

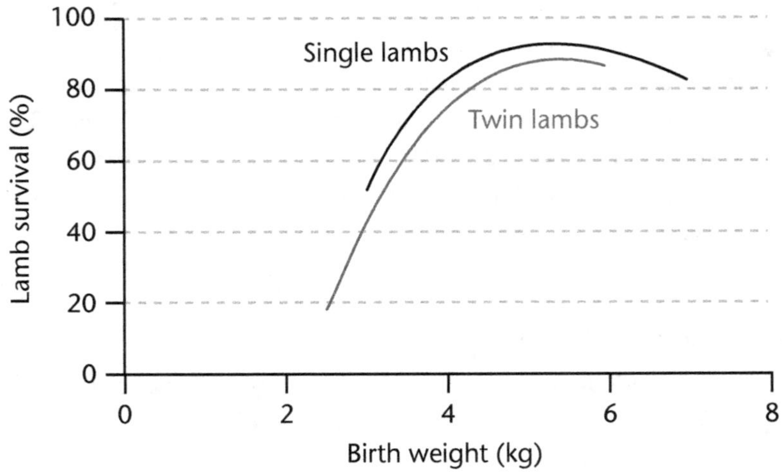

Figure 5.1. Lamb birth weight and survival. (Source: www.lifetimewool.com.au)

Lamb survival and performance

Lamb mortality is a major cause of reproductive wastage in sheep, especially southern regions where weather conditions are more severe. To reduce lamb mortality, ensure that the ewe condition score at lambing is at least 3 and ewes are grazing adequate pasture at lambing.

To ensure good lamb survival, lambs should weigh between 4 and 5.5 kg at birth. Lambs that are less than 4 kg birth weight are very vulnerable to hypothermia. Lambs that are over 5.5 kg at birth are vulnerable to difficult birth or dystocia. Figure 5.1 demonstrates the impact of birth weight on lamb survival.

There is a strong association between ewe condition score and lamb survival. Good nutrition in early pregnancy will result in improved placental development and in the last 50 days of pregnancy good nutrition will result in improved foetal growth. Figure 5.2 shows the relationship between ewe condition score at lambing and lamb survival. Ideally, ewes should be in condition score 3 at lambing or in condition score 3–3.3 if they have twins.

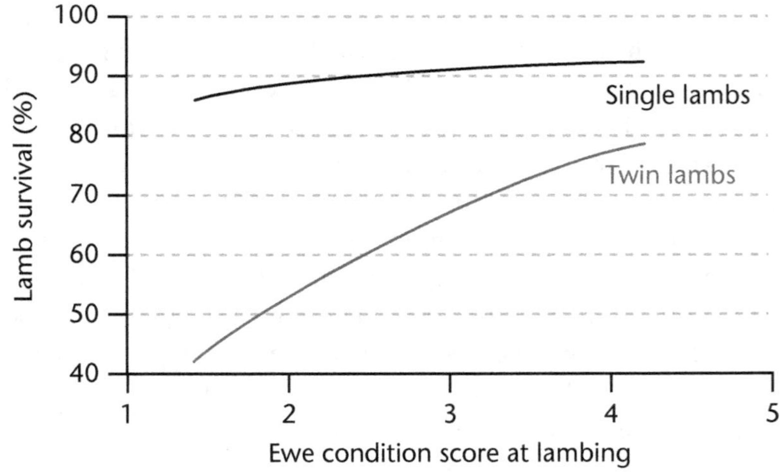

Figure 5.2. Ewe condition score and lamb survival. (Source: www.lifetimewool.com.au)

The most common cause of death in new born lambs is a combination of:

- starvation
- hypothermia
- mismothering.

Other causes are difficult birth owing to:

- difficult birth (dystocia)
- predation, most commonly by foxes.

Occasionally, trace mineral deficiencies – such as iodine deficiency causing goitre, selenium deficiency causing white muscle disease and copper deficiency causing enzootic ataxia – result in serious neonatal lamb losses. If uncertain, seek veterinary advice to determine the cause of death in new born lambs.

Lactation

Lactation will influence lamb survival and performance. The energy requirement for a lactating ewe ranges from 2.5 DSEs for single lamb Merino ewes to as high as 4.5 DSEs for multiple lamb crossbred ewes. The increased energy demand is matched with an increased need for protein. In most cases, it is not possible for ewes to be able to eat enough feed to supply these needs and weight loss during lactation can be high. Autumn lambing ewes, which tend to lamb on low quantity and poor-quality feed, may lose over 14 kg during lactation.

To achieve acceptable production during lactation, it is critical that ewes have adequate pasture. As a guide, ewes with single lambs should be grazing at least 1000 kg DM/ha and ewes with twin lambs should have 1200–1500 kg DM/ha with at least 20% clover. If these pasture benchmarks are not met, lamb growth will be poor, resulting in weaner ill-thrift and higher death rates over summer. For prime lambs, the consequence will be that lambs do not reach weight specification at sale, so achieve lower prices or take longer to reach target markets.

Time of weaning

Weaning time is a critical issue in sheep flocks. As a guide, all lambs can be weaned by 12 to 14 weeks after the start of lambing, assuming the joining period is only 6 weeks. By this time, all lambs will have a functional rumen to graze pasture and milk supply will only contribute 5% of the lamb's total feed requirement. Meat sheep breeds have superior milk production and lambs can remain with their mothers for slightly longer. Early weaning in crossbred flocks is important in poor years.

Delayed weaning will have important implications on the subsequent performance of ewes and lambs. For ewes, delaying weaning will mean that ewes take longer to gain condition before joining, thus reducing reproductive performance in the next season. This impact will be less in good seasons when feed is abundant. In addition, ewe wool production will be penalised. For lambs, delaying weaning will result in a slower lamb growth rate owing to worms and possibly lambs not grazing the best available pastures.

Other influences on reproductive performance

Paddock topography can influence ram–ewe contact. Join maidens in smaller, open paddocks with good feed and a single watering point. Timber belts, gullies, poor feed, multiple watering points and large paddocks promote flock dispersion and may reduce ram–ewe contact.

Paddock shelter can markedly increase lamb survival. Shelter belts must be well planned, with the correct tree species and good fencing. Wind penetrability should be 30 to 50%. Lower wind penetration creates turbulence and decreases the area that is effectively sheltered. Good shelter may improve lamb survival by up to 10%, although responses are variable.

Identifying dry ewes is useful if there is a high percentage (greater than 10%) of dry ewes, or when a lot of supplementary feed is required in a drought situation. For many seasons in southern Australia only 5 to 8% of ewes remain dry, so the cost of identifying dry ewes is often not recovered by savings in feed.

The **mob size** may have an effect on lamb survival. To increase lamb survival, the optimum mob size is up to 400 ewes (depending on the total flock size) or less for mobs with a high proportion of twins.

A **stocking rate at lambing** over 18–20 ewes per hectare should be avoided because the risk of mismothering is too high.

Predators can cause lamb losses of up to 10%. Foxes are the most common predator: baiting can help minimise losses, but must be continued because new foxes will quickly fill territory vacated by the resident population.

Shearing during joining has no effect on ovulation rates but can increase the number of ewes that do not show signs of heat (silent heat). To avoid this phenomenon, avoid shearing during joining.

Shearing maiden ewes 4 weeks before joining may increase the number of lambs born. This is thought to be due to improved embryo survival rather than increasing the number of maidens mated. Shearing ewes before joining can increase ovulation rates by 10%, although the response is unreliable and is not recommended as a reason to set the shearing time before joining.

Pre-lambing shearing aims to encourage ewes to seek shelter. Ewes seek shelter for only 3 to 4 weeks off-shears. Shearing close to lambing to encourage ewes to seek shelter and will increase the risk of pregnancy toxaemia and hypocalcaemia, which will often outweigh the benefits of extra lamb survival associated with encouraging ewes to seek shelter.

Pre-lambing crutching does not increase lamb survival. If pre-lambing crutching is carried out for other reasons – such as the prevention of flystrike, or wool stain – it should be done at least 1 month before lambing.

Ram management

Rams represent a large investment and the standard of management applied should protect this investment. Too often, rams are only looked at the week before joining starts. They are drenched and may be footpared and then expected to work for the next 6 to 8 weeks.

The common problems found in ram flocks include light condition, too much wool at joining, bad feet, disease (scrotal mange or poll infections) and old rams. These problems cannot be fixed in the week before joining. It takes at least 50 days for rams to produce viable sperm after a ram has had a health problem such as elevated body temperature that has temporarily rendered him infertile.

The lambing percentage can be reduced by running infertile rams alongside fertile rams. Infertile rams can serve ewes and keep fertile rams out of the mating.

Characteristics of a good ram team

A condition score 3.5 or 4 is required because rams lose several kilograms during mating and the aim should be to have them in 'good working condition' at the end of the joining period.

The rams' feet and legs should be sound because lameness will reduce a ram's ability to serve.

The testicles should be of adequate size and consistency because size is highly correlated with semen output. Scrotal circumference needs to be at least 34 cm at the start of joining to allow for 0.5 to 1 cm per week decrease in size during joining. Below a scrotal circumference of 30 cm, conception rates start to fall. Lupins will increase testicular size and may be worth feeding if rams are not in good condition 1 month before joining. Firm springy testicles are an indicator of good semen quality. Soft flabby testicles produce semen of poor quality. Flabby testicles can be the result of poor condition, foot abscess, flystrike, scrotal mange and other diseases.

The sheath and penis should be examined before joining for signs of inflammation or injury.

Provide supplementary feed before joining to prevent grain poisoning when rams are introduced to a mob of ewes that are being supplementary fed.

Buy rams from an accredited ovine brucellosis-free source. The testing of the ram flock for ovine brucellosis can be carried out by a veterinarian. Brucellosis is generally not a major cause of infertility in Merino ram flocks.

Provide enough sound rams to join maidens above 1.5% (i.e. two rams per 150 ewes) and ewes at 1% (plus one for the paddock). For example, a mob of 500 ewes requires six rams for joining (five rams plus one for the paddock). Avoid joining maiden rams to maiden ewes. In extensive pastoral environments, a higher joining percentage is recommended to ensure that adequate ewe ram contact is achieved.

Ram health

Rams should receive an annual booster vaccination to protect them from clostridial disease. Newly purchased rams should receive two doses of vaccine to protect them from clostridial disease 4 to 6 weeks apart.

Rams should be included in the worm-control program (and fluke control, if fluke disease occurs on the property). Often rams miss drenches and as a result they tend to have more worm problems.

Poll infection following fighting should be treated promptly. Fighting can be minimised by running young and newly purchased rams separately from adult rams. Poll infection may result in temporary infertility in a ram flock.

Pare the rams' feet if necessary. Paring should be carried out at least 8 weeks before joining; do not use lame rams because their serving ability is likely to be compromised.

Poor nutrition, foot problems, flystrike, scrotal mange and other diseases may cause testicular degeneration and impairment of fertility lasting 2 months or more.

Ram paddock

The ram paddock should not be the small holding paddock used for other purposes. It should have regular fertiliser application, adequate shelter and fresh water. Rams should not be shunted from paddock to paddock and should not be run with killers or culls.

Table 5.5. Target joining weights of maiden ewes of different mature sizes.

Adult weight (kg)	Joining target weight (kg)
45	34–36
50	37–40
55	41–44
60	45–48
70	52–56

Ram shearing and dipping

Avoid mating rams with more than 9 months of wool growth. Do not wet dip rams in the 8 weeks before joining in case they suffer minor infections that render them infertile. Other topical treatments to manage flies and lice can be applied.

Maiden ewe management

The lamb marking percentage of the maidens should be within 10% of the adult flock. Low body weights are often the cause of poor maiden performance.

Table 5.5 outlines the target body weight of maiden ewes at joining to ensure adequate reproductive performance. Maiden ewes should be at least 75%, and preferably 80%, of their mature body weight at joining to achieve adequate conception rates. In southern areas, this weight must be achieved when pastures dry off because the pasture quality will not be good enough to maintain growth over summer when pasture digestibility declines.

Maiden performance can be maximised by attention to:

- the age at weaning (wean 12 to 14 weeks after the start of lambing)
- the worm-control program (weaners are the most susceptible class of sheep to worms)
- nutrition
- ram management (maidens have a shorter, weaker oestrus than adults and are less active when seeking out the ram, so they should be joined separately, using at least 1.5 to 2% of experienced rams.

Joining ewe lambs

In some circumstances joining ewe lambs can be a worthwhile strategy. Generally, this option is only available for meat breeds and only if the live weight at joining is above 45 kg. Ewe lambs less than 45 kg live weight are less likely to have reached puberty and, if they do conceive, the risk of neonatal losses is high owing to poor mothering.

It is very important to provide young ewes with good-quality pasture to ensure that they continue to grow to reach target weights after mating. If young ewes cannot be provided good quality feed, early joining is not recommended.

To prevent lambing difficulties, young ewes should only be mating to rams with low-birth-weight lambs or select rams from within a breed with low-birth-weight breeding values.

Diseases causing reproductive loss in sheep flocks

Infectious causes of abortion in sheep

The infectious agents can cause sporadic outbreaks of abortion and newborn lamb deaths and are probably responsible for low levels of undetected losses on many farms. 'Abortion storms' tend to be associated with high stocking densities.

The main infectious causes of abortion in Australia are:

- vibriosis
- toxoplasmosis
- listeriosis
- salmonellosis.

Vibriosis, also known as campylobacteriosis, is a bacterial infection that causes mid to late term abortions, stillborn or weak non-viable lambs. It is the most common cause of abortion in ewes. Outbreaks of abortion are usually preceded by a period of high stocking density, particularly in winter and spring, but also after periods of hand feeding in summer and autumn. Flocks may experience a few sporadic cases or there may be abortion storms with up to 50% of the flock affected. Abortion storms may follow 2 to 3 weeks after the first sporadic cases. The bacterium is carried asymptomatically in some sheep; if a ewe is infected and aborts the foetus, foetal fluids and after-birth are highly infectious and the infection can spread to other ewes. Ewes may retain their placental membranes and develop metritis (uterine infection). The bacteria survive in the gall bladder and gut of infected sheep and are spread by ingestion of pasture contaminated by sheep faeces. Crows and magpies can become infected for several months and thus become vectors. To minimise spread, aborted ewes should be removed from the lambing paddock. If hand feeding, change to a new clean area each feed. Aborting ewes develop good immunity and are unlikely to abort from this cause again. One control strategy is to run maiden ewes with older ewes well before joining or in paddocks that abortions have occurred, so they build up immunity. A vaccine against campylobacter infection, marketed as *Guardian®*, has recently been released in Australia by Coopers Animal Health.

Toxoplasmosis is a protozoan parasite spread by cats or infected aborted placenta. Toxoplasmosis is a zoonosis; that is, it can infect people. The infection causes stillborn or weak lambs or foetal resorption and mummification, depending on the stage of pregnancy when infection occurs. Prevention involves avoiding contamination of stockfeed and pasture with cat faeces.

Listeriosis is a bacterial infection that can cause late-term abortion in ewes, septicaemia in lambs or, where the bacteria infects the brain, circling disease in sheep of any age. It is unusual to see both the abortion form and the neurological form of listeriosis in the same outbreak. Listeria is often present in the gut of normal sheep and can survive and multiply in faecal material and soil. The feeding of poor-quality silage is often associated with outbreaks of listeriosis. Grazing potato crops can expose sheep to the bacteria. The disease commonly occurs under wet, muddy conditions and its occurrence is sporadic and unpredictable. It is probably a cause of widespread losses from abortion and perinatal death, but generally with a low flock prevalence. Listeriosis can infect humans.

Salmonellosis is well known as a bacterial infection of the gut, resulting in severe scouring, but it can occasionally cause abortion. Outbreaks of salmonellosis in pregnant ewes will usually cause abortion in a large proportion of ewes in addition to signs of gut infection. Signs of salmonellosis in the ewe flock include diarrhoea, smelly dark red vaginal discharge, foetal death, abortion, retained foetal membranes, septicaemia and high fever. Outbreaks of *Salmonella* abortion are usually associated with stressful conditions, including overcrowding, hand-feeding and under nutrition. Enteric salmonellosis in non-pregnant sheep is also associated with the same stressful conditions. Separation of affected ewes from unaffected ewes, and a reduction in stocking rate of the latter, may reduce the incidence of new cases, but is unlikely to stop them altogether. Affected sheep can be treated with antibiotics under veterinary prescription as a life-saving measure.

Samples for the diagnosis of the cause of abortion include aborted foetuses, dead lambs and placenta (afterbirth). Several lambs and placentas should be collected where available, put into clean plastic bags and kept cool. Early detection of any disease – especially new, emerging or exotic diseases – is the key to their effective management.

Clover disease

Sub clover cultivars such as Dwalganup and Yarloop are highly oestrogenic; neither is common in most areas of south-eastern Australia. Others, such as Mount Barker, Trikkala or Leura have low oestrogen levels and rarely cause problems. Clover infertility may be either temporary or permanent.

Temporary infertility – ewes joined on green-clover-dominant pasture in spring become infertile. Ewes regain fertility within 6 weeks of clover drying off or removal from the pasture. Temporary infertility occurs in all age groups of ewes.

Permanent infertility – older ewes are worse affected because of repeated exposure. Ewes mate, but fail to conceive, so they will be marked with ram harnesses but a high percentage (more than 25%) will be remarked at the next cycle.

Under extreme circumstances, wethers can appear to lactate if they graze in highly oestrogenic pastures.

Longer term prevention involves replacing highly oestrogenic cultivars of sub clover with safe cultivars. Ongoing management includes avoid grazing ewes on high-risk pastures, especially when they are green. Young ewes should take priority. In addition, increasing the ram percentage and extending the joining period is sometimes adopted, but is not an ideal solution. Higher soil fertility (increased phosphorus) appears to reduce the level of oestrogen in clover.

Brucellosis

Ovine brucellosis (OB) is caused by a bacterium called *Brucella ovis*. It is characterised by infertility in rams and causes losses due to greater ram turnover, lower lambing percentages (especially if more than 10% of rams are affected), extended lambing and ill-thrifty newborn lambs. *Brucella ovis* does not infect humans.

Infection is spread by sodomy, resulting in infected semen being deposited into the rectum of other rams. This behaviour is more pronounced in younger rams and can be seen in up to 30% of all rams. Infection can also be spread by 'clean' rams servicing ewes that have been exposed to infected semen from a previous mating. Ewes do not remain infected from season the season, but the infection is persistent in rams.

Brucella ovis causes inflammation of the epididymis (the duct that conveys sperm from each testicle), which blocks the movement of sperm from the testicle. Rams may be sub-fertile or infertile, depending on the extent and severity of the infection.

Although ewes rarely become infected, OB can establish in the uterus of a pregnant ewe, but the infection is only carried for the term of that pregnancy. The main result of infection is placentitis (inflammation of the placenta), which interferes with foetal nutrition. Lambs born from infected ewes are usually full term, but of significantly lower birth weight.

A reduction in lambing percentage is likely in flocks joined for a restricted period using OB-infected rams. For, example, in a 6 to 8 week joining, a reduction of lambing percentage of up to 30% is possible if more than 10% of the rams have OB. Flocks joined for extended periods may experience normal lambing percentages, but, due to the lower semen quality, it may take ewes longer to get into lamb. This will result in an extended lambing period. The effects of OB are often insidious and unrecognised, especially in areas where marked fluctuations in fertility occur due to variation in ewe nutrition, clover disease, predation, or other circumstances.

There are a number of controls that can be implemented to assist in the prevention and control of OB. Firstly, all boundary fences should be secure to prevent sheep straying from neighbouring properties. Secondly, rams should be monitored for OB on a regular basis. One way of doing this is to palpate or feel the testes of a ram for any lumps. Ram's testes should be checked before joining or purchase. When palpating rams' testes, start as high up in the scrotum as possible and gently palpate each testicle and epididymis. Both testicles should be the same size and feel smooth and firm; there should be no lumps in the epididymis. Experience in palpation of normal testicles, and guidance in the procedure from a veterinarian, will help to develop the skill of detecting abnormalities.

If OB has been introduced into a flock, the disease can be eradicated under the guidance of a veterinarian by a program of repeat blood testing, palpation and culling of infected rams. The prevalence of OB in Australia has been reduced by OB accreditation programs for flocks, and other regulatory and voluntary control measures.

CHAPTER 6

Disease management

There are hundreds of known diseases of sheep. Some of the more commonly occurring diseases of sheep in southern Australia are described in this chapter. It is important that people working in the sheep meat and wool industries manage their flocks to prevent commonly occurring diseases. It is also important to recognise signs of disease and distress and take immediate action to diagnose the cause and implement control and prevention strategies.

Diseases that result in very high costs to the sheep industries, such as internal parasites, flystrike, lice and footrot, have been dealt with in detail. Common diseases associated with abortion and reproductive failure are described in Chapter 5.

The possibility of an 'emergency animal disease' should always be in the minds of livestock producers and it is vital that livestock managers report any unusual signs of disease immediately to their veterinarian or animal health staff of state Departments of Primary Industries or equivalent.

Internal parasites

Internal parasites – roundworms, tapeworms and liver fluke – are found in the in the gut or some other internal organs.

Roundworms

Roundworm-related losses are due to:

- death of sheep
- weight loss and/or reduced weight gain/scouring/dags
- reduced quantity and quality of wool
- reduced milk production by ewes resulting in poor lamb survival and growth
- depressed appetite and poor feed utilisation
- predisposition to other diseases
- costs of drenches, labour and associated management practices.

Worm control is a compromise between maximising sheep health and production and minimising selection for drench resistance.

A roundworm control program is based on three critical practices:

- grazing management (required to ensure that the most susceptible sheep – weaners and lambing ewes – are grazed on paddocks with adequate feed and low numbers of worm larvae)
- a drenching program, especially the strategic drenches (needed to kill worms in the sheep's gut and to minimise pasture contamination by worm eggs)
- monitoring (to determine when drenching is required and to test for drench resistance).

The program will vary from property to property depending on stocking rate, flock structure, feed availability, effectiveness of the drench groups and seasonal conditions.

These three critical practices will be discussed after basic facts relating to roundworms are presented. Sheep industry web sites, such as WormBoss (http://www.wormboss.com.au), provide extensive information on most aspects of worm control.

The most important roundworms of sheep in southern Australia are the brown stomach worm (*Ostertagia circumcinta*) and the black scour worm (*Trichostrongylus* spp). In the summer rainfall areas of New South Wales and Queensland, the most important roundworms are barber's pole worm (*Haemonchus contortus*) and black scour worm (*Trichostrongylus* spp). The thin-necked intestinal worm (*Nematodirus* spp) sometimes causes losses in lambs and weaners. Brown stomach worms, black scour worms and thin neck intestinal worms cause scouring ('scour worms'), ill-thrift and weight loss. Barber's pole worms suck blood and cause anaemia and weakness, but there is no scour; barber's pole worm can cause losses in moist summers in southern Australia. The large lungworm (*Dictyocaulus filaria*) causes coughing and pneumonia and is sometimes a problem in young sheep in spring.

Tapeworms (*Moniezia*) are common in lambs and are easily seen, but there is little evidence that they cause any damage and most losses blamed on tapeworms are caused by the brown stomach worm and black scour worm, which are very small and not so easily seen.

In most seasons, the scour worms are more important than the other roundworms except in northern New South Wales and Queensland where barber's pole worm is the predominant worm. The principles of roundworm control apply equally to lung, barber's pole and scour worms, but separate sections are included to cover the peculiarities of the lung and barber's pole worms.

To devise an effective worm-control program it is necessary to understand the life cycle of roundworms. Sheep carrying worms pass worm eggs into their faeces; given warmth and moisture, these eggs hatch and larvae move onto pasture. If larvae are eaten by susceptible sheep, they mature in 2 to 3 weeks, then mate and the females start laying eggs, and so the cycle continues (see Figure 6.1). Larval worms survive on pasture for up to 6 months in all seasons other than summer. Few larvae on pasture survive hot, dry summer conditions. Most worms are carried over summer in sheep. The worm population on a property is therefore found in two places: on pasture and in sheep.

Susceptibility ranking

Classes of sheep vary in their susceptibility to roundworms. Weaners and lambs are the most susceptible, followed by lambing ewes, especially maidens. Dry ewes and wethers usually have good immunity to the roundworms that cause scour. Susceptible sheep develop large worm burdens and

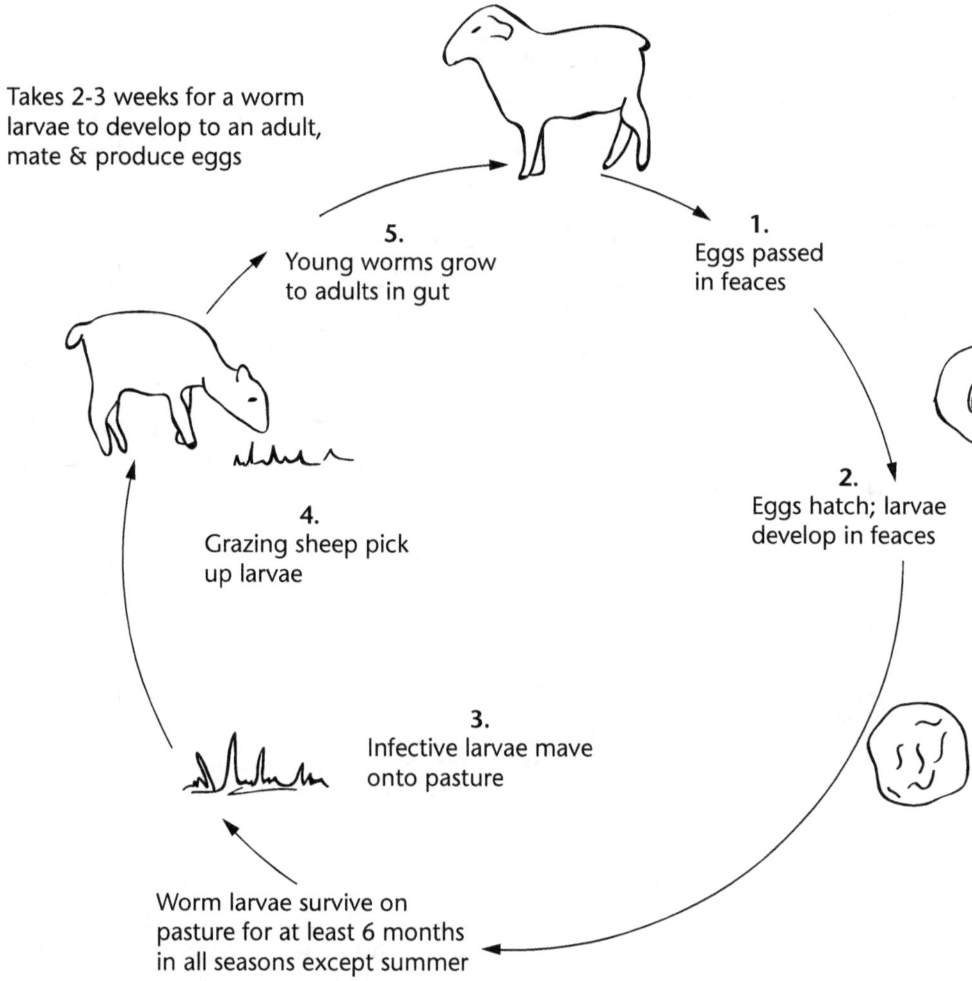

Figure 6.1. The life cycle of roundworms.

pass out large numbers of worm eggs in their faeces. Resistant sheep develop smaller worm burdens and pass out fewer worm eggs. Sheep are not infected by most cattle roundworms and vice versa, the exception being *Trichostrongylus axei* or stomach hair worm.

Roundworm drenches

Drenches are divided into three major groups according to their mode of action. Group 1 drenches are commonly known as benzimidazole (BZ or white) drenches, Group 2 as levamisole (LEV or clear) drenches and Group 3 as macrocyclic lactone (ML or mectin) drenches. Group 1, 2 and 3 drenches are effective against all the important stomach and intestinal roundworms, providing drench resistance has not developed. Narrow-spectrum drenches are effective against barber's pole worm and/or liver fluke. Recently, Zolvix® containing monepantel – the first member of the new class of anthelmintics, the amino-acetonitrile derivatives (AADs) – was launched in New Zealand. It has broad-spectrum activity against all roundworms and will become available in Australia. The drenches currently available for sheep are shown in Table 6.1.

Table 6.1. Sheep drenches available in Australia.

Chemical components	Example products	Meat WHP[a]
BENZIMIDAZOLE (BZ or white drenches) (Broad-spectrum)		
Albendazole	Alben, Rycoben, Strategik, Valbazen, Albendazole	10
Albendazole + Pyraclofos	Colleague	14
Albendazole capsules	Extender	0
Albendazole + Ivermectin capsules	Optamax	126
Fenbendazole	Fenbendazole, Panacur, Fencare	14
Mebendazole	Benzicare	7
Oxfendazole	Oxfen	10
LEVAMISOLE/MORANTEL (LEV or clear drenches) (Broad-spectrum)		
Levamisole	Levamisole, Nilverm, Ripercol, Rycolzole, Big L, Levamisole gold, Levimax	3
Levamisole + Praziquantel	First drench	3
Morantel	Oraject	7
MACROCYCLIC LACTONES (ML or mectin drenches) (Broad-spectrum)		
Abamectin	Abamectin, Combot abamec, Paramectin, Rycomectin, Vetmec, Virbamec, Absolute, Abamax, Baymec, Ovimectin, Zoomec, Avomec Genesis injection (35 day WHP)	14
Abamectin + Praziquantel	First Mectindrench, Genesis tape	14
Abamectin + Albendazole capsules	Bionic	125
Ivermectin	Ivomec, Genesis drench, Imax SC, Noromectin, Paramax concentrate, Virbamax, Austmectin	10–14
Ivermectin capsules	Ivomec maximiser	126
Moxidectin	Cydectin	7
Moxidectin + Praziquantel	Cydectin plus tape	7
Moxidectin (injection)	Eweguard, Weanerguard	28
Moxidectin (injection)	Cydectin Long Acting	91
COMBINATIONS: BZ + LEV (Broad-spectrum)		
Albendazole + Levamisole	Combi, Leviben, Rotate, Combinator, Vetmec dual	10
Fenbendazole + Levamisole	Combo, Combat duo, Combination, Duocare, Nucombo, Combimax, Switch	14
Oxfendazole + Levamisole	Scanda, Coalition	10
COMBINATIONS: ML + other broad-spectrums		
Ivermectin + Levamisole + Albendazole	Triton	14
Ivermectin + Levamisole + Oxfendazole	Scanda Max	14

Chemical components	Example products	Meat WHP[a]
Abamectin + Levamisole + Albendazole	Pyrimide	14
Abamectin + Levamisole + Albendazole + Closantel	Q-drench	28
Abamectin + Levamisole +Oxfendazole	Hat-trick, Avomec triple	14
COMBINATIONS: ORGANOPHOSPHATE (OP) + Broad-spectrum(s)		
Naphthalophos + BZ or LEV or BZ/LEV or ML	Rametin, Combat or Pole vault can mixed with some BZs, LEVs or MLs. Check labels	Various
Naphthalophos + abamectin	Rametin ML	14
Naphthalophos + levamisole + fenbendazole	Rametin combo	14
ORGANOPHOSPHATE (OP) (Mid-spectrum)		
Naphthalophos	Rametin, Combat, Pole vault	7
LIVER FLUKE DRENCHES		
Closantel	Closantel, Closicare, Seponver, Sustain, Closamax	28
Triclabendazole	Fasinex, Flukare, Flukeguard, Tremacide, Tricla, Triclabendazole, Exifluke	21
LIVER FLUKE DRENCHES + Broad-spectrum		
Closantel + Albendazole	Closal	28
Closantel + Oxfendazole	Closicomb, Rotafluke	28
Closantel + Abamectin	Genesis xtra, Avomec dual	49
Levamisole + Oxyclozanide	Nilzan LV	14
Triclabendazole + Abamectin	Flukamec	21
Triclabendazole + Ivermectin	Fasimec, Paramax-F, Triclamec	21
Triclabendazole + Moxidectin	Cydectin plus fluke	21
Triclabendazole + Oxfendazole	Flukazole	21

[a]WHP = withholding period (days). Producers should also be aware of export slaughter intervals (ESI).

Notes

Different chemicals within the same group act in a similar manner to kill worms. Worms resistant to one chemical in a group often have side resistance to the other chemicals in the same group. Some worms will have multiple resistance to chemicals from more than one group.

Several brands of drenches are prepared in a range of concentrations, formulations and mineral additives.

See product labels for full claims, withholding periods and restraints.

Always read and follow the label instructions.

Sources, Infopest CD July 2009 © DPIF Queensland 2009 (www.dpi.qld.gov.au/infopest) and APVMA March 2010 (www.apvma.gov.au).

Drench resistance

Drench resistance occurs when a significant proportion of a worm population survives a dose of drench that would normally kill them. This is defined as a less than 95% reduction in the worm egg count (WEC) after drenching. Drench resistance is widespread in Australia. Currently, drench resistance occurs on approximately 90% of properties to benzimidazole (BZ or white) drenches, approximately 90% to levamisole (LEV or clear) drenches, approximately 80% to combination (BZ + LEV) drenches and at least 30% to macrocyclic lactone (ML or mectin) drenches. Many properties have worm populations resistant to more than one drench group. On many farms with drench resistance, drenches or combinations of drenches will be partially effective (e.g. 70–95%) and provide a reasonable clinical response, though research shows use of such drenches will result in substantial production losses and economic loss in the order of $3.00 to $7.00 per weaner. Continued use of these drenches will tend to increase the level of resistance.

Resistance is not 'all-or-nothing' in the early stages: drenching still kills the majority of worms present. As resistance develops through repeated drenching, a greater proportion of the worm population survives the drench. For example, instead of killing 98% of the worms in the sheep, a drench may only kill 70%.

Drench 'failure' occurs when sheep do not respond to drenching. On the majority of properties on which drench resistance is confirmed by laboratory testing, there is no evidence of drench 'failure'. Drench resistance may go unrecognised in many flocks. If resistance is present, the efficiency of drenches is reduced and pasture contamination with roundworm larvae is increased, resulting in a need for more frequent drenching to control worms. The first indication of drench resistance may be the need to drench more often, not the failure of animals to respond to drenching. Frequent drenching with a failing chemical rapidly leads to selection of more strongly resistant worms. Eventually, sheep will fail to respond to drenching.

How drench resistance develops

A very small proportion of worms are naturally resistant to the drench. When a drench group is first introduced and used at the correct dose rate, it is highly effective: killing the vast majority of the population in the sheep (99% plus), but some naturally resistant worms survive.

The surviving resistant worms reproduce and pass on the ability to survive the drench to their offspring. With continued use of the drench, the susceptible worms are slowly eliminated and the proportion of resistant parasites in the population increases. With frequent and continued use of the drench, most of the population will be resistant worms. Even in the early stages of detectable drench resistance, a substantial proportion of the worm population carry resistant genes.

Drench-resistant worms are also likely to be introduced in purchased sheep and may have different patterns of resistance to the worms in the flock.

To delay the development of drench resistance:

- weigh the sheep and give the full dose for the heaviest sheep in the mob (use a volume measure to check that the correct dose is delivered by the drench gun)
- test the drench efficacy and use effective drenches at critical times, such as during summer
- monitor the worm egg counts and use a minimum number of extra drenches
- limit drenching when conditions are dry and there are few worms in refugia on pasture to dilute resistant worms that survive drenching
- use narrow-spectrum drenches instead of broad-spectrum drenches whenever possible; for example, for barber's pole in summer, where barber's pole only is the only worm present and the drench does not coincide with a strategic summer drench
- limit the use of combination roundworm and liver fluke products to when treatment of both roundworm and liver fluke is required
- use quarantine drenching to minimise importing resistant worms with purchased sheep by drenching them on arrival with a highly effective combination of drenches such as BZ + LEV + ML (consult your adviser to select appropriate drenches and quarantine procedures); if

possible, release the sheep into a paddock that will not be grazed by other sheep for at least 3 months; consider conducting a 'DrenchCheck' 10 to 14 days after treatment to confirm that the treatment has been effective.

Tests for drench resistance

Test for drench resistance at least every 3 years to determine which drenches are effective. Useful tests are the worm egg count reduction trial (WECRT), DrenchRite™ test or a 'DrenchCheck' (monitoring worm egg counts at drenching, then again 10 to 14 days after drenching a mob). WECRT measures the efficacy of several drenches in the same test and should be conducted every 3 years. 'DrenchCheck' is easier to conduct; it measures the efficacy of the current drench and should be done at least once a year. Lambs at weaning time are the ideal group for WECRT. A 'DrenchCheck' can be conducted at any time that WECs are high enough.

Grazing management

The worm population on a grazing property is not confined to stock. The largest proportion of the worm population (the worm larvae) is on the pasture. The aim of grazing management is to provide young sheep that are most vulnerable to worms with the pastures with adequate feed and lowest worm contamination. Sheep maintained in good body condition are more resilient and can better withstand the ill effects of worms.

Low-risk paddocks have few worm eggs and larvae. Sheep carrying a worm burden will contaminate low-risk pasture. A contaminated pasture harbours worm eggs and larvae. Susceptible sheep grazing contaminated pasture will pick up a worm burden.

The principles outlined below can be used to prepare low-risk pastures.

Principles of grazing management:

- drenching only kills worms in the sheep; limiting worm pick-up depends on grazing management
- worm eggs and larvae survive for over 6 months on pasture, except in summer when survival is shorter; it is not practical to spell pastures to clean them; grazing management requires forward planning
- paddocks can be classed as low-, moderate- or high-risk worm paddocks (few paddocks are worm free)
- use cattle to prepare low-risk paddocks for sheep because the major cattle worms do not infect in sheep and vice-versa
- not all classes of stock are equally susceptible to roundworms; weaners are most susceptible, dry sheep are least susceptible
- adult wethers and dry ewes in good condition usually have the lowest egg output
- prepare low-risk pastures by cropping or hay making
- if sheep are being moved onto a low-risk paddock, drenching will be beneficial, especially in young sheep.

Low- to moderate-risk paddocks include:

- pasture grazed by cattle
- new pasture or crop
- crop stubble (usually poor feed value)

- dry pasture not grazed by sheep since summer
- hay aftermath
- paddocks grazed by adult dry sheep (including using Smart Grazing), which have received two effective summer drenches.

High-risk paddocks include:

- lambing paddocks
- paddocks grazed by weaners
- paddocks grazed by wormy sheep.

Preparing weaner paddocks

There are two types of weaner paddocks: firstly those that are used for weaners immediately after separation from ewes in spring and into summer and, secondly, in high rainfall regions, those that are used for 8- to 12-month-old weaners in winter.

The first weaning paddock should be prepared using the list above (low- to moderate-risk paddock). Ideally new pasture, fodder crops or paddocks grazed by cattle for at least 6 months are the lowest risk. Paddocks grazed by sheep are less preferable and avoid lambing paddocks or paddocks previously grazed by weaners. Pasture quality is critical for weaned lambs, so pastures must be good quality or, if not, adequate supplementary feed must be provided.

In high rainfall regions, the first weaning paddocks often become heavily contaminated with worms in winter. In this situation, a second weaning area must be prepared so that weaners can be drenched and moved in late autumn and winter to ensure they continue to graze low-risk paddocks over the critical winter period. This strategy is not usually necessary in regions below 600 mm rainfall if worm control is good. Ideally, prepare the winter weaning paddocks by grazing with cattle for at least 6 months, or with Smart Grazing (see WormBoss http://www.wormboss.com.au).

Drenching

Use an effective drench at the correct dose rate. Test for drench resistance at least every 3 years.

Strategic drenching: time drenching to minimise drench frequency

Either one or two summer drenches should be given. These drenches control worms for the rest of the year. Give the first drench when pastures dry out, but before late December, regardless of the region. In lower rainfall areas (<500 mm rainfall), monitor the worm egg count (WEC) because the first summer drench may not be required. Always monitor WECs of all mobs to see if a second drench is required in February.

The other 'must-do drench' is the weaning drench for lambs. Drench and wean all lambs off Merino ewes at 12 to 14 weeks. In some circumstances, meat breed lambs benefit from staying on meat breed ewes slightly longer, but all lambs benefit from drenching at this time. It is not necessary to drench at lamb marking if ewes are lactating well and lambs are weaned at 12 to 14 weeks.

Pre-lambing drench for ewes

If ewes are in good condition and grazing recommended pasture, then pre-lambing drenching is not necessary, especially with spring-lambing flocks. If ewes are less than condition score 2.5 at lambing and grazing short feed on heavily contaminated pasture, pre-lamb drenching on the point of lambing may be justified. The need for pre-lambing drenching will depend on the level of worm infection, the likely pasture contamination, the amount of pasture available and the condition of the ewes. WEC monitoring, 2 to 4 weeks before lambing, is useful to determine if drenching is necessary. Prime lamb ewes, where lambs remain on the ewes for more than 14 weeks, may benefit from pre-lamb drenching, partly to reduce pasture contamination for lambs.

Extra drenching

Additional drenching should only be done on information provided by WEC monitoring. Ideally, adult sheep need only one to two drenches per year and weaners two to three. Sheep producers exceeding this frequency should seek expert advice.

Introduced sheep

Give a quarantine drench as described under 'Drench resistance'.

Choice of drench group

All broad-spectrum worm drenches on the market can be classified by the way in which they kill worms. For example, the benzimidazole (BZ or white) drenches act by interfering with the respiratory pathways and structural protein synthesis of worms. Levamisole (LEV or clear) and macrocyclic lactone (ML or mectin) drenches act by paralysing the worms. Although there are some differences between the active chemicals within drench groups, changing from one drench to another within the same group is effectively only changing the trade name used.

The decision of which drench group to use is important because drench resistance now occurs on most properties. Determine which drenches are effective on an individual farm using WECRT or 'DrenchCheck'. Consult an animal health adviser on which drench groups or combinations are effective on a property and their use in an integrated worm-control program. Using the same drench every year is not recommended; generally, a rotation of effective drench groups is suggested – the rotation depending on individual property drench efficacy.

Monitoring

WEC monitoring is done to decide on non-strategic drenching and summer drenching. Monitoring is also carried out pre and post drenching to check on drench efficacy.

Monitor worm egg counts in weaners about 4 weeks after the autumn break and at 4- to 6-week intervals through winter and early spring.

Integrated parasite management (IPM) strategies for worm control

In southern Australia, larval populations of the common worms (*Ostertagia* and *Trichostrongylus*) are highest in the cold, wet months of winter and lowest during the hot, dry months of summer. Eggs deposited during late summer and autumn contribute the largest number of larvae available to sheep during next winter (June to August). These months are cold and wet, with limited pasture growth. The additive effects of low pasture availability, high nutritional demands and high numbers of infective larvae cause ill-thrift, clinical worm infections and deaths. In Merinos, this is more severe in weaners and maiden ewes.

Consequently, worm-control programs in the high winter rainfall areas are based on reducing pasture contamination with worm eggs during summer and autumn. The simplest way of achieving this is a strategic drench in early and mid-summer to all classes of sheep, using a highly effective combination of drench compounds.

Each farming system is quite different and so a 'general' worm control program must be tailored to each farm.

Key messages – roundworms

1. **Testing for drench resistance every 3 years** – this will show which drenches are effective.
   - Useful tests are the worm egg count reduction trial (WECRT), DrenchRite™ test or a 'DrenchCheck' (monitoring worm egg counts at drenching, then again 10 to 14 days after drenching a mob).
2. **Monitoring worm egg counts at critical times** – weaners should be monitored every 4 to 6 weeks after the autumn break. Other classes of sheep should be monitored as follows:
   - All mobs should be monitored in late January early February to see if a second summer drench is needed in February, or if it can be delayed. This will also indicate which paddocks may be more 'high-risk' the following winter.
   - All mobs should be monitored in July/August to check on peak winter egg counts and whether or not a pre-lambing treatment of ewes may be needed (there is a big interaction with ewe condition score, available pasture and the drenching decisions taken at this time)
   - All ewe mobs should be monitored at marking in October – this is a check on how effective the control program has been in ewes, and can highlight some mobs that need an additional drench (only needed in high worm risk areas).
   - All sheep should be monitored at the first summer treatment in December – this gives an indication of any higher risk paddocks from the 'spring' generation of worms that might occur next winter (only needed in high worm risk areas).
3. **Giving introduced sheep a quarantine drench** – this consists of at least two effective drench groups, but preferably four.
4. Use **grazing management to provide low risk pastures for weaners in their first winter:**
   - Smart Grazing
   - alternate grazing between sheep and cattle, preferably rotating every 6 months.
5. **Breeding sheep for increased resistance to worms** (low worm egg count) and less scouring (a low dag score).
6. **Having targets for ewe condition score at lambing** (for example, condition score 3) and weaner body weight over summer-autumn-winter, and monitoring progress against these targets.

Further details of IPM worm control programs for winter rainfall, summer rainfall and Mediterranean regions are provided on IPMs, AWI and MLA web sites, listed in References and further reading at the end of the book.

Lungworm

Lungworms are also roundworms, but the life cycle is different to other roundworms. Lung worm live in the airways of the lungs; adult worms lay eggs that are coughed up and swallowed. These eggs hatch to release larvae while they are still in the sheep's intestine, and larvae are passed onto pasture in faeces. Sheep are infected by grazing pasture contaminated by these larvae. After ingestion by sheep, the larvae burrow through the wall of the intestine and migrate to the lungs via the bloodstream and the lymphatics. The development to adults is completed in the lungs.

Larvae survive best under cool, wet conditions, so numbers build up on pasture in winter and early spring.

Adult sheep usually have a strong immunity to lungworm. Lambs and weaners are most susceptible; lungworm is commonly found in young sheep but usually in small numbers. The main symptom is coughing. Heavy infections can cause difficulty in breathing and death. Complications include pneumonia. Lungworm is usually found along with heavy scour worm burdens, which do more damage to the host. All broad-spectrum drenches are effective against lungworm.

Barber's pole worm

Barber's pole worm *(Haemonchus contortus)* is the most important worm in summer rainfall regions, but can also occur in summers in the rest of New South Wales and parts of Victoria. The life cycle is similar to that of the other roundworms. Adult barber's pole worms live in the fourth stomach (abomasum). Barber's pole worm can become a problem in warm wet summers. Barber's pole worms suck blood from the lining of the fourth stomach and in the acute form of the disease causes severe anaemia and a heavy mortality. Affected sheep may be found dead; others may lag behind if the mob is driven and may collapse and die. There is no scour. In the chronic form, barber's pole worm causes mild anaemia with loss of condition and often a 'bottle-jaw'.

The eggs and larvae of barber's pole are very sensitive to heat, cold and dry conditions. A single female barber's pole worm may lay 5000 eggs per day. A heavily infested sheep harbouring 3000 females may give a daily output of 15 million eggs. Infections can build up rapidly under the right conditions for egg and larval survival.

Under outbreak conditions, pastures usually contain large numbers of larvae before the flock shows any signs of disease. Re-infection will be rapid unless the sheep are shifted to a fresh pasture.

Drench resistance is now common in barber's pole worm. Consult an adviser to select effective products and control programs.

In summer rainfall regions where barber's pole is the major worm, the control program will vary between season, drench resistance status and grazing management. Monitoring worm egg counts is critical to determine when drenching is required. Both barber's pole and scour worms need to be considered as part of the control program.

The program required will depend on the resistance status. The barber's pole season usually starts in October and extends to April on the tablelands. Time the first drench for late October to coincide with infective larvae becoming available on pasture. Additional drenches may be required in late December and late February. Closantel used to be the preferred drench but it is no longer effective on more than 60% of properties, so alternative drenches must be used to control barber's pole over this time. Other options include combining Closantel with a broad-spectrum drench or Naphthalophos 3 weeks later, or using Moxidectin (if effective). Another option includes using an effective broad-spectrum drench and/or Naphthalophos, with a move to clean pasture. Always monitor to see if these drenches are required and consult a veterinary or other animal health adviser to fine-tune plans appropriate for a property. Relying solely on long-acting products such as capsules will be a very short-term option for barber's pole worm control because drench resistance is likely to develop.

In a barber's pole worm control program, additional drenches that may be required for broad-spectrum control of brown stomach worm and black scour worm include pre-lamb drenching to ewes based on need and weaning drench to lambs at 12–14 weeks after the start of lambing. Additional drenches for lambs should always be based on worm test over the following late autumn and winter. WEC monitoring is done to decide on non strategic drenching and summer drenching. Monitoring is also carried out pre and post drenching to check on drench efficacy.

The principle of grazing management to provide young sheep that are most vulnerable to worms with the pastures with lowest contamination also applies to barber's worm control. Prepare weaning paddocks with cattle for at least 6 months or with smart grazing (see WormBoss: www.wormboss.com.au). Alternatively, use wethers to prepare weaning areas for at least 6 months before lambs are weaned. Ideally, graze weaners on a second 'safe' paddock the following winter. Research indicates that barber's pole can be controlled effectively by rotational grazing where the rotation length extends to at least 50 days. This is useful in summer, when barber's pole larvae survive less than 50 days.

Research shows that long-term selection for sheep with low worm egg counts can reduce pasture contamination and possibly drench frequency. Integrate the selection pressure with selection for all important wool and meat traits. This strategy is most useful for control of barber's pole.

Sheep with good nutrition manage worm burdens better. Maintain sheep above condition score targets. Research in the New England region indicates that protein supplementation on poor-quality pasture can have a significant effect on worm control in lambing ewes.

Tapeworm

Infections with the sheep tapeworm *Moniezia* are common and obvious in lambs. When infected lambs are yarded, they often pass large quantities of tapeworm segments. As sheep mature they develop a strong immunity to this parasite.

There is no evidence that tapeworm causes production loss. Losses attributed to tapeworm are often caused by the scour worms, which are not easily seen.

Treatment for tapeworm alone is not warranted. Broad-spectrum drenches containing a benzimidazole or praziquantel will be effective against tapeworm.

Sheep tapeworms have an interesting life cycle. Tapeworm segments are passed in the dung of infected sheep. The segments release tapeworm eggs onto pasture, and the eggs are eaten by pasture mites and undergo the next stage of development in the mite. Sheep then eat the infected pasture mites while grazing and immature tapeworms are released in the sheep's intestine. The head of the tapeworm attaches to the lining of the intestine and the tape worm grows, shedding segments in dung to repeat the cycle.

Sheep measles

'Sheep measles' are small, white, oval shaped cysts about three to 10 millimetres long found in sheep muscle. They are the intermediate or 'larva cystic' stage of a tapeworm called *Taenia ovis*. Sheep measles occur in Australia wherever there are sheep (or goats) and dogs.

The life cycle of *Taenia ovis* is similar to that of the hydatid tapeworm, *Echinococcus granulosus* (see Chapter 9). The dog is the final or definitive host and harbours the adults of both tapeworm species in its intestines. The adult *Taenia ovis* tapeworm is about 2 metres long; the adult hydatid tapeworm is tiny: about four to six millimetres long.

Tapeworm eggs are shed in the dog's droppings; sheep ingest the eggs when eating pasture. These eggs hatch and the tiny larvae develop into cysts in muscles of the sheep, in the case of sheep measles (or in organs such as the liver and lungs, in the case of hydatids). Dogs get infected by eating viable tapeworm cysts in sheep or goat tissues.

The fox is not an important definitive host for *Taenia ovis* in Australia. Field evidence indicates that dingoes and wild dogs are only very occasionally infected with *Taenia ovis* in certain areas. Domestic dogs are the most important definitive host for *Taenia ovis* in Australia.

Unlike hydatids, sheep measles is not a human health hazard. The parasite does not develop in humans, but infection is nonetheless considered unacceptable and results in carcase condemnations.

Like hydatids, the key to control of sheep measles is breaking the sheep–dog life cycle by:

- controlling dog movements
- preventing dogs from eating sheep or goat meat or offal
- treating dogs regularly with an efficient tape-wormer.

The occurrence of sheep measles in sheep carcases is an important issue, with partial or full condemnation of affected carcasses and producers being penalised with discounts for affected carcasses.

Sheep measles can cost producers a lot of money, but its control (and that of hydatids) is relatively easy: seek advice from a veterinarian.

Liver fluke

In Australia the life cycle of the liver fluke, *Fasciola hepatica*, involves a fresh water snail *Lymnaea tomentosa* (plus *L. columella* in some areas) as the intermediate host, and sheep, cattle, goats and other animals as final hosts; see Figure 6.2.

The time scale of events described in Figure 6.2 is as follows:

- Stage 1 to Stage 2 takes about 3 weeks in summer and longer in winter
- Stage 2 to Stage 3 takes about 1 month at 25°C and 3 months at 15°C
- Stage 5 to Stage 1 takes between 8 and 12 weeks.

Adult liver flukes can live for several years in the liver of sheep (6 to 12 months in cattle) and daily egg output can be as high as 5000 eggs per fluke. The development stages of the fluke also multiply within the snail.

The snail prefers slow flowing fresh water; its main habitats are temporary or permanent springs. Big rivers and dams are not very suitable, but backwaters, swamps, irrigation channels, slow moving creeks, drains and billabongs provide good habitats.

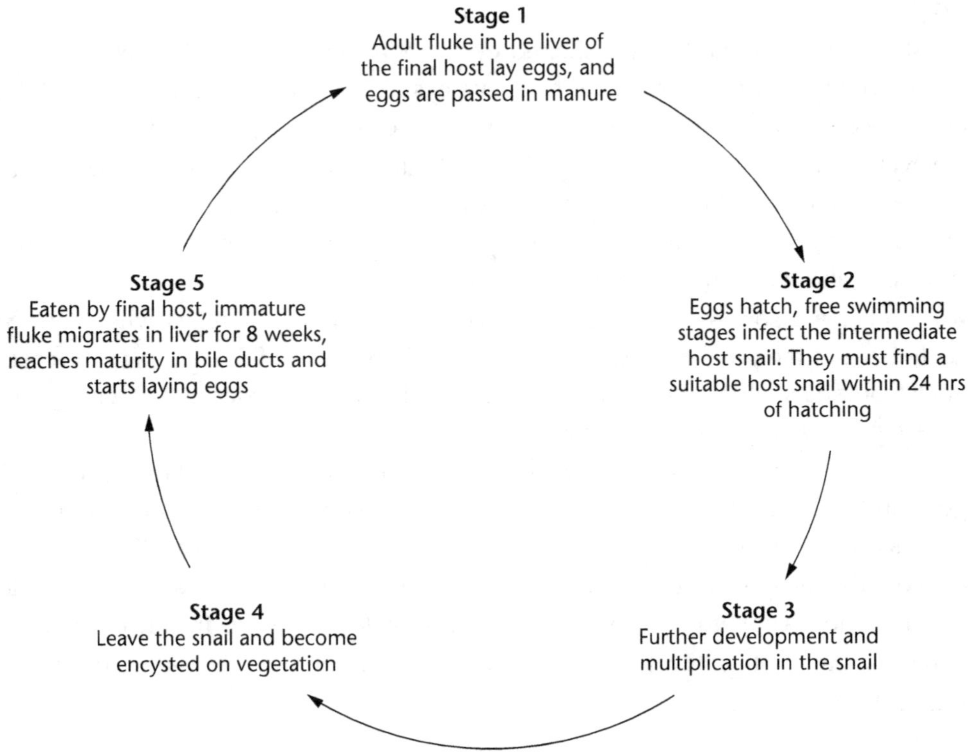

Figure 6.2. The life cycle of liver fluke.

Snails have a large reproductive capacity, laying up to 1000 eggs per month. Temperatures need to be above 10°C for reproduction and for the progression of the development stages of flukes in the snails. Snails multiply from early spring to early summer and, if the temperature and moisture levels are right, some multiplication can occur at the beginning of autumn. There is usually little snail activity in the winter or summer. When the environment becomes dry, snails survive by burying themselves and aestivating (or hibernating) in the drying mud. Snails can migrate long distances assisted by rain, flooding and irrigation.

The important features of the fluke life cycle are:

- its dependence on the intermediate host snail which has limited habitats
- both snails and flukes are very prolific breeders
- the snail is temperature and moisture dependent.

There are two types of fluke disease:

- **Acute fluke disease –** Acute fluke disease usually occurs 6 to 8 weeks after the intake of a large number of the developmental stages of flukes, which are found as cysts on pasture. This tends to occur from mid-summer when pastures dry off and sheep graze swampy areas. Acute fluke disease is a form of acute hepatitis due to massive damage to the liver caused by immature flukes. Sheep may be found dead; there is rapid loss of condition, pale membranes and a loss of appetite and abdominal pain. Damage to the liver caused by immature flukes can also predispose sheep to black disease (one of the clostridial diseases).

- **Chronic fluke disease** – Chronic fluke disease is more common than acute fluke disease, and tends to occur in autumn/winter. It is caused by the activity of adult flukes in the bile ducts of the liver.

Signs develop slowly and consist of ill-thrift, anaemia and bottle-jaw. Chronic fluke disease causes severe loss of wool growth and body weight in proportion with the number of flukes present. Loss of production without any other signs can also occur with light fluke infestations.

All classes of sheep are susceptible because mature sheep do not develop immunity to flukes.

Diagnosis of acute fluke disease is by post mortem examination. The liver is usually swollen and congested, there are numerous burrows, either filled with blood or starting to fibrose. On slicing the liver, numerous immature flukes about 5 to 10 mm long can be squeezed out. Liver fluke ELISA blood tests detect immature as well as adult liver flukes in the live animal.

In chronic fluke disease, fluke eggs will be present in faeces or there will be a positive ELISA blood test for flukes. Adult flukes in the bile ducts of the liver cause thickening and fibrosis of these ducts. When the ducts are opened they are usually packed with adult flukes.

Control of liver fluke

Liver fluke can be controlled by controlling snails:

- Drain wet areas and improve irrigation channels to reduce the snails' habitat.
- Using chemicals to control snail habitats is not recommended because, in most situations, snails quickly recolonise treated areas.

Liver fluke can be controlled by grazing management:

- Fence off infected areas.
- Graze cattle; mature cattle are more resistant to fluke than sheep, so cattle could be used to graze fluke-prone areas during summer.
- Use alternate grazing: divide paddocks into fluke-infected and fluke-free; graze the infected paddocks for 8 weeks, then move the sheep to fluke-free paddocks for 8 to 12 weeks, treat before returning them to fluke-free paddocks (see Table 6.2). This is theoretically possible, but difficult to implement on many farms.

This breaks the fluke life cycle because when sheep are on infected pastures no eggs are being excreted by the sheep; when they are on clean pasture eggs are being excreted but no snails are available to complete the life cycle.

Liver fluke can also be controlled by drenching:

- Control should aim to reduce the number of parasites in the final host and thus the consequent pasture contamination by strategically timed drenches. Drenches for liver fluke are included in Table 6.1.
- Common times to drench are late winter (July) and late summer (February). The late winter drench is to remove adult flukes, and hence remove the source of fluke eggs before the snails become active. The summer drench is to remove immature flukes picked up over the spring and early summer. For summer drenching use drenches containing triclabendazole (effective against early immature, as well as late immature and adult flukes) or containing closantel (effective against late immature and adult flukes).

Table 6.2. Alternate grazing for liver fluke control.

Month	N	D	J	F	M	A	M	J	J	A	S	O
Pasture infectivity	High				Medium				Low			
Pasture rotation	Graze fluke-free pasture			Graze infected pasture			Graze fluke-free pasture			Graze infected pasture		
Treatment			▲						▲			

- On heavily infected farms or in wet years, additional drenches may be required. Consult a veterinarian or animal health adviser.

Drench resistance

Liver flukes resistant to triclabendazole occur on some farms in NSW and Victoria. Consider rotating from one flukicide to another flukicide containing a different active ingredient every 1 to 2 years. Seek veterinary advice for diagnosis and delaying drench resistance. Combination products effective against both liver fluke and roundworms are available. Limit the use of these products to when treatment of both liver fluke and roundworms is required.

External parasites

Blowfly strike, lice, itchmite, ked and scrotal mange affect the skin of sheep.

Blowfly strike

Blowfly strike is a formidable challenge to the sheep industry.

Primary flies – the green Australian sheep blowfly, *Lucilia cuprina* and the brown *Calliphora species* blowflies – are capable of initiating flystrike by laying eggs on unstruck sheep. They also extend strikes by laying further eggs on already struck sheep. Secondary flies, such as the 'hairy maggot fly' *Chrysomya rufifacies* only attack sheep already struck.

Lucilia cuprina is the major problem fly: it is involved in about 95% of strikes in Australia.

Flies are attracted by odours from smelly moist wool on susceptible sheep. Eggs laid on moist wool hatch and the larvae (maggots) move down to the skin. As they develop, they become capable of damaging the skin and extending the strike making conditions suitable for more flies to lay eggs.

Maggots feed for about 4 days on the sheep before dropping off and burrowing into the soil. During warmer times of the year, they pupate within 2 to 3 days and adult flies emerge about a week later. Maggots leaving the sheep in late autumn do not pupate until the following spring: the fly over-winters in the soil as a maggot.

Adult female flies require a protein feed and must mate before seeking out a suitable site for egg laying. Females lay batches of 150 to 250 eggs and may lay two or three batches in their lifetime. Under optimum conditions, a complete life cycle (egg to egg) can be completed in less than 3 weeks.

Flystrike is most likely to occur between November and May and is dependent on rainfall. Two factors operate to encourage the development of the disease: the presence of susceptible sheep and a large fly population. Susceptible sheep include those with a breech soiled with urine or faeces, a wet pizzle, or fleece rot, lumpy wool, wounds or footrot. Clean wounds are not usually attractive to flies.

Breech strike is the term for strikes occurring on the crutch and tail region and is the most common site for strike to occur. Tail docking to the correct length – to allow sheep to elevate their tail during defecation to reduce faecal soiling of the breech wool – jetting and crutching are the recommended control methods for breech strike. Good worm control to minimise scouring is essential for effective control of breech strike.

Pizzle strike occurs in wethers and rams when long wool around the pizzle becomes wet with urine. Removal of wool around the pizzle by shearing or ringing is the most common control method.

Poll strike is mainly confined to rams and is very common. Strike can develop in fight wounds or in the skin folds at the base of the horns. The use of chemicals and shearing are the major forms of control.

Body strike involves any part of the body other than the breech, head and pizzle. Most commonly affected sites are the shoulder and back regions. Outbreaks of body strike or 'flywaves' can develop rapidly after periods of prolonged and heavy rainfall in the warmer months. Young sheep are the most susceptible. The two major factors influencing the likelihood of a sheep developing body strike are fleece rot and lumpy wool.

Prevention and control by management

Most forms of flystrike – in particular breech strike – can be prevented by management practices such as mulesing (or its replacement), correct tail docking, crutching, shearing and internal parasite control. However, the problem of body strike can be serious because the costs associated with prevention, control and treatment can be high.

Prevention and control by chemicals

The chemicals available for fly prevention and control include cyromazine, dicyclanil, spinosad and ivermectin. Product details are shown in Table 6.3.

Application is by spray-on, hand jetting, automatic jetting or dipping. Automatic jetting is the quickest method of application and gives reasonable results when cyromazine is used.

Hand jetting is more effective than automatic jetting races if done thoroughly, but is relatively slow. Dipping is also thorough but slow and only used for sheep with 2 to 6 weeks wool.

Treatment of flystrike

Ideally, the affected area should be clipped and treated with an insecticidal solution of, for example, spinosad, ivermectin or cyromazine. Jetting can be used if rapid treatment of large numbers of struck sheep is required, but healing is quicker if the wool is clipped from the struck area and wounds are able to dry out. Treat the wool clipped from the sheep to kill the maggots.

Blowfly danger periods can be predicted under extended conditions of high temperature and humidity, which favour sheep susceptibility and fly reproduction rates. The flock should be inspected at least three times per week for evidence of strike in these periods and preventative measures should be undertaken.

Integrated Parasite Management (IPM) strategies for flystrike control

The following strategies are used for flystrike control:

1. Ensuring the correct tail length (the third palpable joint should cover the tip of the vulva).
2. Selecting sheep with reduced susceptibility to body strike (less fleece rot and body wrinkle) and breech strike (plainer breech, larger bare area, less scouring).
3. Controlling scouring:
   - in young sheep through effective worm control
   - in older sheep by genetic selection against 'hypersensitivity scouring'; use rams that have a low dag score and cull daggy ewes.
4. Strategic timing of crutching and shearing before periods of high risk for flystrike.
5. Other methods that may be applicable:
   - choosing low-risk paddocks with respect to worms for the most susceptible sheep (weaners)
   - using flytraps to monitor fly populations (this will define low- and high-risk paddocks and identify patterns of fly activity)
   - strategic application of pesticides with regard to fly biology, flystrike risk and wool residues (the timing of strategic treatments can be early in the season (September to October), to control the first generation of flies, or late in the season (April to May) to limit the population of over-wintering larvae).

Table 6.3 lists veterinary chemical preparations available for control of flystrike in Australia.

Lice

Three types of lice may be found on sheep:

- body louse (*Bovicola ovis, formerly Damalinia ovis*) – a biting louse found in the fleece
- face louse (*Linognathus ovillus)* – a sucking louse found in the hair
- foot louse (*Linognathus pedalis)* – a sucking louse found in the hair.

Body lice are found in approximately 10 to 30% of sheep flocks in Australia. Foot lice are found occasionally. Face lice, which are very sensitive to high temperature, are not common in Australia.

The body louse

The body louse (*Bovicola ovis*) is a small (1 to 2 mm) banded insect, pale yellow to brown in colour, which is usually found close to skin in fleece covered body regions, with the heaviest concentrations on the upper half of the body and neck.

All stages occur on the sheep. A generation from egg to nymphal stages to egg-laying adult takes 34 days.

Spread occurs through direct contact between sheep. Body lice usually cannot survive off sheep for more than a few days.

Shearing reduces the lice population by up to 50%. This occurs through the removal of some lice in the shorn wool and the exposure of the remaining lice to rain and extremes of temperature.

Sheep rubbing against fixed objects, biting or scratching their fleece with their hind feet are indicators that lice may be present. Other causes of irritation include flystrike, grass seeds, lumpy wool and itchmite.

Table 6.3. Flystrike control products available in Australia.

Time of application	Application method	Chemical group	Example products
Off-shears	Spray-on	Fly IGR[a]	Clik
		Fly IGR + Lice IGR	Magik, Clik plus
Short wool (up to 6 weeks wool)	Spray-on	Fly IGR	Clik
	Dip	Fly IGR	Vetrazin, Venus, Virbazine, Cyromazine, KFM, Cyro-Fly, Cy-guard, Cyrazin, Cyromazine 500, Lucifly, Proguard, Clout
Long wool (6 weeks to 9 months wool)	Spray-on	Fly IGR	Clik
Long wool (6 weeks to 10 months wool)	Spray-on	Fly IGR	Vetrazin SO[b], Venus SO, Virbazine SO, Cyromazine SO, Cyro-Fly SO, Cy-guard SO, Cyrazin SO, Lucifly SO, Proguard SO, KFM SO, Clout SO
	Jet[c]	Fly IGR	Vetrazin, Venus, Virbazine, Cyromazine, KFM, Cyro-Fly, Cy-guard, Cyrazin, Cyromazine 500, Lucifly, Proguard
	Spray-on	Synthetic Pyrethroid	Vanquish
Long wool (6 weeks to 10½ mths wool)	Hand jet	Macrocyclic Lactone	Blowfly and Lice Jetting Fluid, Paramax Concentrate, Zinjet
Long wool (6 weeks to 12 months wool)	Jet	Spinosyn	Extinosad

[a]IGR = Insect growth regulator
[b]SO = Spray-on
[c]Hand jet or jetting race
See product labels for full claims, withholding periods and restraints. Do not apply later then the specified number of days before slaughter or harvesting wool.
Always read and follow the label instructions.
Sources: Infopest CD July 2009 © DPIF Queensland 2009 (www.dpi.qld.gov.au/infopest) and APVMA March 2010 (www.apvma.gov.au).

To examine sheep for lice, part the wool to skin level in a number of places along the back line and sides and over the neck. Lice are readily seen in good light, but they avoid light by moving away through the fleece when exposed to it.

With a moderate to heavy infestation, the clean fleece weight is reduced by up to 0.5 kg per sheep. With a light infestation, the clean fleece is reduced by up to 0.2 kg per sheep. Wool loss during processing is also increased. Lice have no effect on live weight, but heavier infestations may occur on malnourished, lighter sheep in the mob.

The main chemical groups used for treating lice are insect growth regulators, organophosphates, synthetic pyrethroids, ivermectin and spinosad.

A list of veterinary chemical preparations currently available for use as sheep dips, backline or jetting treatments in Australia can be found in Table 6.4.

When to treat body lice

Treatment should only be undertaken if an infestation exists or there is reasonable suspicion that one exists. If lice are present, the aim is to eradicate lice at the next shearing by effective short wool

Table 6.4. Sheep-lice-control products available in Australia.

Time of application	Application method	Chemical group	Example products
Off-shears	Backliner	Lice IGR[a]	Zapp, Magnum, Stampede, Clipguard, IGR Pour-On, Epic, Triflumuron 25, Command Pour-On, Triffik, Exilice, Exit, Cannon
		Fly IGR + Lice IGR	Magik, Clik plus
		Spinosyn	Extinosad Pour-On
		Neonicotinoid	Avenge
		Synthetic Pyrethroid	Clout-S, Cypercare, Spurt, Cypermethrin 25, Outflank
	Spray-on	Organophosphate	Eureka Gold
Short wool (up to 6 weeks wool)	Dip	Lice IGR	Fleececare, Strike
		Spinosyn	Extinosad
		Magnesium Fluorosilicate	Flockmaster Mk 2
		Organophosphate	Assassin, Wham
Long wool (6 weeks to 6 months wool)	Hand jet	Lice IGR	Strike, Fleececare
	Spray-on	Lice IGR	Magnum
Long wool (6 weeks to 10 months wool)	Spray-on	Synthetic Pyrethroid	Vanquish
Long wool (6 weeks to 10½ mths wool)	Hand jet	Macrocyclic Lactone	Blowfly and Lice Jetting Fluid, Paramax Concentrate, Zinjet
Long wool (6 weeks to 12 months wool)	Jet[b]	Spinosyn	Extinosad
Long wool (6 months to 12 months wool)	Backliner	Spinosyn	Extinosad Pour-On

[a]IGR = Insect growth regulator
[b]Hand jet or jetting race
See product labels for full claims, withholding periods and restraints. Do not apply later then the specified number of days before slaughter or harvesting wool.
Always read and follow the label instructions.
Sources: Infopest CD July 2009 © DPIF Queensland 2009 (www.dpi.qld.gov.au/infopest) and APVMA March 2010 (www.apvma.gov.au).

treatment. Lice in long wool (between 6 and 44 weeks wool growth) can be controlled, if necessary, by jetting or backline application. These treatments will not eradicate the lice completely and these sheep should be effectively treated at the next shearing with a pesticide from a different chemical group.

It is unnecessary, and to some extent counter-productive, to treat for lice if they are not present. Treatment at shearing gives only short-term protection against re-infestation, which varies depending on the product used and how well it is applied.

Eradication

Lice live their entire life cycle on sheep. This fact, coupled with the availability of effective pesticides, should make eradication possible. However eradication is complicated by:

- failure to treat all sheep including rams, lambs, killers or incomplete muster
- sheep not being clean shorn (due to lumpy wool or neck skin folds)
- insufficient chemical being applied to sheep
- split shearing (if sheep on the property are shorn and treated for lice at different times of the year, re-infestation can occur)
- re-infestation from neighbouring sheep or purchased sheep
- shearing and using pour-on lice control preparations within 6 weeks of lambing
- lice resistance to the chemical.

If an outbreak of lice occurs in sheep with more than 6 to 8 weeks wool, the alternative courses of action are:

- control by midline jetting or long wool pour-on and retreat to eradicate after the next shearing with a product from a different chemical group
- if a very light infestation is detected within 2 or 3 months of the next shearing, isolate the infested mob/s but do not treat; long-wool treatment is unlikely to be warranted
- re-shear and treat to eradicate (this is expensive and disrupts farm management and is not recommended).

Application is by dipping or pour-on (see Table 6.5). The basic aim with dipping is to thoroughly wet every sheep all over with dip solution of the correct strength. With pour-ons, all sheep must be treated with the specified dose, and in the recommended manner, within 24 hours of shearing. The likelihood of eradication is the same with both methods.

Mistakes are often made in the dipping or pour-on treatment program. Some important points to remember are:

Dip preparation

- ensure there is no risk of arsenic residues in old dips
- plan to dip all sheep between 2 and 6 weeks off shears
- clean and measure the exact capacity of dips and sumps
- install a pressure gauge to a shower dip pump
- trial run a shower dip a week before planned dipping
- yard or shed sheep overnight before dipping

Table 6.5. Comparison of application methods for lice control.

Pour-on or Spray-on	Versus	Plunge/shower dip
High chemical cost, low capital cost		Generally lower chemical cost, high initial capital cost
Low labour requirement		High labour requirement
No need for re-muster after shearing		Re-muster necessary (unless dipping off board)
Decreased health risk for sheep		Increased health risk for sheep (lumpy wool, cheesy gland, post dipping lameness, etc.)

- choose a fine day
- start early and finish early to allow sheep to dry.

Plunge dipping

- do not overcrowd
- sheep should spend at least 10 seconds in the dip
- keep dip well agitated
- check the first few sheep for saturation to the skin.

Shower dipping

- do not mix classes of sheep in the dip
- pump pressure should be 140 kPa (20 psi) operating
- overhead spray assembly should do 5 to 10 rpm
- check the first pen for saturation and adjust time in shower accordingly.

An alternative to setting up a plunge or shower dip is to employ a professional trained dipping contractor.

Pour-on application

- sheep should be clean shorn
- set dose to the weight of the heaviest sheep in mob
- apply from poll to tail along the centre of the backline
- lumpy wool may reduce efficacy
- treat lambs at foot or lambs due within 6 weeks of treatment of their mothers.

LiceBoss website

LiceBoss (http://www.liceboss.com.au) contains detailed information on all aspects lice control and several tools to assist growers assess the risk of lice infestation and select appropriate treatments.

Lice resistance to chemicals

Some lice have become resistant to some sheep-lice-control chemicals. Resistance is associated with administration of sub-lethal doses of chemical, continuous use of the same chemical and/or poor application. This results in populations of lice that can withstand previously lethal doses of a chemical to survive and breed. The survivors pass on the genes that confer resistance to the next generation, which gradually increases the level of resistance in the overall lice population.

Resistance to synthetic pyrethroids became evident in the 1980s and is well documented. Although the extent is unknown, evidence of resistance and reduced susceptibility to insect growth regulators (IGRs) is now emerging. Resistance to other chemicals is rare.

Currently there is no quick, convenient test to confirm resistance. When treatment fails, it is important to explore all possible causes because resistance usually only accounts for a small proportion of treatment failures. Management practices, poor application and under dosing are the usual causes of failure to eradicate lice.

Prevention of resistance can be achieved through strategic use of chemicals, dosing for the correct live weight and precise application techniques. Only treat for lice when they are present. If lice treatment is required in consecutive years, rotate the chemical groups used and ensure that the dose rate is set for the heaviest animal. Backlines application requires a straight line from the poll to the rump: strips off to the side can leave an area of low chemical concentration where lice can survive. If shower or plunge dips are used, it is important that they are operating properly.

Integrated Parasite Management (IPM) strategies for lice control

1. Monitor for lice before shearing, crutching and at other times when sheep are yarded. Pay particular attention to high-risk mobs such as those containing newly purchased sheep or where strays have been found.
2. Institute a lice biosecurity plan, which includes:
   - checking lice history and quarantining purchased or agisted sheep
   - preventing strays with stock-proof fences
   - guarding against the introduction of lice by contractors or shearers who have recently been on a lice-infested property
   - discussing lice management with neighbours to coordinate where possible.
3. If lice are detected, treat with an effective chemical at next shearing, regardless of whether or not a long wool treatment has been used (**all** sheep must be mustered and treated).
4. Choose a chemical that will enable the wool to meet market specifications. Minimising chemical residues is critical to maintaining the 'pure and natural' marketing image of Australian wool. New environmental legislation in major wool markets will limit the level of chemical residues in wool able to be processed by many mills and most Ecolabels place strict limits on the level of chemical residues allowable.

Table 6.4 lists the veterinary chemical preparations available for control of lice in Australia.

Face and foot lice

Foot lice (*Linognathus pedalis*) can survive for 2 weeks off the sheep, so clean paddocks are required following treatment. Face lice (*Linognathus ovillus*) can extend from the face into the wool on the neck and occasionally on the lower body. The foot louse may extend to the belly and scrotum if infestation is severe. Infestations are rare. Seek veterinary assistance for diagnosis and treatment.

Itchmite

Before the introduction of the macrocyclic lactone (ML or mectin) drenches, itchmite (*Psorobia ovis*) was present in about 30% of sheep flocks, although itchmite was much less important than body lice. Mectin drenches are effective against itchmite and it is now very rare. Itchmites are very small eight-legged mites that live on and in the superficial layers of skin. They are about 0.2 mm long and are not visible to the naked eye.

Mites do not survive off the sheep. Shearing causes a temporary decrease in numbers. Spread occurs by direct contact between sheep. The spread is greatest between recently shorn sheep and to lambs when they are suckling. Usually only 5 to 10% of a mob is affected and a build up in the percentage of affected sheep occurs slowly.

Sheep often have tags of wool hanging from their fleeces where they have been biting. Their fleeces may also have the appearance of being rubbed. In severe cases, the fleeces may be cotted. There may be an increased amount of scurf on the skin. Microscopic examination of the skin scrapings from sheep with severe fleece derangement and scurf is needed for diagnosis. Badly affected sheep are allergic to the mite. Low numbers of mites in these allergic sheep can cause severe irritation and fleece derangement.

Itchmites have no significant effect on fleece weights, but the quality of fleece from sheep with moderate to severe fleece derangement is affected. Fleeces from the small number of sheep with severe fleece derangement can be discounted 20% because of cotting and discolouration.

Treat individually affected sheep with an ML drench. The use of ML drenches to control worms or dipping in magnesium fluorosilicate (Flockmaster) will control itchmite incidentally.

Minimise the spread of itchmite by:

- not shearing ewes with lambs at foot (to minimise spread from ewe to lambs)
- reducing the time sheep spend in counting out pens after shearing because this is the greatest opportunity for spread to occur between sheep.

Ked

A ked (*Melophagus ovinus*) is a brown wingless blood-sucking fly about 7 mm long. Keds live their entire life cycle on the host and are spread by direct contact. They are now rare.

They cause intense itching and dark staining of wool. Registered treatments include short wool dipping with magnesium fluorosilicate (Flockmaster) and ivermectin capsules.

Scrotal mange

Chorioptic mites (*Chorioptes bovis*) cause leg and scrotal mange. The mites are spread by direct contact. The damage caused by the mites results in crusting of affected areas. Scrotal mange is important because it may cause a rise in temperature of the scrotal contents, followed by testicular degeneration. Seek veterinary assistance for diagnosis and treatment.

Lameness

A number of diseases that affect sheep cause lameness. The specific diseases of the feet of sheep including footrot, ovine interdigital dermatitis, foot abscess, shelly hoof and strawberry footrot will be discussed here. Lameness may also be part of a generalised disease such as blue tongue, foot and mouth disease (both 'emergency' or 'exotic' diseases) or post-dipping lameness.

Footrot

The progression and rapid spread of footrot is favoured by warm and moist seasonal conditions.

Footrot is caused by the bacterium *Dichelobacter nodosus* and effects sheep (and goats) of all classes and ages. The bacterium causes a separation, or under-run, of the horn of the claw from the underlying soft tissues, which results in lameness. There is often a greyish scum between the under-run horn and the soft tissue and a bad odour, but never any pus.

There are three essential factors for the rapid spread of footrot:

1. **Initial damage to the skin between the toes**

   The skin between the toes must be damaged by continual wetting before bacteria can gain entry. Initially the skin will have a moist, reddened appearance and is infected by a bacterium called *Fusobacterium necrophorum*. This condition is called ovine interdigital dermatitis (OID). OID is commonly seen in sheep grazing on lush pastures where the damage caused by water and bacteria will allow *Dichelobacter nodosus*, the footrot bacterium, to enter.

2. **Presence of footrot bacteria**
   D. nodosus must be present to result in footrot. For the disease to occur, uninfected sheep need to come in contact with infected sheep or goats, carrier cattle, or contaminated pasture during a period suitable for footrot spread. The footrot bacterium depends on living tissue for its survival. An important feature of the disease is that while the bacterium can survive in the feet of infected sheep for many years, it cannot survive on pasture for more than 7 days, even in warm moist conditions. Virulent footrot also occurs in goats, while cattle can act as passive carriers of virulent footrot. These species are a source of infection for sheep and should be considered in any control or eradication program.
3. **Warm and moist conditions**
   Footrot does not spread from sheep to sheep in hot dry conditions. Transmission is slow in very cold conditions: it **mainly** occurs when conditions are warm and moist. Exposure to moist conditions for at least 3 to 4 weeks when the mean daily temperature exceeds 10°C will lead to an outbreak of footrot, if footrot bacteria are present in the flock. Summer storms do not provide suitable conditions for the spread of footrot, unless there is enough moisture to promote significant pasture growth. During warm and moist seasonal conditions, a very high percentage of a flock can become rapidly infected, often in more than one foot. Affected sheep are usually severely lame, especially when the under-running is extensive.

Virulence of footrot

Different strains of footrot vary in 'virulence'. From the practical point of view, footrot is recognised as benign, intermediate or virulent. Virulence is the ability of the footrot bacteria involved to under-run horn, and to affect multiple toes and feet in a high proportion of the flock.

The stages of footrot

A four-stage scoring system has been developed to describe the progression of the disease, with score 1 being the earliest stage of infection and score 4 being the most advanced stage of footrot.

The first sign of infection is on the skin between the toes, in the interdigital skin. This skin becomes red and hair on the skin falls out. There is no separation of the horn from the foot (score 1 and 2).

Given warm, moist conditions, the disease can rapidly develop into score 3 or 4 footrot if a virulent strain is involved.

Score 3 footrot involves the separation of the horn, starting at the inside edge of the heel at the skin–horn junction (score 3a) and continuing at least half way across the heel (score 3b). A grey, paste-like scum will form between the horn and the underlying soft tissues of the feet. This scum often has a distinctive, foul smell.

If only mild strains are involved, then this will be as far as the disease progresses, regardless of the environmental conditions. However, with more virulent strains the disease is just starting to develop.

In a warm wet environment, if a more virulent strain of footrot is present, separation of the horn progresses across the heel to the outside wall of the hoof (score 3c). This is a characteristic sign of more aggressive strains of footrot. No other disease causes this sign.

In very badly affected sheep, the whole of the horny layer of the sole of the hoof can be separated from the foot (score 4). This causes extreme animal welfare problems in the affected sheep. Sheep with both front feet affected will often kneel to graze because of the severe pain associated with standing.

Benign footrot in the early stages is indistinguishable from ovine interdigital dermatitis (OID). There may be slight separation at the skin–horn junction between the claws and some individuals may be under-run further (fewer than 1% have score 4 lesions).

Benign footrot is a common disease of sheep and is evident in all moderate to high rainfall areas of south eastern Australia. It is present in many flocks and is apparent and harmless except in some very wet years. Spontaneous recovery when pastures dry off is a characteristic of benign footrot.

The problem of differentiating benign footrot from more severe footrot occurs when lesions are seen in the saleyards or when lesions are seen at the beginning of a spread period. With virulent footrot, complete under-running of the horn takes 2 to 3 weeks and, in some situations, several inspections a week apart may be necessary to distinguish benign footrot from early virulent footrot.

Intermediate footrot is characterised by interdigital dermatitis with under-running of the heel and sole only. Intermediate footrot cannot be readily distinguished from virulent footrot in individual sheep. On a flock basis, intermediate footrot is a milder disease with fewer severely affected feet and with less tendency for chronic lesions to develop. Note that from an eradication perspective, intermediate footrot should be considered eradicable.

Virulent footrot is characterised by interdigital dermatitis with under-running involving the outside wall of the claw in more sheep. It causes severe weight loss in affected sheep and the disease becomes persistent and chronic if left untreated. Sheep are commonly affected in more than one foot. Both claws of each infected foot are usually involved. Sheep affected in both forelimbs commonly support their weight on their knees. More time is spent lying down. Affected feet and the brisket may become flystruck.

Some or all of the following features are seen:

- inflammation and damage to the interdigital skin
- under-running and damage to the soft tissues under both the soft horn of the heel and sole, and the hard horn of the toe and wall of the claw.

However, there is no evidence of draining abscesses or pus (as in foot abscess).

Some characteristics of the three clinically distinguishable forms of footrot are summarised in Table 6.6.

It is not possible, nor is it economically justifiable, to attempt eradication of benign footrot from flocks. For the purposes of this discussion, 'footrot' will refer to the virulent and intermediate disease. The disease can come from footrot-affected straying livestock or introductions. It is virtually impossible however for boots, tyres and wildlife to spread footrot.

Merinos are more susceptible to footrot and have more severe under-running of the horn than British breeds.

It is important to understand how footrot progresses in the host and the times of the year when the different stages of the virulent disease will be found (see Figure 6.3).

It takes 1 to 3 weeks for virulent footrot to progress from stage 1 to stage 4.

Not all sheep that become infected with footrot in spring will become 'chronic carriers'. Even if left untreated, the hot dry conditions of summer will result in the 'self-cure' of up to 50% of infected sheep.

Table 6.6. Characteristics of benign, intermediate and virulent footrot.

	Virulence		
	Benign	**Intermediate**	**Virulent**
Ability to under-run horn	Very low	Moderate	Severe
Ability to heal when dry	High	Moderate	Low
Lesions	Most score 1–3a	Score 2–4	Score 3–4
Percentage of sheep in a mob with score 4 lesions	<1%	<1–3%	>3%
Proportion of total score 2–4 lesions > score 3c	<10% score 3		>20% score 3c–4
Degree of lameness	Moderate	High	High
Eradication possible?	No	Yes	Yes

The control and eradication of virulent footrot

The principle of footrot eradication is simple, but in practice it is difficult to achieve on many properties. The importance of a systematic, planned approach to footrot control and eradication cannot be over-emphasised. As discussed, footrot can survive in the feet of infected sheep for months, or even years, but cannot survive on pasture for more than 7 days. Accordingly, if all infected sheep are identified and culled from a flock when footrot is not spreading, footrot will be eradicated.

The essential steps are to minimise footrot spread during footrot transmission periods, and to identify and cull remaining infected sheep in non-transmission periods (such as dry summer conditions).

Footrot spread is minimised during the transmission periods (usually spring but sometimes during autumn and winter or even wet summers) by using a range of footrot treatments without necessarily segregating infected sheep or spelling paddocks or foot-paring. The remaining infected sheep are identified by repeat inspection of all feet of the entire flock in a non-transmission period (usually during dry summer conditions). All infected sheep are then culled.

Vaccination and foot-bathing are the primary methods used to control the spread of footrot in transmission periods. They do not, however, eliminate the need to examine all sheep in non-transmission periods and cull all remaining infected sheep to eradicate the disease.

The combinations of control options available for footrot allow flexibility in the design of programs for individual flocks. Professional advice is essential to ensure that an effective program is developed at minimum cost.

Foot bathing for footrot control

Foot bathing is a short-term measure to prevent footrot spread, but does not provide long-term protection. Foot bathing can enhance the effectiveness of a footrot vaccination program and is a useful 'precautionary' treatment for purchased sheep and mobs that may have contacted infected sheep.

The choice of foot bathing solution and the duration of foot bathing will depend on the number and severity of footrot infections, the number of sheep, the quality of facilities and the time available.

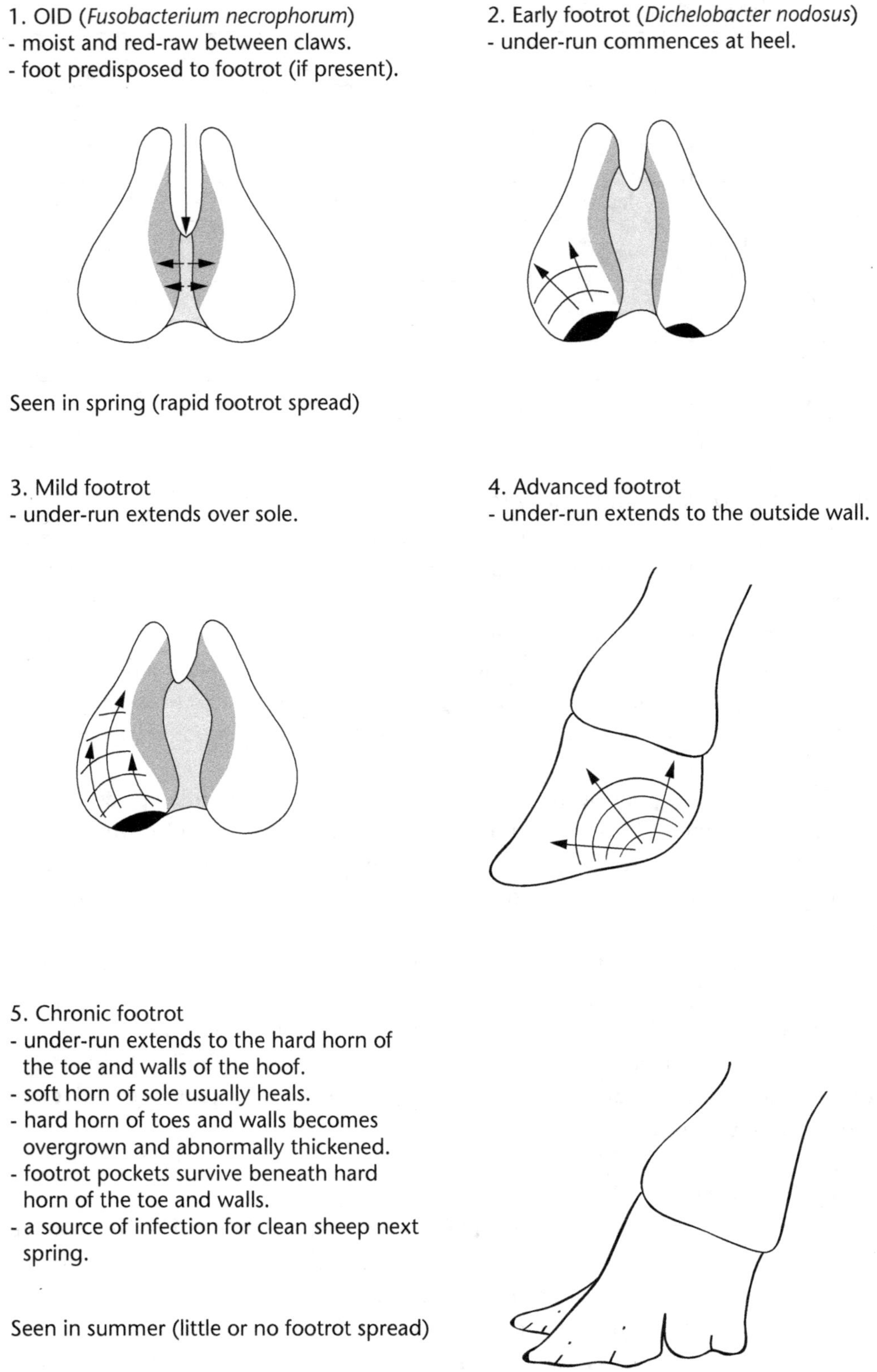

Figure 6.3. The progression of footrot.

Short-duration foot bathing uses a 10% zinc sulphate solution to treat sheep every 7 days using a footbath at least 8 metres long. Sheep are walked slowly through the bath or stood in the bath for up to 5 minutes. Walk through foot bathing is effective against mild lesions and must be done weekly.

Long-duration foot bathing uses a 20% zinc sulphate solution only when conditions are wet. 'Radicate' is a solution of copper salts used for foot bathing every 2 weeks. Consult a veterinarian or other animal health adviser for further information. Sheep are stood in the footbath for between 15 and 60 minutes, depending on the product used.

Vaccination for footrot control

Vaccines are an effective tool in the control of virulent footrot. Although vaccines can cure a percentage of infected sheep (as high as 60%), they are most effective in preventing footrot spread. A vaccination program consists of two injections given at a minimum of 4 to 6 weeks apart. The protection rate achieved is very high for the period of protection.

It is important to note that vaccines will not eliminate the need for summer inspections and culling to eradicate footrot in non-transmission periods.

Vaccines have no role on properties where virulent footrot does not exist. They are not effective against benign footrot.

Timing of vaccination

The timing of vaccination is extremely important because the vaccine gives good protection 2 weeks after the second injection and protection may only last for 8 to 12 weeks. If the transmission period extends beyond the period of protection given by the vaccine, then further preventative measures such as foot bathing or re-vaccination are necessary until the pasture dries off. A single dose of vaccine up to 12 months after the initial two doses can re-establish immunity for 8 to 12 weeks.

The first 'sensitising' vaccination should be given 4 to 6 weeks before the second vaccination. For maximum effectiveness, the second vaccination must be given about the time the seasonal conditions are suitable for footrot spread. Sheep vaccinated in September should have reasonable protection until the end of spring. Increase the frequency of foot bathing or revaccinate from late November onwards if the pasture has not dried off.

Vaccination alone is not effective if the program is commenced when footrot is already spreading and a high percentage of feet are under-run. The sheep must be protected by an appropriate foot-bathing treatment until at least 2 weeks after the second injection to prevent footrot spread while the vaccine is taking effect.

If footrot is not eradicated over the following summer and autumn, a booster vaccination will be needed in early spring of the following year.

Using vaccines

If a vaccine is to be used as part of the control program, all mature sheep and weaners should be vaccinated. Foot bathe young lambs whenever they are mustered.

It is very important to make sure the vaccinating gun is accurately calibrated and that extra care is taken when vaccinating woolly sheep. Do not use the vaccine beyond the expiry date and follow the manufacturer's recommendations on storage. The site of vaccination is important. Follow the manufacturer's instructions carefully.

Vaccination, if timed correctly, is an effective and labour efficient technique for controlling footrot. It should be the cornerstone of footrot control programs where virulent footrot is diagnosed and regular and frequent foot bathing cannot be carried out, such as during lambing.

Phases in eradicating footrot

The phases in eradicating footrot include diagnosis, planning, control, eradication, surveillance and prevention of re-infection.

Diagnosis

Obtain a correct diagnosis from a veterinarian, because other diseases of the feet of sheep can be confused with virulent footrot.

Planning

The importance of a systematic, planned approach to footrot control and eradication, using professional advice and guidance, cannot be over-emphasised. Footrot eradication is not easy nor is it cheap. Success requires firm commitment and good management.

Planning should occur before the start of the spread conditions to ensure that sheep are protected during spread.

Consider management practices, financial resources, facilities for handling stock, labour, motivation and management skills when planning a program.

Ensure that boundary fencing is sheep-proof. If necessary, install or improve existing facilities for handling and foot bathing sheep.

Learn to recognise the disease in all its stages and acquire a sound knowledge of footrot and the likely transmission and non-transmission periods in the locality of interest.

One option when planning a program is de-stocking. This should be seriously considered when:

- only one mob of sheep on the property is affected
- the footrot affected sheep are close to the end of their economically useful lifespan
- the change-over cost for clean sheep is low
- the risk of buying in footrot is low.

Control

The aim of the control phase is to contain footrot to fewer than 5% of the flock during spring. Realistically, in many situations this is difficult to achieve. However, good control reduces summer inspection costs and increases the chances of eradication.

During the control period, the best management strategy is the one which keeps the level of infection low enough to allow for culling, not treatment of infected sheep during summer. Control should start as soon as conditions are suitable for footrot spread and continue until the end of the spread period. The length of the control phase is determined by the length of the spread period.

Ensure that any mobs known to be clean are kept strictly isolated from infected mobs and away from anywhere infected mobs have been within the last 7 days. Avoid mustering clean sheep so they do not become infected. If shearing or other husbandry procedures are necessary during spread periods, bring clean mobs in first and ensure they do not walk over ground where infected mobs have been for the past 7 days.

Control methods for infected mobs include vaccination, foot bathing, or a combination of both. Paring of feet, spelling of paddocks and segregation of 'clean' from 'infected' sheep (within known infected mobs) are not necessary in the control phase.

If environmental conditions are suitable for footrot spread, weekly foot bathing in zinc sulphate or two weekly foot bathing in 'Radicate' may be necessary to keep the level of infection down. Few farmers are able to keep up weekly walk-through foot bathing for more than 6 to 8 weeks during transmission periods, let alone long-duration foot bathing. If shearing, lambing, haymaking or other commitments are likely to interrupt the foot bathing schedule, the use of footrot vaccine is recommended.

Vaccination is the control option of choice where adequate foot bathing cannot be undertaken, or the cost of foot bathing exceeds the cost of vaccination.

Eradication

The eradication phase aims to identify the remaining infected sheep for immediate culling. Eradication is accomplished only by the repeated inspections of all feet on all sheep. Ideally, two clean inspections, without finding any footrot affected feet, are required. This means inspecting all feet on all sheep at least three times before the autumn break.

This eradication phase should commence when conditions are no longer suitable for footrot spread. Success is more likely if eradication is attempted after a successful control phase and the level of infection is low. Low levels of infection make eradication by culling the remaining infected sheep a realistic target.

Method: The eradication phase consists of repeated inspections of all sheep, at 4-weekly intervals. The first inspection should take place as soon as possible after the end of the transmission period before pockets of footrot dry up and become difficult to detect.

The first inspection: Carefully isolate all sheep that are going to be sold for slaughter and for which inspection of feet is therefore not necessary (for example, cast-for-age ewes and prime lambs).

Prepare 'clean' paddocks for the 'clean' sheep after inspection. These paddocks will need to be spelled for at least 7 days beforehand.

Look for sheep with four normal feet, and only keep those sheep. All remaining sheep with footrot or doubtful feet should be culled for slaughter. Some may have conditions other than footrot, but keeping them can lead to eradication failure.

Individually tip sheep and inspect all feet; 'diagnostic' pare where necessary to establish whether or not footrot is present. As soon as footrot is found, paring should stop and the sheep should be identified and isolated. If in doubt, cull the sheep.

'Infected sheep' should be isolated and culled immediately after the completion of the inspection. As the inspection proceeds, the clean sheep should be placed in separate spelled paddocks.

Treatment during the eradication phase is not recommended.

If the prevalence is still very high, the control over the 'spread' period has failed. In this case, options are a very high cull rate or an antibiotic treatment program.

Antibiotic treatment: If fewer than 10% of sheep are still infected with footrot at the first eradication inspection, they should be culled. If more than 10% are infected, it maybe worth trying to salvage them with antibiotic treatment.

There is a risk associated with treating infected sheep: the longer they remain on the property, the greater the risk of failure of the footrot eradication program. The decision to use antibiotics in an eradication program must be made in consultation with a veterinarian.

Antibiotics will not cure all footrot-affected sheep. Sheep that do not respond to the antibiotic treatment must be culled. To be effective, the antibiotic must only be injected if feet can be kept dry for 24 hours after treatment (for example, on a slatted floor in a shearing shed). Foot paring has a minimal effect on the cure rate when injectable antibiotics are used, though it is advantageous if sheep have normal 'shaped' feet to ensure they are easier to inspect at the next inspection. Antibiotic-treated sheep must be clearly identified and not sold for slaughter until the withholding period for the antibiotic has expired.

Examination of feet: The feet of all sheep should be carefully examined to establish whether or not all four feet are normal. Minimal paring may be done to make inspection easier or to cosmetically reshape the foot. If diagnostic paring is necessary, avoid drawing blood because it causes pain and lameness, obscures the inspection of the foot and is not humane. Mutilation of the claw by paring does not increase the chances of eradication. Paring only needs to be done to ensure that the feet are normal.

Accuracy in footrot diagnosis at these inspections is vital. Do not work too fast and inspect each foot. Use sheep-handling equipment to increase efficiency. The owner or manager of the sheep should closely supervise the eradication inspections.

The old adage 'if in doubt, chuck it out' applies. If the claw needs to be trimmed excessively to make a decision, treat the sheep as infected. If there are large numbers of sheep suspected to be infected with footrot and a decision is made to treat and attempt salvage, then it is safer, cheaper, faster and more humane to treat these sheep with antibiotics rather than waste time paring their claws.

Salvaged sheep are treated with antibiotics at the time of inspection, identified, foot bathed and held overnight on slatted floors in the shearing shed and run as a separate mob.

Identification of sheep: Identification of all sheep is essential to ensure that a thorough inspection is carried out. A good system is:

- Red mark: infected or 'abnormal'
- Green mark: clean or 'normal'

All sheep should be accounted for at each inspection. Any unmarked sheep has not been inspected.

Foot bathing: Sheep with 'normal' feet can be foot bathed and returned to a paddock that has been spelled for at least 7 days.

The second inspection: All sheep should be given a second inspection 4 weeks after the start of the first inspection. Again, mark all sheep as they are inspected. Any unmarked sheep has not been inspected and may be responsible for the failure of the eradication program. The inspection process is the same as the first inspection: use a minimum level of paring to ensure that all four feet are normal.

All feet on all sheep must be inspected, with no distinction being made between treated 'salvaged' sheep and previously 'clean' sheep. Cull all sheep that do not have four perfectly normal, healthy feet at the second inspection. No sheep should be treated for salvage at this inspection, or later.

Brand all culls with clear distinguishing marks and send them for slaughter immediately.

The third and subsequent inspections: The third inspection is carried out 4 weeks later. The procedure is the same as the second inspection and all suspect sheep are culled.

As a guide, up to 1 to 2% (10 to 20 per thousand) of suspects can be found at the second inspection (or 5 to 10% if they are in a salvage mob) and about 0.5% (five per thousand) at the third inspection.

Regular inspections must continue until there have been two successive, totally clean inspections of each mob. The number of inspections may vary between each mob, so good records are essential. The last clean inspection may be delayed until after the autumn break, but must be completed before conditions are suitable for footrot spread.

Surveillance

The aims of the surveillance phase are to detect and remove footrot breakdowns early and to treat in-contact sheep to reduce chances of footrot spread and to ensure that the flock is free of the disease (no evidence of footrot during spring).

Surveillance should start as soon as eradication inspections are completed and consists of regular surveillance of the flock until the end of the following transmission period.

Method: Inspect each mob in the paddock at least once a week. Any sheep that is lame or has abnormal feet **must** be examined. This procedure should also be followed whenever the sheep are yarded. Lame sheep with foot abscess or arthritis should be clearly marked to avoid catching them again. Lame sheep are more likely to be seen if they are walked quietly.

Carefully examine the suspect hoof. If footrot is not present, raddle the sheep and let it go. If footrot is confirmed, or suspected, immediately remove the sheep from the mob and humanely destroy it.

Sheep that have been in contact with a footrot case should be tipped, examined and foot bathed into a 'reserved' spelled paddock. If a spelled paddock is not available, repeat the foot bathing 5 to 7 days later.

It is a good idea to split mobs to use all available paddocks and hence reduce the number of sheep per mob. This makes it easier to deal with breakdowns as the number of 'in contact' sheep are reduced.

Do not cross any other mobs behind the suspect mob in the yards, lanes or shed.

Success: Footrot can be considered eradicated if all sheep pass through the next spring without breakdown.

Footrot breakdowns: Footrot breakdowns may occur because it is possible that some infected sheep will be missed at the eradication inspections. These sheep will have small, deep pockets of infection. Over late autumn and winter they will eventually become lame as the footrot pockets progress within the hoof.

Control procedures during the surveillance phase: Footrot control procedures during the surveillance phase make breakdowns difficult to detect. For this reason, frequent foot bathing or further vaccination is not desirable.

Eradication failure: If footrot breakdowns are found beyond early spring and, in particular, if there is evidence of footrot spread, the eradication attempt has failed. Select and commence an appropriate footrot control program as soon as possible and ensure the infected mob does not come in contact with clean mobs.

Prevention of footrot

By following a few simple rules, the risk of introducing or reintroducing footrot can be considerably reduced.

Saleyards

At public saleyards, examine as many sheep as possible in the mob intended for purchase. Footrot is most likely to be detected in poor conditioned or lame sheep, and in sheep with grossly deformed hooves. The only way to check for footrot is to tip the sheep up, part the toes and look closely. Question the vendor about the history of footrot and, if in doubt, do not buy the sheep.

Property of origin

Whenever possible, the property of origin should be checked for a history of footrot. Do not be embarrassed to inquire about the footrot status of the flock of origin. Ask the vendor if he or she is willing to sign a vendor declaration to the effect that, to the best of their knowledge, the sheep are not infected with footrot.

Sheep purchased on farm

A similar procedure should be followed for sheep purchased on farm. If sheep are bought from a property known to be free of footrot, the risk of buying infected sheep is low.

Trucking hygiene

Ensure that the truck has been hosed out before loading.

Foot bathing on arrival

All sheep should be foot bathed as they come onto a clean property. This ensures that any footrot bacteria that have been picked up in transit are killed. Five to 10 minutes in a 10% zinc sulphate footbath is recommended. Foot bathing will not cure chronic footrot lesions.

Isolate purchased sheep

All purchased sheep should be considered suspect. Hold them in isolation from the rest of the flock until they have passed through a season suitable for footrot spread without breakdown. It may not be practical to isolate a small number of introduced sheep such as rams. In this case, they must be carefully examined before introducing them to the flock.

Secure fencing

Secure boundary fences are essential. Stray sheep are all too often the source of footrot. If an infected stray sheep is found in a flock, remove it and immediately tip, examine and foot bathe the sheep that have been in contact with it. Do not start footrot eradication until the boundary fences are 100% stock proof.

Other species

Virulent footrot also occurs in goats, and cattle can act as passive carriers of virulent footrot. These species are a source of infection for sheep and should be considered in any control or eradication program.

Recent treatments

Beware of vaccinated sheep or sheep with evidence of recent foot paring or foot bathing.

Other causes of lameness

Ovine interdigital dermatitis (OID)

OID occurs when the skin between the claws becomes 'water damaged'. This happens when sheep are kept on wet pasture. Water-damaged skin allows the entry of bacteria that are commonly found in the environment, especially *Fusobacterium necrophorum*. OID heals spontaneously with the advent of dry conditions. The significance of OID is that OID precedes infection with footrot or foot abscess. OID cannot be distinguished from benign footrot or early virulent footrot except with laboratory tests.

Foot abscess

Like footrot, foot abscess is preceded by OID. Abscess results from a deep invasion into the tissues of the foot by bacteria always found in the environment of sheep. It is usually seen in heavy sheep, such as rams and pregnant ewes, in wet conditions. It can become a serious problem in wet seasons. Usually fewer than 10% of sheep are affected.

Generally only one foot and one claw is involved. The foot becomes very swollen and lameness is acute. Rupture of the swelling produces a discharge of pus and a sinus forms, usually at the coronary band. If the infection involves the joint, the sheep may become permanently lame.

Foot abscess is readily distinguishable from footrot. There is no pus, swelling or sinus formation associated with footrot.

Treatment of early cases with intramuscular antibiotic, as recommended by a veterinarian, may be worthwhile, but usually cases are found when the abscess has progressed too far.

Prevention by foot bathing in 10% zinc sulphate is successful, but often impractical, particularly if sheep are late pregnant or have to walk through muddy yards, which can predispose to OID.

Toe abscess

This is an infection in the wall of the claw due to breaks in the horn. Front feet are the most commonly affected, usually in one claw only. Pus is often discharged at the top of the claw on the coronet. Treatment is by paring to expose underlying infection.

Shelly hoof

Shelly hoof is over-growth of the horn of the claw and separation of the outer wall of the claw to form cavities that become filled with mud, manure and other organic matter. Lameness may occur. The condition will predispose sheep to toe abscess.

Strawberry footrot

When strawberry footrot occurs, extensive scab formation is found around the coronary band and on the lower leg. The causative agents include the scabby mouth virus and the bacterium that causes lumpy wool, either alone or in combination. Outbreaks are seen in conditions that cause damage to the skin of the lower limbs such as wet conditions or grass seed penetration. Strawberry footrot heals in 5 to 6 weeks and treatment is usually not necessary.

Lameness can be very severe, resulting in rapid weight loss and even death of weaners already in poor condition.

Foot bathing in 10% zinc sulphate and/or the use of an appropriate antibiotic may aid healing in severe cases.

Other scouring/wasting diseases

There are a number of causes of scour (diarrhoea) in sheep. The more common causes include:

- worms
- yersiniosis
- campylobacteriosis
- coccidia
- nutritional causes
  - sudden changes in feed
  - lush green feed
  - nitrate, nitrite in the diet (for example, capeweed and some grasses).

When dealing with a group of scouring sheep, the usual reaction is to reach for the drenching gun. Worms are a common cause, but they are not the only cause of scouring. Treating scouring sheep with a worm drench may be unwarranted in some situations.

In the face of an outbreak of scouring sheep, call a veterinarian. Successful treatment depends on obtaining the correct diagnosis. A detailed flock history, clinical examination and appropriate laboratory backup tests will assist diagnosis.

Yersiniosis

Yersiniosis is caused by bacteria known as *Yersinia. Yersinia* can cause ill-thrift, with or without scouring, and death. Scouring and ill-thrift result from damage to the gut wall. *Yersinia* is found in the gut of most sheep, with most sheep carrying *Yersinia* showing no sign of disease. The disease is generally seen in young weaned sheep up to 12 months of age, often in winter and spring, and is usually associated with a concurrent problem such as worms, nutritional or cold stress, and a change in diet or stressful management procedures, such as transportation.

Yersiniosis is spread by susceptible sheep eating feed or drinking water contaminated by the faeces of infected sheep. Diagnosis is based on clinical signs, histopathology of the gut wall and culture of *Yersinia* from the gut of affected sheep. Antibiotic treatment prescribed by a veterinarian is the preferred treatment. Care should be taken to prevent breech strike by flies during the fly season.

Avoid feeding young sheep on ground heavily contaminated with sheep droppings, especially in wet conditions. Supplementary feed and water supply should be high enough off the ground to avoid faecal contamination.

Campylobacteriosis

Campylobacteriosis, also known as weaner colitis or sulpha-responsive scours is caused by bacteria *(Campylobacter)* that attach to the bowel and produce scouring. It occurs exclusively in weaners 3 to 9 months of age during summer months, especially in wet conditions, and typically affects about one-quarter of a flock at any one time. Weaner colitis should be suspected when weaners continue to scour after drenching. It is a common cause of 'apparent drench failure'.

Healthy and ill-thrifty sheep are equally affected. If nutrition is good, deaths are uncommon and sheep recover without treatment.

In stressed sheep, however, especially in small weaners suffering from malnutrition, higher death rates can occur. In this case, the cost and inconvenience of treatment can also be considerable, and increased susceptibility to flystrike can be a further impact of the disease.

Predisposing factors are not known, although the disease does appear to occur more commonly in wet summers. The bacteria that cause weaner colitis are found in most normal sheep and only occasionally cause disease. Weaner colitis occurs repeatedly each year on some farms, while others remain free from the disease.

Diagnosis of campylobacteriosis relies on ruling out worms as a potential cause of enteritis and seeking veterinary advice. Post mortem examination and histopathology of the gut wall usually provides the most useful information for diagnosis, and appropriate antibiotic therapy can then be prescribed.

Coccidiosis

Coccidiosis is an infection of the gut wall with a protozoan parasite. Coccidiosis causes diarrhoea that may contain blood. Affected animals may strain when they pass faeces and may have a broad, black, triangular stain on the breech. Coccidiosis usually occurs in lambs between 1 and 3 months of age.

Coccidia 'eggs' (oocysts) are shed in faeces. The disease is spread by susceptible sheep eating feed or drinking water contaminated by the faeces of infected sheep. Coccidia are common in lambs, but often not in high enough numbers to cause disease. Overcrowding causes high levels of pasture contamination. Overcrowding, combined with wet conditions, stress, and poor feed, leads to outbreaks of coccidiosis. The disease occurs commonly in lambs at pasture with their dams where ewes are producing insufficient milk (due to poor body condition and shortage of feed) forcing lambs to graze early. In this situation, lambs ingest large numbers of coccidia 'eggs'. Some lambs may receive only a slight setback: scouring a little but continuing to eat. Immunity develops after infection. Coccidiosis does not respond to worm drenches. Most lambs recover without treatment.

Coccidiosis often occurs with other problems such as roundworms, yersiniosis or stress. When coccidiosis occurs concurrently with other diseases, severely affected sheep may die.

Diagnosis is based on clinical signs, histopathology of the gut wall and identification of oocytes in the faeces of affected sheep.

Treatment of coccidiosis that occurs without a concurrent infection is usually not warranted. Improved feed and a clean paddock will generally solve the problem. If treatment is needed, discuss the options with a veterinarian.

As for yersiniosis, avoid feeding young sheep on ground heavily contaminated with sheep droppings, especially in wet conditions. Supplementary feed and water supply should be high enough off the ground to avoid faecal contamination.

Ovine Johne's disease

Johne's disease is an incurable, infectious intestinal disease characterised by wasting, which is caused by the bacterium *Mycobacterium paratuberculosis*. There is no treatment for this disease.

Mycobacterium paratuberculosis can persist in the environment for up to 1 year, but does not multiply in the environment. Low temperature, moisture and protection from sunlight favour survival outside the sheep. In cool, wet, shaded areas the bacterium may survive for more than 12 months.

The disease has a long incubation period (the time between establishment of *Mycobacterium paratuberculosis* in the intestine of the sheep and the appearance of the first signs of the disease). The incubation period may be 2 years or longer. Clinical disease (wasting) generally does not occur in sheep less than 2 years of age.

Subclinically infected sheep do not show any outward signs of disease. These sheep carry the bacteria in their intestines and may spread bacteria via their faeces. As the infection progresses, subclinical disease can develop into clinical (visible) disease. Some subclinically infected sheep may never develop clinical disease.

Clinically infected sheep show progressive weight loss leading to emaciation (wasting). Chronic scour may be present, but is not a common feature of the disease in sheep. The average time from the onset of signs of disease to death is about 6 weeks to 3 months.

There is often a distinct 'tail' of sheep in poor condition in OJD infected flocks. These sheep eventually die.

Different cattle and sheep strains of *Mycobacterium paratuberculosis* are recognised. Potential for cross infection between sheep and cattle is a subject for further research. Sheep strains are known to infect sheep, goats, alpaca and deer, and these species must be included in control and eradication programs on sheep properties.

Sheep become infected by eating feed or drinking water contaminated by the dung of infected animals. Unlike cattle, sheep can become infected at any age. Clinically infected sheep shed large numbers of bacteria in their faeces, resulting in heavy environmental contamination.

Johne's disease is most commonly introduced into a previously disease-free flock through the introduction of one or more infected sheep.

Contaminated drainage or surface water from an infected farm may be a source of infection for non-infected flocks. Dilution of faeces from infected animals in large volumes of water may diminish the risk of environmental spread of the disease because dilution diminishes the chance of any one animal taking in an infective dose (sufficient bacteria to become infected). Faeces from infected animals may also be carried onto non-infected properties on footwear, clothing or equipment or blown between properties by wind.

The annual death rates attributed to Johne's disease in flocks where the disease has been present for some time are on average 2.5%, with some flocks reporting stock losses of up to 10–15%.

The diagnosis of clinical (visible) Johne's disease is straightforward and based on:

- the appearance of the sheep
- the exclusion of other diseases that may cause weight loss (such as parasitism, poor nutrition or broken mouth
- post mortem examination, with the detection of characteristic changes in the intestinal tissue and associated lymph nodes. The lining of the intestine may be thickened and corrugated and the lymph nodes may be enlarged. Microscopic examination following fixing and staining of the tissue (histopathology) is conducted to confirm the diagnosis.

The characteristic changes in the tissues resulting from infection with *Mycobacterium paratuberculosis* include the presence of the causative bacteria, which can be seen in tissues when special stains are used, and characteristic giant and epithelioid cells.

The diagnosis of subclinical Johne's disease is much more complex and challenging.

The currently available blood tests include the AGID and ELISA tests to detect antibodies in blood samples. The blood tests are flock screening tests only and are used to select sheep for confirmatory testing by microscopic examination of tissues, or sometimes by culture.

Faecal culture is used to detect *Mycobacterium paratuberculosis* in faeces. Because infected sheep shed *Mycobacterium paratuberculosis* in their faeces, the disease must be present in the flock if the OJD bacteria can be grown from faecal samples. However, faecal culture will not detect every infected sheep. Faecal culture takes at least 12 weeks because the OJD bacteria grow very slowly in the laboratory.

Pooled faecal culture (PFC) involves the collection of faecal pellets from up to 50 sheep and pooling these to present a single sample for culture at the laboratory.

In most circumstances, faecal culture tests are preferred to blood tests because it is possible to detect infection in the flock at an early stage. Testing seven pools of 50 sheep (350 sheep per flock) is considered equivalent to blood testing 875 sheep.

Diagnostic tests are used to classify individual animals and/or flocks according to their disease status. No diagnostic test is perfect. Many tests, including those for OJD, have limited sensitivity (the ability to identify all infected animals as test positive). Some infected animals return a negative test and are described as 'false negatives'. Some non-infected animals will return a positive test and are described as 'false positives'. This is a feature of biological tests.

The immune system of a sheep can respond to infection with *Mycobacterium paratuberculosis* by producing antibodies that can be detected in blood samples. Generally the antibody response does not occur until the disease is advanced (a sheep tested soon after infection is likely to return a negative result). Antibody levels may fluctuate (an infected sheep tested at monthly intervals may be positive one month and negative the next), and some animals may not produce antibody at all (an infected sheep may never test positive).

Test results must be interpreted in the light of knowledge of the capabilities of the test and of the disease in the individual or flock in which the test is applied.

The AGID and ELISA blood tests will, on average, detect 30% of the subclinically infected sheep in the group under test at any one test (for example, if there are 10 infected sheep in the group under test, the AGID or ELISA tests will detect three of them). Therefore, 70% of infected sheep are not detected resulting in a high false negative rate. This means the blood tests have limitations when applied to individuals or small groups of sheep, or for test and cull programs. However, they are very useful tools when applied to large groups of sheep for flock diagnosis and market assurance programs.

A vaccine is available to control OJD. The vaccine can reduce the number of deaths due to OJD by 90% and reduce the amount of OJD bacteria excreted by sheep in their dung by 90%. Vaccination is recommended for most infected flocks. Other flocks may use vaccination to assist in the prevention of the introduction of OJD into an unaffected flock and to provide extra assurance to purchasers of sheep that they have a low risk of being infected. Sheep only require a single vaccination in their lifetime, preferably given between 4 and 16 weeks of age. The only site for vaccination is under the skin high on the neck behind the ear. Accidental self injection may result in a severe local inflammatory reaction. The product leaflet should be read carefully and safety precautions followed.

Other methods of control can be used in combination with vaccination for infected flocks including:

- removing sheep showing obvious signs of OJD as soon as possible to reduce contamination of pastures with OJD bacteria
- controlling internal parasites, ensure adequate nutrition and preventing other forms of stress
- identifying and considering culling high-risk mobs to reduce contamination of pastures with OJD bacteria
- avoiding high stocking rates on very short pasture
- cropping or grazing cattle on paddocks contaminated with OJD bacteria (to reduce contamination with OJD bacteria)
- avoiding lambing on heavily contaminated areas
- crutching or shearing before lambing to reduce the amount of faecal build-up on the udders and legs of infected ewes to reduce the risk of lambs ingesting faeces containing OJD bacteria
- weaning early to reduce the time of contact between infected ewes and lambs
- avoiding the contamination of water supplies with faeces.

The introduction of live animals poses the greatest risk to the biological security of the flock. OJD is most commonly introduced into a flock by purchasing sheep that are subclinically infected. It is very difficult to breed all replacement sheep and the majority of sheep operations buy rams. However, it is important to keep purchases to a minimum and always buy from a low-risk flock. Is it also valuable to know if the sheep were bred by the vendor and if the vendor's flock is closed.

Individual animal testing will not give adequate protection from the introduction of ovine Johne's disease for the reasons described above.

Absence of signs of clinical ovine Johne's disease can be a poor guide to a flock's disease status. Subclinical ovine Johne's disease can exist in a flock for years before the clinical disease is recognised. For this reason, seeking vendor advice that there is no history or suspicion of disease in the flock should not be the only precaution taken. The Assurance Based Credits (ABC) Scheme forms the basis of a national approach to the management of OJD. The ABC scheme involves providing the ABC score for consigned sheep up to a score of 10 credit points based on:

- location risk
- OJD testing
- vaccination against OJD
- risk assessment.

The higher the ABC score, the lower risk that sheep are infected, and the greater the assurance provided to the purchaser. The sheep health statement (SHS) provides the basis for the ABC Scheme. The vendor provides to potential purchasers an SHS showing the ABC score for the consigned sheep and how the score was derived.

As the faeces of infected animals can spread OJD, do not allow dirty boots or equipment to be moved between farms. If possible, insist that all people moving on and off a farm disinfect their boots and equipment.

Boundary fences must be kept in sheep-proof condition to stop neighbours' sheep and strays entering a farm. If necessary double-fence high-risk boundaries or exclude sheep and other susceptible species from the high-risk areas on the farm. Avoid sharing rams and avoid agistment unless the OJD status of the animals that have grazed the agistment property for the period including the two most recent summers is known.

If necessary, fence off waterways and provide drinking water from a source that cannot be contaminated by sheep faeces.

Trace element and vitamin deficiency diseases

Trace elements deficiencies are not a major cause of poor production in flocks in southern Australia. However, in some instances, trace element supplementation of either sheep or pasture has brought about dramatic improvements.

Sheep have a requirement for a range of trace elements. The most common deficiencies include copper, cobalt, selenium and, occasionally, iodine.

The trace element status of livestock is lowest in late winter and spring. Consequently this is the ideal time to assess the trace element status of sheep. Measurements of copper, cobalt and selenium can be made from the blood samples of ten animals. Liver samples can be more valuable to assess copper deficiencies. A veterinarian is able to collect samples, interpret the results and advise on appropriate action.

Few cases of ill-thrift are solely due to trace element deficiencies. However, poor performance is often attributed to trace element deficiency. Before a supplementation program is started, it is essential to confirm that a deficiency exists.

Cobalt deficiency

Cobalt is required by rumen microbes for the synthesis of vitamin B_{12}. Vitamin B_{12} is essential for energy metabolism in sheep and deficiency can lead to wasting even when good feed is available. A cobalt deficiency causes 'starvation in the midst of plenty'. The signs of cobalt deficiency are similar to those of malnutrition and include emaciation, poor appetite, weeping eyes, anaemia and decreased milk production. Cobalt deficiency predisposes sheep to white liver disease, where sheep develop photosensitisation or 'sunburn' associated with the damage to the liver.

Cobalt deficiency occurs on lush pasture in spring and may be associated with a reduction in soil intake by stock when grazing. The deficiency may also occur in coastal and other areas where the soil is sandy or well drained. The application of lime reduces the availability of cobalt.

Vitamin B_{12} injections are recommended for the immediate treatment of cobalt deficiency. An injection provides protection for 3 months. Cobalt pellets may also be used for immediate treatment and last 12 months. The choice between using pellets and vitamin B_{12} injections is dependent on the cost and the time period for which animals require supplementation.

Cobalt deficiency can be prevented by a range of cobalt supplements added to drenches, licks, pasture sprays and topdressing, though vitamin B_{12} injections and pellets are the most commonly used and effective option.

Selenium deficiency

Selenium is required for growth and for the prevention of selenium-responsive white muscle disease. Selenium also appears to have a role in the resistance to disease (immunity). Many of the pastures grazed by sheep in southern Australia are deficient in selenium.

White muscle disease is the most severe form of deficiency. It is usually seen in lambs, but it can occur in sheep up to 1 year of age. Muscle damage caused by the disease causes stiffness, lameness and sometimes an inability to stand. Damage to heart muscle can lead to sudden death.

Selenium-responsive ill-thrift has been reported in southern Australia. From some trials conducted in flocks with very low selenium nutrition, the increased body weight gains have been in the order of 2 kilograms in 10 weeks, and wool gains have been in the order of 0.4 kg in 12 months. However, responses of this order are not always obtained.

Selenium deficiency occurs on acid soils and is associated with high rainfall, high superphosphate usage and clover-dominant pasture. Deficiency is usually seen in spring and early summer.

It is important that an assessment of the selenium nutrition be made before animals are treated because selenium is toxic if overdosed.

Immediate treatment of deficiencies can be supplemented with injections or oral drenches of selenium. For long-term treatment and prevention of selenium deficiency, lambs can be treated with selenium at 4 to 6 weeks of age and again at 3 months of age. Lambs can receive further doses of selenium at intervals of no less than 3 months until 15 to 18 months of age.

Selenium can be administered with clostridial vaccines and worm drenches. Other products such as selenium pellets, long-term injections and selenium in fertiliser are also available.

There is less need to supplement adult animals, with the exception of pregnant ewes in situations where white muscle disease may occur in lambs in the first few weeks after birth.

Vitamin E deficiency

Selenium works in conjunction with Vitamin E to repair cells and body tissues. Any green feed is a rich source of Vitamin E and even a small green pick will provide an adequate amount. Stored grain or hay contain little Vitamin E, depending upon the age and storage conditions. During a drought, lambs may not have access to enough green feed before summer and so have low Vitamin E reserves. In this case, treatment of lambs with Vitamin E as an oral drench is generally the most cost-effective solution, with retreatment depending on availability of green feed.

Copper deficiency

In sheep, copper is required for body, bone and wool growth, for pigmentation, and for nerve fibre and white blood cell function.

Many deficient areas have now been top-dressed with copper. In some situations, a single copper top-dressing has provided adequate copper for animal and pasture production for at least 13 years.

Copper is more available from dry herbage than green herbage, and is in higher concentration in clovers than in grasses. Therefore, copper deficiency is more likely on lush, green grass-dominated pastures during winter in southern Australia.

High intakes of molybdenum and sulphur have been shown to decrease the availability of dietary copper to animals and to induce a copper deficiency. Many areas that are marginal or deficient in copper are also deficient in molybdenum. Therefore, molybdenum applications to these areas should be accompanied by an application of copper.

Loss of wool crimp is one of the first signs of copper deficiency in sheep. The wool lacks character and develops a lustre ('steely wool'), with crimps three to four times normal width. Grey banding on black-woolled sheep is also a sign of copper deficiency.

Sway back (enzootic ataxia) – seen as paralysis or as a hind limb staggering gait in lambs – is a clinical condition caused by copper deficiency.

Immediate treatment of copper deficiency may be given by injection or oral drenches. Pasture top-dressing with copper is the preferred long-term treatment. The standard recommendation has been to apply 2 kilograms of copper per hectare every 5 to 7 years. However, copper topdressing may only be necessary every 15 years. If copper levels are marginal, copper should be applied to pasture when molybdenum is applied.

Iodine deficiency

Adequate iodine intake by pregnant ewes is essential to enable lambs to produce sufficient thyroid hormone for growth and development and to survive after birth. Signs of iodine deficiency include 'goitre' (enlargement of thyroid glands in the neck) and poor lamb survival.

Outbreaks of goitre have occurred in lambs on or adjoining the Great Dividing Range and in small pockets throughout south-eastern Australia. Seasonal conditions favouring iodine deficiency include an early autumn break and a wet winter. Iodine deficiency is diagnosed by high thyroid weight to body weight ratio in lambs.

Flocks lambing between August and October in high rainfall areas where goitre has occurred require iodine supplementation. Ewes may be drenched with potassium iodide in late pregnancy or be given access to iodised salt licks. Lambs may be drenched with potassium iodide or injected with iodised oil.

Trace element toxicity

Copper, molybdenum, cobalt and selenium toxicities have occurred due to inappropriate supplementation or overdosing.

Other common diseases

Lumpy wool

Lumpy wool is a type of dermatitis caused by infection with the bacterium *Dermatophilus congolensis*. The disease is also known as mycotic dermatitis. Continued wetting of the skin increases the chance of infection. The bacterial infection causes weeping from the damaged skin, which mats the wool fibres together and forms a scab. The raw surface beneath the scab of an active infection is a prime site for flystrike. Lumpy wool occurs mainly on the back of sheep and is more common in wet seasons. Infection on the ears, nose and legs produces thin flat scabs. Weaners are usually the most severely affected class of sheep. Infected young sheep have a decreased growth rate.

The spread of lumpy wool occurs when bacteria from an infected sheep contact susceptible skin on another sheep. Any sheep with moist skin is susceptible to lumpy wool, so infection may be spread while dipping or handling wet sheep.

There are minimal benefits from culling affected sheep, because the heritability is thought to be very low. It is not possible to identify susceptible sheep in a dry season and many infected young sheep may not be affected again in their lifetime due to good immunity after recovery from infection.

If dipping must be carried out for lice control, pay attention to hygiene by daily cleaning and recharging the dip, adding disinfectant, or by adding a 1% solution of zinc sulphate (1 kilogram in 100 litres of water). Sheep affected by lumpy wool should be shorn and dipped last.

On most sheep, the scabs will grow out without treatment, but a few sheep develop severe lumpy wool that does not grow out. A single treatment of antibiotic prescribed by a veterinarian will allow the skin under the scab to heal. Subsequent wool growth will lift the scab away from the skin to allow shearing about 6 to 8 weeks after treatment. Affected sheep should be shorn last. The treatment of sheep with lumpy wool is essential when lice are present to guarantee clean shearing to allow easier eradication.

Most of the scab is removed from the wool during processing, but any that remains results in uneven dye uptake, causing prices paid for affected wool to be significantly reduced.

Fleece rot

Fleece rot is a type of dermatitis caused by prolonged wetting of the skin, resulting in matting of the wool by serum exudate. Fleece discolouration is common, with colours ranging between red, blue, yellow, grey, brown and green. The important bacterium involved in the development of fleece rot is *Pseudomonas aeruginosa*.

Fleece rot is by far the most important predisposing factor to body strike by *Lucilia cuprina*, the green Australian sheep blowfly. Weaners and hoggets are the most commonly affected. The prevalence of infection falls significantly in a mob as the sheep age.

Fleece rot is probably related to the ease with which the skin can be wetted. There is a variation in fibre diameter in every fleece. A wide variation of fibre diameter in a fleece leads to an increased susceptibility to fleece rot. Coarse long secondary fibres produce a tipped end and uneven crimp, and the staple becomes difficult to separate because fibres become disorganised. The end result is increased water trapping and decreased ventilation of the fleece, preventing the skin from drying quickly after wetting.

A thick staple and dense blocky tip are associated with an increased resistance to fleece rot. Fine woolled sheep are generally less susceptible to fleece rot than medium and strong woolled sheep.

Heritability of fleece rot is high but it is difficult to consistently identify susceptible sheep for culling, especially in a dry environment.

Pizzle rot

Pizzle rot is an infection of the prepuce and penis of wethers. The disease occurs in most wether flocks, occasionally in rams and very occasionally as an inflammation of the vulva in ewes. The bacteria are spread in sheep camps by sexual contact and by flies.

Pizzle rot occurs when wethers graze improved pastures containing legumes. The causative bacterium, *Corynebacterium renale*, can hydrolyse urea in urine to produce ammonia, which causes irritation and ulceration of the pizzle. If this spreads to the rest of the penis and sheath, it can result in urine blockage and death.

Testosterone treatments will help to cure mild cases and prevent the development of new cases. Testosterone preventive treatment should be timed to coincide with growth of clover-dominant pastures. If a good autumn break occurs, a treatment should be carried out in autumn, but in most years a late winter/early spring treatment will be all that is required. Testosterone treatment results in increased horn growth, masculine behaviour, increased growth rate (1 to 2 kilograms), increased wool growth (0.15 kilograms) and an increase in fibre diameter.

Severe cases may require surgical treatment. Surgical treatment involves opening the sheath to relieve the obstruction. Seek veterinary assistance with this procedure.

The surgical procedure of pizzle dropping can also reduce the risk of pizzle rot. This procedure is normally done on wether hoggets before they are 12 months of age, but not on young lambs because of too much variation in pizzle length.

Pink eye

Pink eye is an infection of the conjunctiva and cornea, usually caused by *Rickettsia conjunctivae*, but other organisms can also be the cause. Initially, a pink eye infection induces a watery discharge and reddening of the conjunctiva followed by the cornea becoming cloudy. Occasionally the eyeball ruptures.

The disease may be spread either directly by immediate contact, or indirectly by flies, long grass and dust contaminated by the tears of infected sheep. The disease is usually seen in summer. Outbreaks may occur in either young or older sheep because immunity following infection is short lived. After infection and recovery, the bacteria persist in the eye for long periods. Carrier cases are the source of infection for subsequent years.

Most sheep recover within 2 to 3 weeks without treatment. Avoid mustering because it increases the spread of pink eye infection in the flock. Severely affected sheep, or those with both eyes involved, can be put in a hospital mob and treated with a suitable topical antibiotic as advised by a veterinarian.

Grass seeds trapped in the eyes under the eyelids, including the third eyelid, cause signs similar to pinkeye. Carefully examine the eyes: checking under the eyelids, to ensure that there are no grass seeds present.

Cheesy gland

Cheesy gland is caused by the bacterium *Corynebacterium pseudotuberculosis* and is characterised by abscess formation in the lungs and lymph nodes. Infection is widespread in Merino sheep in

Australia. Export restrictions are placed on infected meat destined for several countries. The annual cost to Australia from carcase rejection and condemnation due to cheesy gland is very high.

The bacteria penetrate intact skin with less than 3 weeks wool growth. Bacteria may also gain entry through broken skin from shearing cuts or marking wounds. A skin abscess forms at the entry site and from there bacteria spread to lymph nodes and lungs. Infected sheep with discharging lung abscesses are the main source of bacteria for new infections because these sheep spread bacteria when they cough. Abscesses that discharge through the skin are also a source of infection. The bacteria survive for up to 6 months in the environment and all sheep are susceptible to infection. Maximum spread occurs at shearing and dipping.

Avoid mixing cheesy-gland-affected sheep with clean sheep. Young sheep usually have a lower incidence of infection. Older sheep are more likely to be infected because they are more likely to have had exposure to the bacteria that cause the disease.

The management procedures that should delay the spread of cheesy gland are to shear or dip the youngest sheep first, avoid dipping within 4 weeks of shearing and not mixing age groups. These procedures delay the spread of cheesy gland within a flock. However, given time, a large proportion of adult sheep become infected.

Vaccination provides effective control of cheesy gland. A vaccine is available in combination with the clostridial vaccines. Sheep require two doses as lambs (at marking and 4 to 6 weeks later) and an annual booster before shearing. The vaccination will not eradicate the disease, but will reduce the number of abscesses per sheep and the number of sheep affected.

'5 IN 1' vaccination and the clostridial diseases

5 in 1 vaccination provides protection against the clostridial diseases. These diseases include blackleg, pulpy kidney (enterotoxaemia), malignant oedema, black disease and tetanus.

To be fully effective, two doses of '5 in 1' vaccine must be given. For lambs, the first dose should be given at marking and the second dose 4 to 6 weeks later. A booster dose should be given each year.

The lamb is protected from the clostridial diseases for its first few weeks of life by antibodies obtained from its mother's colostrum. Pre-lambing vaccination increases the level of antibodies in the ewes' colostrum.

When a lamb is first vaccinated with '5 in 1', it becomes sensitised to the vaccine and produces some antibody: perhaps enough to protect it for the next couple of months. When the second dose is given, there is a rapid response and the protection lasts for at least 12 months. A third 'booster' dose given one year later will, in most cases, give lifelong protection against all the above diseases with the exception of pulpy kidney (enterotoxaemia).

In flocks that are not properly vaccinated, the clostridial diseases can claim the lives of many sheep. Sheep found dead in the paddock are often cases of blackleg, black disease or pulpy kidney. Ewes dying after lambing may be cases of malignant oedema.

Wethers are as susceptible to the clostridial diseases as ewes.

In summary, vaccinate all sheep including rams:

- first dose at marking
- second dose 4 to 6 weeks later
- booster vaccination 12 months later.

Ewes should receive additional boosters given yearly or bi-annually within a month of lambing.

Scabby mouth

Scabby mouth is caused by a pox virus that can survive for many years in the environment. The virus enters the skin through abrasions. Outbreaks are usually seen in the summer when dry stalky feed and grass seeds abrade the skin of the lips or in spring when young sheep are grazing lush pasture when the feet form is most common.

Scabby mouth sores are most often found on the lips, or feet, and occasionally on the udder. The scabby mouth virus is often associated with strawberry footrot. In the early stages of the disease the scabs are firmly attached. Scabs drop off in 4 to 5 weeks. Early manual removal of scabs will delay healing. Following initial infection, sheep develop a strong lifelong immunity.

Once infected sheep have been on a property, outbreaks will occur in later years in susceptible sheep (those which have not been previously infected or vaccinated).

A scabby mouth vaccine is available; vaccine should not be used unless the disease has occurred on the property because the vaccine contains live virus.

The usual vaccination program on a property with a continuing scabby mouth problem is to vaccinate all lambs. Only one vaccine is required for lifelong immunity. The vaccine is given at marking or earlier. The bare skin on the inside of the thigh or ear is scratched with a forked needle that has been dipped in vaccine. A line of pustules appearing along the scratch 1 week later indicates a good response to the vaccine.

Scabby mouth virus can also infect humans. Sores commonly occur on the arms and lifelong immunity develops after infection.

Ryegrass staggers

Most perennial ryegrasses are infested with fungi known as endophytes. The endophytes are important for the survival and persistence of the plant by helping protect them from insect pests and improving their environmental tolerance. However, the endophyte does this by producing chemical compounds called alkaloids, which can be harmful to livestock.

Ryegrass staggers usually occurs in livestock grazing perennial ryegrass in late summer and early autumn when pastures are dry, short and growing slowly. Alkaloids produced by the endophyte are concentrated in the crown of the plant and are ingested under these conditions. In addition to staggers, these compounds can cause a range of animal problems, including lower growth rates and lower milk production. In some circumstances death can occur. The occurrence of perennial ryegrass staggers is extremely variable.

Affected sheep are uncoordinated, have a stiff, proppy gait, may fall on their sides and show body tremors and convulsions. Sheep usually recover within a few minutes. Disturbing, yarding or driving sheep induces attacks making them difficult or impossible to muster. Deaths can also occur owing to misadventure should stock become caught in fences or fall into dams or troughs. Ryegrass staggers can also contribute to weaner ill-thrift because sheep graze poorly while affected.

The removal of stock from infected paddocks is difficult because affected animals cannot be driven. Stock can be drifted into safe paddocks using supplementary feed as a lure. It is difficult to assess the safety of a paddock because, when faced with an outbreak, perennial ryegrass is so widespread. Supplementary feeding to reduce the intake of fungal toxin is unlikely to be effective because the green pick is very palatable.

Sheep will recover in about 3 days after removal to safe pasture. Outbreaks should cease after the autumn break because the disease normally disappears following a fall of rain and increased pasture growth.

Managing endophyte in ryegrass:

- The use of endophyte-free pasture species is possible, but this may result in poor pasture persistence (may need to budget for more frequent pasture re-sowing, depending on the district).
- Dilute the endophyte effect by adding some toxin-free species such as cocksfoot, phalaris and fescues.
- Grow special-purpose paddocks containing endophyte-free species such as lucerne or fodder crops to use during the summer/early autumn danger period.
- Use cultivars with safer 'novel 'endophyte strains: some are already available, and many are in the trial stage and will hopefully become more available as research continues.
- Cultivars with safer endophytes may be more prone to certain insect damage and less resistant than other cultivars. Safer endophytes can also improve the palatability of ryegrasses and may be more prone to damage from overgrazing.

Seed from the ryegrass plants remain in the ground in the seed bank and may last several years. This needs to be considered when sowing seed either without endophyte or with one of the novel endophytes.

Novel endophytes

Different endophytes produce different alkaloids. The traditional 'wild type' endophyte produces lolitrem B, ergovaline and peramine. Peramine is not believed to cause any side effects, but lolitrem B causes staggers and ergovaline is believed to cause heat stress and reduce animal performance.

A number of novel endophytes are now available that produce different alkaloids and therefore reduce risks to stock. The alkaloids produced by a range of the novel endophytes are summarised in Table 6.7, for reference and decision making as different cultivars become available.

NEA2 produces lolitrem B, ergovaline and peramine – like wild-type endophyte – but in much lower quantities, making it safer. AR1 produces peramine only, which makes it much safer, but protection from insect pests is also reduced as a result.

Endo 5 produces peramine and a reduced amount of ergovaline. This effectively results in greater insect protection than by not producing this alkaloid at all, but minimises side effects – and avoids staggers by not producing lolitrem B. AR6 is similar.

AR37 does not produce any of these compounds, but produces janthitrems instead. Janthitrems are believed to provide broad protection from insect pests with a minimal risk of staggers – but field research is still continuing.

It is important to note that different endophyte–grass combinations can produce different ranges and concentrations of alkaloids. This can also vary with environmental conditions. The most appropriate endophyte for a particular enterprise may therefore vary, and new information will continue

Table 6.7. Alkaloids produced by a range of novel endophytes.

	Alkaloids			
Endophytes	**Lolitrem B**	**Ergovaline**	**Peramine**	**Janthitrems**
Nil	X	X	X	X
Standard (Wild)	✓	✓	✓	X
AR1	X	X	✓	X
Endo 5	X	✓	✓	X
AR6	X	✓	✓	X
AR37	X	X	X	✓
NEA2	✓	✓	✓	X

to become available. Likewise, new endophytes may be found that minimise the risk of staggers while maximising plant persistence.

Annual ryegrass toxicity

Annual ryegrass toxicity is a fatal neurological disease caused by the ingestion of a toxin that can be present in the seed heads of annual ryegrass. The toxin is not produced by the plant, but by a bacterium known as *Clavibacter toxicus* in association with a nematode that parasitises the seed heads of the grass.

Sheep affected by annual ryegrass toxicity show signs of incoordination and may collapse or convulse, especially when driven. Eventually they become recumbent and die in 1 to 2 days. To date, this disease has only occurred in South Australia and Western Australia.

Diagnosis is made on clinical signs and history of recent access to annual ryegrass in pasture or hay.

Phalaris poisoning

Phalaris poisoning can result in staggers, with nervous signs resulting directly from ingestion of alkaloids produced by the plant. Phalaris can also cause sudden death and a chronic staggers syndrome, which is often fatal. Autumn and early winter are the high-risk periods and care must be taken when introducing hungry sheep to phalaris pastures, especially if sheep are not accustomed to it.

Sudden death occurs soon after hungry sheep graze fresh phalaris pick and is caused by resulting high blood ammonia levels causing brain damage. Compounds in the plant interfere with normal nitrogen metabolism in the sheep.

In contrast, chronic phalaris staggers caused by alkaloids can develop months after grazing toxic plants, especially when stock are handled. Sheep generally do not recover from chronic phalaris staggers. The staggers can sometimes be prevented by treating sheep that are grazing at-risk pasture with intraruminal cobalt pellets: the cobalt helps to break down the alkaloids from the plant.

There is no prevention for the sudden death syndrome apart from avoiding moving hungry sheep to short fresh phalaris at, or just after, the autumn break. In particular, avoid moving sheep that have been off feed for more than 12 hours. Sheep set stocked on phalaris pastures are usually at lower risk because they adapt to the compounds in the plant that interfere with their normal nitrogen metabolism.

Nitrate/nitrite poisoning

Nitrate poisoning occurs when ruminants (including cattle and sheep) eat crops, fodder or pasture containing high levels of nitrate. Nitrates are converted into nitrites in the rumen and then into ammonia, which is belched out or detoxified by the liver. If the intake of nitrate is high, some nitrate passes through the rumen without being converted to nitrite and causes irritation to the lining of the gut, which results in scour. When nitrites are not converted to ammonia rapidly enough, the nitrites are absorbed into the bloodstream where they bond with haemoglobin in the blood. The haemoglobin loses its ability to transport oxygen and the affected animals die from oxygen starvation.

Nitrate poisoning should be considered as a possible cause of scouring – and nitrite poisoning should be considered as a possible cause of sudden death – in sheep grazing pastures or fodder high in nitrates.

Nitrite poisoning occurs very quickly. Symptoms are often not observed and animals are found dead. The symptoms are that the affected animals stand apart, have an unsteady gait, have rapid and shallow breathing and eventually collapse. The whites of the eyes and membranes in the vagina, mouth and nostrils will appear initially bluish in colour and then muddy brown.

Death usually occurs suddenly and treatment is not possible. If stock are observed in time, intravenous administration of methylene blue by a veterinarian can save them.

Plants containing more than 1.5% nitrate are dangerous. Nitrate concentration in forages increases greatly during drought conditions. Other conditions frequently associated with losses are cool temperatures, cloudy days and heavy or recent applications of nitrogen fertilisers. As well as annual ryegrass, some plant species are more likely to accumulate toxic concentrations of nitrate. These include forage crops such as millet, corn, rape, sudan, oats, weeds (such as cape weed) and some pasture species.

Management practices can reduce and prevent losses from nitrate poisoning. Awareness of high-risk pastures or crops following dry conditions is important. High-risk pastures or crops can be measured for nitrate levels via laboratory analysis or by using quick paddock-side nitrate test strips. These strips can give an indication of nitrate levels in plants, but the pasture sample must accurately reflect what the stock are eating. Stock grazing a high-risk crop should be observed carefully and **quietly** removed at the first sign of toxicity.

Photosensitisation

Photosensitisation is the sensitisation of lightly pigmented skin to sunlight. Dermatitis develops when sensitised skin is exposed to sunlight. Areas not protected by wool, such as the face, ears and tail, can be affected. Primary photosensitisation can be caused by eating plants containing primary photodynamic agents. Secondary photosensitisation is the result of faulty excretion of the breakdown product of chlorophyll (the green pigment in plants) due to liver damage.

In some cases of photosensitisation, is not easy to determine if the cause is primary or secondary, especially in clinical cases of photosensitivity occurring sporadically in sheep grazing lush pasture.

Primary cases have skin signs only, including redness, swelling, and the death of skin. Secondary cases may also have signs of liver dysfunction including jaundice. Photosensitisation is painful: affected sheep are restless, they will shake their heads and seek shade. The affected skin eventually peels off leaving a raw area that can be slow to heal.

St John's Wort, a plant containing a primary photodynamic agent called hypericin, can cause primary photosensitisation. Secondary photosensitisation can follow liver damage. The breakdown product of chlorophyll is called phylloerythrin. Phylloerythrin is excreted in the bile, and when biliary secretion is blocked by liver damage or bile duct obstruction, phylloerythrin accumulates in the body and may reach levels in the skin, that make the skin sensitive to light.

There are many plants that contain substances that are toxic to the liver, including ragwort, cyanobacteria associated with blue green algae in water, and lupins with the accompanying fungus phomopsis. The ingestion of these plants can result in photosensitisation.

Facial eczema is a secondary photosensitisation caused by the ingestion of spores of the fungus *Pithomyces chartarum*. The fungus proliferates on litter in the base of pasture in warm, humid weather. In these cases, other clinical signs such as jaundice may be seen and production losses occur. Depending on the extent of liver damage, sheep may die or not recover condition and have to be culled.

Photosensitised sheep should be given shade and access to water and good-quality feed. Supportive treatments are available, so always consult a veterinarian.

Metabolic diseases

Pregnancy toxaemia (twin lamb disease)

Pregnancy toxaemia occurs in pregnant ewes: often those carrying twins or triplets and receiving insufficient feed in the last 4 to 6 weeks of pregnancy. Pregnancy toxaemia can also be induced in late pregnancy by poor appetite in fat ewes, bad weather causing reduced grazing, disease such as footrot, foot abscess or perennial ryegrass staggers, and handling, mustering and shearing resulting in time off feed.

Pregnancy toxaemia is caused by insufficient energy intake at the time of high-energy demand due to pregnancy. If the ewe is not offered enough quality feed to provide sufficient energy for herself and the unborn lamb, she will run short of glucose and will break down her own body tissues, resulting in accumulation of high levels of ketones and a loss of body condition. Often there may be an underlying disease problem, such as foot abscess, that reduces the ewe's ability to graze.

Initial signs of pregnancy toxaemia include drowsiness, loss of appetite, incoordination and weakness, progressing to blindness, fine muscle tremor and grinding of teeth. Later signs include recumbency, convulsions and coma. These signs progress over a course of 2 to 8 days. Death is very common unless treated very early in the course of the disease. If the ewe is in the later stages of the disease then humane destruction is recommended because the treatment is not often successful.

For ewes showing early signs of pregnancy toxaemia, provide good shelter and nursing care and administer glycerol or propylene glycol orally twice daily as an energy source, as well as administering calcium borogluconate under the skin. Repeat treatments with glycerol or propylene glycol twice daily until the ewe's appetite returns.

If pregnancy toxaemia is occurring in the flock, other ewes may be marginally affected, but not showing immediate outward signs of the disease. Starting or increasing supplementary feeding will often be essential to reduce further problems. The longer term impact may be low-birth-weight lambs, poor lactation in the ewes and lower lamb survival.

Avoid under nutrition and stress in late pregnancy (see Chapter 3).

Hypocalcaemia

Hypocalcaemia is a fall in serum calcium levels occurring in the last 6 weeks of pregnancy and the first 2 weeks of lactation. Older ewes are more susceptible.

Circumstances such as a sudden change to lush pasture, prolonged feeding of rations of cereal grain only, bad weather causing stress, management stress such as shearing or crutching, or grazing plants containing oxalate can all also cause hypocalcaemia. In these situations, younger ewes, wethers and rams may also be affected.

Signs include a stiff, staggery gait and eventual recumbency with an inability to rise. If left untreated, sheep lapse into a coma and die. Response to treatment is good, especially if the sheep are treated early.

Treat with calcium borogluconate. Oral treatment with Unmix can give 24-hour protection in high-risk situations. Administer as advised on the label or as recommended by a veterinarian. If secondary pregnancy toxaemia occurs, give glycerol or propylene glycol as for pregnancy toxaemia.. Provide shelter and warmth and repeat the treatment with calcium borogluconate and glycerol or propylene glycol if necessary.

Environmentally induced diseases

Cold stress

Acute hypothermia or cold stress occurs in cold wet, windy weather in recently shorn sheep. It can occur in sheep shorn within the previous month, but is most common when sheep are only a few days off-shears.

Shorn sheep succumb to cold stress when there is a sudden change to cold, wet, windy weather, creating high chill factors. Following the loss of insulation provided by the fleece, the sheep's metabolic processes need time to adapt to higher heat production requirements. In cold, wet, windy weather, the body temperature of recently shorn sheep falls to critical levels unless the heat loss can be reduced.

Cold stress occurs in all classes of sheep, although losses are generally higher in weaners. This is because weaners have a higher skin surface area per kilogram of body weight.

Other factors that are important in cold stress and may affect management decisions are:

- Sheep that are losing condition at shearing are more susceptible to cold stress than sheep maintaining or gaining weight.
- The actual body weight of adult sheep at shearing has little effect on mortality.
- Autumn-shorn sheep are more susceptible to cold stress than spring- and early summer-shorn sheep, probably because sheep in autumn are losing weight and cold, wet, windy weather often occurs at this time.

- Cold stress in off-shears ewes in late pregnancy may predispose them to hypocalcaemia and pregnancy toxaemia.
- Paddock shelter may or may not be of much use in preventing cold stress (timber belts and plantations need to be large in area and dense in order to provide protection for enough sheep). Sheep tend to be driven with the wind in very bad weather, so the plantations need to be on the edge of the paddock, although the direction of wind in bad weather cannot always be predicted with accuracy.
- Sheep yarded for 1 day are at less at risk than those yarded for 2 or more days.

Handling cold-stress-affected sheep is extremely difficult and attempts to save recumbent animals are often unsuccessful. The principles involved in prevention and treatment are as follows.

Prevent sheep becoming affected by:

- stopping shearing if a sheep weather warning is issued
- minimising the fasting period during shearing
- changing time of shearing
- feeding supplements to reduce weight loss during shearing (see Chapter 3)
- using sheds, timber belts, plantations and any other shelter to reduce the chill factor on the sheep; place the most recently shorn and youngest sheep in the shed
- consider use of plastic coats (see below).

Treat affected sheep by:

- reducing heat loss and/or replacing the heat lost by applying plastic coats (these can be made from garbage bags by cutting triangular pieces out of both closed corners to create holes for the head and front legs – place on the sheep and tie in place with baling twine around the belly. This is a good method for treating affected sheep and preventing cold stress in sheep with no shelter.)
- treating with calcium borogluconate injections is not usually of much value except perhaps in ewes in late pregnancy and weaners.

Livestock management during periods of high fire risk

Preparing a fire plan for livestock may help protect livestock assets and welfare, and financial and emotional wellbeing of livestock owners. Have a workable plan that can be implemented quickly and safely when fire threatens.

Livestock fire plan

The basis of the plan is to move stock to designated low-risk areas (those with a lower risk of burning) when fire threatens. Decisions on when to enact the fire plan will be based on the weather, or the immediate threat of fire. Listen to the radio and observe the environment to decide when to put your plan into action.

Low-risk areas should be centrally located and easy to access and can include:

- recently cultivated paddocks
- bared-out paddocks, provided they are well defended by fire breaks
- irrigated paddocks or paddocks containing green summer crops: green feed does not burn easily

- stock yards: cattle and sheep can be held in yards with bare or ploughed surrounds, which can be wetted in advance; however, the yards must be well defended as the fire front passes.

Cattle are adept at moving to avoid fire, so may be kept in low-risk paddocks with relative safety. The mobbing instincts of sheep may prevent movement away from a fire if they are kept in the open. Using sheep yards as a safe area can be useful if time allows, but extra care should be taken in preparation for the coming fire front. Wetting down the surrounds and extensive fire breaks are a must. If sheep are to be moved, do it early in the day. Do not wait until the hot part of the day because sheep can be difficult to move in very hot conditions.

All low-risk areas should be further protected by firebreaks and be free of litter build up. Thought should be put into the preferred location of low-risk areas, and should take into account terrain, accessibility, direction that the fire threat is most likely to come from, prevailing winds in the summer months, and the location of nearby forests and bushland. Farms with a large number of stock may need to designate and manage several low-risk areas on the property.

Managing burnt sheep

Many burned sheep are destroyed when they could have returned to normal productivity. Immediately after a fire, burns often look worse than they really are.

The urgent, initial need is to destroy burnt livestock that are down and unable to rise. Such stock can be shot in the paddock. Burnt stock that are able to walk must be contained for assessment and carefully examined individually, preferably by a veterinarian. It is inefficient and dangerous to try to assess and shoot mobile, affected sheep in a paddock. Skilled assistance is needed for mustering stock, which can be very difficult if any fencing has been damaged or destroyed. If permanent yards are not available, portable yards can be erected or temporary yards constructed with steel posts and any fencing materials on hand. If possible, temporary yards should be sited adjacent to burial sites.

When assessing individual sheep, areas to be examined closely include the face, ears, lips, legs, hooves, anus, vulva, penis, prepuce, scrotum, udder and other non-woolled areas including axilla (under forelimbs) and inguinal (between hind limbs) areas. Sheep often need to be tipped in order to make a thorough assessment.

Singed wool is not always a good indicator of the severity of a sheep's burns. Wool is a good insulator and therefore the degree of charring of the fleece is not necessarily related to the chances of survival.

Examine the non-woolled areas closely. Sheep with more than 15% of their body surface burnt are unlikely to survive, even with intensive treatment and nursing.

Mobility is a good indicator of chances of survival. If the coronary band (the area immediately above the hoof) cracks and leaks fluid it is an indication that the hoof will be shed within the next few days. Stock without hooves will not survive and should be destroyed.

Sheep should be classified into the following groups:

1. **Destroy immediately:** any downers (sheep that are unable to get up), sheep with extensive burns, sheep with swollen limbs and sheep having difficulty in breathing (due to lung damage). Stock with extensive burns to the legs below the knee and hock, with swelling and a dry leathery appearance of the skin have a poor prognosis for survival.
2. **Reassess within 1 to 3 days or salvage slaughter:** sheep with minor burns to the non-woolled areas. Hospital mobs should have good access to water, feed and shade. Stock should be

destroyed if their legs become swollen, the hooves begin to separate from the underlying tissues of the foot, or if the sheep are not able to feed or go down.

3. **Reassess within 5 to 7 days:** animals without apparent damage. Often burns to the axilla and inguinal areas, which have been overlooked at earlier inspections, become apparent.

The salvage slaughter of burnt stock is not an option if they have to travel long distances to an abattoir or if they have more severe burns that would cause pain during transportation and likely condemnation at abattoirs.

Recovering sheep should be kept under close observation in a paddock with good fed and water and access to shade. Check for sheep that may become flystruck, have pizzle obstruction or have difficulty in walking. Treat flystrike to kill maggots. Scabs on the pizzle causing urinary obstruction should be manually removed.

Farm biosecurity

Protection of animal health by the prevention of spread of disease

For commercial success, stock health is managed by good husbandry practices including feeding, vaccination and worming programs and the prompt treatment of sick animals. Often too little attention is paid to the risks associated with bringing in stock from other properties, and particularly from livestock markets. The greatest disease risk to animals is other animals. Livestock from markets pose the greatest risk because these animals have been mixed with many others of unknown disease status and they may have been stressed by transportation and time off feed.

When buying stock:

- consider the diseases known to occur in the district or region or in areas from which animals will be purchased
- know the status of animals that are being purchasing before introducing them to a new property
- source low-risk animals (vendor declarations and sheep health statements help with animal disease assessment)
- inspect animals carefully for disease, including parasitism and footrot
- drench with a triple combination drench, vaccinate and foot bathe incoming animals
- isolate (preferably so that sheep run separately until the end of the next footrot spread period) and observe introduced stock closely
- ensure that vehicles used in the transportation of animals are cleaned and disinfected before use.

Four important diseases of sheep are commonly introduced onto farms by purchased, agisted or stray sheep. These are:

- lice
- drench-resistant worms
- footrot
- ovine Johne's disease.

Even with careful inspection, it is impossible to be absolutely sure that sheep are free of these diseases at the point of sale.

Minimise the chances of introducing lice

Examine a large number of sheep by parting the wool to skin level at a large number of sites along the back line and upper sides and over the neck. Isolation of the introduced mob may not effectively contain lice because transmission may occur through fences. If sheep have only a very light infestation at the time of introduction, the infection may go undetected for several months. Any sheep with fleece derangement should be examined for lice. If sheep are purchased off shears and there is doubt as to whether sheep have lice, then treat the mob.

Minimise the chances of introducing drench resistant worms

Minimise the risk of importing drench-resistant worms with purchased sheep by drenching on arrival with a highly effective triple combination of drenches. Consult an adviser to select the most appropriate drenches. Consider holding imported sheep in a small paddock and conducting a worm egg count 10 days after treatment to confirm that the treatment has been effective.

Minimise the chances of introducing footrot

The footrot causing bacterium *Dichelobacter nodosus* can not survive for more than 24 hours on an exposed dry surface. If the bacteria are protected by mud or manure, they can survive for up to 7 away from the sheep's hooves. *D. nodosus* can survive for at least 4 years in an infected hoof.

Clean flocks are generally exposed to footrot in one of two ways:

- by the introduction of infected sheep or
- by sheep walking over an area that has been contaminated by infected sheep during the preceding week.

It is virtually impossible for car tyres, gumboots, birds, scavengers, the wind or water to spread footrot.

Table 6.8 shows the number of sheep that need to be inspected, if picked out of the mob randomly, to be 95% sure of detecting a 2% prevalence of footrot.

Footrot is more likely to be detected in lame sheep and in sheep with deformed, overgrown hooves. Sheep must be tipped and all four feet must be examined by parting the claws.

Where possible, the property of origin should also be checked for a history of footrot. Obtain a sheep health statement and carefully check the declaration regarding footrot.

A clean flock can pick up footrot by walking along a road, or standing in a market or truck that has been contaminated by infected sheep during the previous week.

Places where sheep of unknown footrot status have been during the previous week should be avoided.

Foot bathe all introduced sheep with a 10% zinc sulphate solution as they are unloaded to ensure that any footrot bacteria picked up in transit are killed before they establish in the hoof.

All introduced sheep should be isolated until they have passed through at least one spread period (spring) without any sign of footrot. The introduced sheep should be yarded last at shearing, drenching or other husbandry procedures.

Table 6.8. The number of sheep that need to be examined to be 95% sure of detecting a 2% prevalence of footrot.

No. of sheep in mob	20	50	100	200	500
No. of sheep to be examined	20	48	78	105	129

Secure all boundary fences. Stray sheep all too often introduce footrot or other diseases to otherwise clean flocks. Cattle may be a source of benign footrot for sheep. Goats are a source of benign, intermediate and virulent footrot for sheep.

Minimise the chances of introducing ovine Johne's disease

Under the Assurance Based Credits (ABC) Scheme, every mob of sheep in Australia has an ABC score. The higher the ABC score, the lower the risk. Ask for a sheep health statement (SHS) when purchasing sheep. An SHS requires vendors to disclose the ABC score of their flock. These statements are mandatory for mobs of store sheep from flocks that are infected, or suspected to be infected, with OJD, and mobs of sheep that have an ABC score of two or less.

To minimise the chances of introducing ovine Johne's disease (OJD), maintain the sheep-proof boundary fencing.

The movement of faeces between properties in run-off water or on boots and equipment is a risk. Double-fenced tree-lined boundaries will filter dung. Boots and equipment should be cleaned before moving to another property.

OJD bacteria may survive in the environment for 12 months or more. Consider the risk that land may be contaminated with OJD bacteria before leasing or purchasing land or using it for agistment.

Owners of all sheep flocks should consider whether OJD vaccination would be appropriate for their flock. Factors to be considered include the enterprise type, the sources of purchased sheep and the risk posed by neighbours. It is important to know that vaccination does not provide complete protection from infection.

Emergency animal disease

Reporting and preparedness

Thanks to strict quarantine measures regarding the importation of animals and animal products, Australia is free of many emergency animal diseases such as foot and mouth disease and scrapie that could devastate our rural environment and economy.

Prompt reporting of any unusual signs or symptoms of disease in livestock is vital for industry protection. All livestock producers have an ethical, legal and economic responsibility to report any unusual signs of disease in livestock promptly, so an early diagnosis can be made.

Emergency or serious animal diseases can prove enormously difficult and costly to control or eradicate if they are not reported and acted on quickly. The total direct cost of the foot and mouth disease outbreak to taxpayers in the United Kingdom in 2001 was A$21.54 billion for animals destroyed, other seized and destroyed items, cleaning and disinfection, disposal and transport of carcasses and veterinary costs.

An outbreak of an emergency animal disease can cripple a livestock industry and cause individual and widespread economic hardship. It can cause anxiety in livestock industries, withdrawal of privileged trade conditions and affect tourism. It can affect the ability of animals to be moved freely and the movement of the public in rural areas. Export markets may not be restored until the disease is eradicated. Some emergency animal diseases may also affect people.

A property owner that does not report unusual signs of disease takes a serious risk that the disease will extend to neighbours, the livestock industries and the community. It is vital that livestock managers contact their veterinarian if any unusual signs or symptoms are seen in their stock. Animal Health staff of the state Departments of Primary Industries, or equivalent, investigate suspect emergency disease cases at no cost. The national hotline number for reporting Emergency Animal Disease is 1800 675 888.

Foot and mouth disease

Foot and mouth disease (FMD) is one of the most significant potential emergency disease threats to the Australian livestock industries. FMD remains a serious problem in many parts of the world, including Africa, Asia and parts of South America. It is a highly contagious viral disease that affects cattle, sheep, goats, deer, buffalo and other cloven-hoofed animals. The disease spreads rapidly by contact with infected animals, animal products or on contaminated feed, equipment, vehicles or people and, in some circumstances, it can be carried by the wind. The disease is rarely fatal in adult animals but causes severe production losses and is a major barrier to international trade in livestock and livestock products.

Animals with FMD are feverish, depressed, off feed and lame, and may show excessive salivation. They lose weight, and milk production in dairy animals falls. Common signs include vesicles (blisters), erosions and ulcers in the mouth, on the feet, and on the udder of affected animals. However, not all of these signs may occur in any single animal. In sheep, FMD can be much milder than in other species. It often presents as lameness and the classical vesicles (blisters) may not be present. Mouth lesions are not prominent in sheep.

Australia is vulnerable to an expanding global trade in animals and animal products if adequate safeguards are not in place. Previous freedom from FMD is no guarantee that it will not occur and Australia needs to maintain disease surveillance. Farmers and veterinarians are at the front line of this surveillance.

Foot and mouth disease and swill feeding

Experience suggests that the most likely way for FMD, and some other exotic diseases, to get started in Australia is through scraps of illegally imported meat and dairy product being used as a cheap source of feed for pigs. This could result in infection, rapid multiplication of the virus and subsequent spread to other pigs and other livestock, such as cattle and sheep. This is the reason for the Australia-wide laws banning the feeding of meat and other animal products to pigs.

All Australians should be vigilant against the practice of feeding meat scraps to pigs. This practice, if it is suspected, must be reported immediately to a local animal health official. An outbreak of FMD would be disastrous for Australia's economy: losing billions of dollars in trade and employment.

Neurological diseases

The function of the nervous system (brain and spinal cord and nerves) is to maintain normal posture and gait, to control the internal environment of the body, the senses of sight, smell hearing, and so on, and the animal's mental state.

There are many diseases of sheep occurring in Australia that affect the nervous system, including listeriosis, poliocecephalomalacia, brain abscess, pregnancy toxaemia, *Clostridium perfringens type D*, enterotoxaemia, focal symmetrical encephalomalacia and lead poisoning. There are also a number of emergency animal diseases that are not known to occur in Australia that can affect the nervous system including scrapie, louping ill, rabies and pseudorabies. Scrapie is described below because it is one of the subjects of a national surveillance program.

Transmissible spongiform encephalopathy (scrapie)

Scrapie is a chronic, fatal degenerative disease affecting the central nervous system of sheep. It is characterised by itching and scratching, an abnormal gait and by a very long incubation period.

Scrapie is one of many transmissible spongiform encephalopathies (TSEs), or brain diseases, affecting a number of animal species. The other widely known TSE is bovine spongiform encephalopathy (BSE), which is also known as mad cow disease.

The spread of TSEs has been associated with the feeding of infected meat and bone meal.

Australia is free of scrapie and BSE. Australian industry and government have had programs in place for many years that ensure our disease-free status is maintained. These measures include :

- a ban on the import of cattle and sheep from scrapie and BSE affected countries
- a ban on the import of meat and bone meal from all countries (except New Zealand, which has a similar disease-free status) since 1966
- a legislated ban on the feeding of meat and bone meal to ruminant animals
- a national surveillance and monitoring program that checks Australian freedom from scrapie and BSE.

The continuing effectiveness of these measures is confirmed through a structured disease surveillance program in Australia and is known as the National Transmissible Spongiform Encephalopathy Surveillance Program (NTSESP). This surveillance testing involves the submission of brains from sheep and cattle that meet a defined clinical profile.

The Australian surveillance system uses sheep (and cattle) with clinical nervous signs associated with a change in mental state (including behaviour change or temperament change), sensation (including hypersensitive or hyposensitive to sound or touch, head pressing and itching) or posture and movement (including circling, ataxia and abnormal head carriage). Providing such cattle are 30 months old or older, and sheep are 18 months old or older, they are eligible for the program. A private veterinarian or state government veterinary officer should be contacted where sheep (18 months of age or older) or cattle (30 months of age or older), are displaying nervous signs. A producer incentive is available and all laboratory costs are covered, which provides a great opportunity to have a thorough post mortem investigation carried out in sheep (or cattle) showing nervous signs.

The information obtained from these eligible cases is recorded in a national database and used to present quantitative evidence to Australia's international trading partners to demonstrate our freedom from transmissible spongiform encephalopathy.

Do not waste eligible animals with neurological disease: use them to help assure Australia's domestic and export markets.

Ruminant feed ban

The ruminant feed ban is a legislated ban on the feeding of meat and bone meal to ruminant animals. A ruminant is an animal that chews its cud, including cattle, sheep, goats, deer, alpaca, llamas and buffalo. The spread of TSEs has been associated with the feeding of infected meat and bone meal. A legislated ban on the feeding of meat and bone meal to ruminant animals is to ensure Australia continues to remain free from transmissible spongiform encephalopathies. It is very important for Australians, and for the protection of our overseas markets, that the ban on the feeding of animal materials to ruminants is strictly observed.

It is an offence to feed any animal material to a ruminant, with the exception of tallow, gelatine or milk and milk products. 'Animal material' means any material, tissue or blood taken from an animal and includes meal obtained from rendering tissues and blood from animals.

In addition, compounded feeds and meals containing animal materials – and/or the documentation, invoices, tags and labels accompanying these products – must contain a statement that the product contains animal material and must not be fed to ruminants.

CHAPTER 7

Genetic improvement

Genetic improvement has the potential to significantly improve most traits that contribute to the profitability and sustainability of any sheep enterprise. Determining the proportion of animal performance due to genetics, rather than the environment, is possible with some understanding of genetic principles.

The process of genetic improvement in a commercial flock seems simple – define the objectives and then find the animals that will achieve these objectives. This chapter aims to assist producers by outlining the process for setting breeding objectives and the genetic principles that help to identify the genetic merit of animals.

Genetic potential

Chapter 1 outlined the key profit drivers and how to identify weaknesses and strengths in the farm business. Many traits that affect farm profit can be improved through genetics, including the quantity of product (e.g. kilograms of meat or wool per head) as well as the quality, which influences value (e.g. fibre diameter and leanness of meat). Other factors that have less direct effect on profit, but can have an impact on future viability and profitability, are traits such as resistance to disease and constitution.

Wool

Merinos have been the main sheep breed in Australia, contributing to both the wool and meat industries. Changing demands over time (e.g. quantity of wool, quality of wool or meat) have led to cyclical changes in the breeding objectives of the Merino. This, coupled with the diverse climatic and geographic areas in which Merinos are run, has provided the sheep industry with a large and genetically diverse population to select from.

A number of resource and demonstration flocks have been developed over time to illustrate the potential for genetic change. One such trial was the QPLUS$ demonstration flock run at Trangie. Ten years of selection showed changes in fleece weight and fibre diameter for two objectives, as

Table 7.1. QPLUS$ – the impact of 10 years of selection on fleece weight and fibre diameter.

Objective	Change in clean fleece weight	Change in fibre diameter
Increase clean fleece weight (CFW) and maintain fibre diameter (FD)	+0.84 kg ***(+19.2%)***	–0.3 μ
Reduce fibre diameter and maintain CFW	+0.31 kg ***(+7%)***	–2.4 μ

shown in Table 7.1. A 19% increase in clean fleece weight (with some reduction in fibre diameter) was achieved after 10 years selection. Similarly selection for fibre diameter resulted in a reduction of 2.4 micron, with some increase in clean fleece weight. Selection for these traits will considerably increase returns for any flock when multiplied over large numbers of animals.

Lamb

In response to consumer demand for larger, leaner lamb, the lamb industry has made significant changes to lamb production through the selection of animals that grow faster with less fat, as illustrated in Figure 7.1. The lamb industry has had this consistent objective for some time and so the overall genetic change made by the industry is easier to measure.

Disease resistance

Selection of disease-resistant animals has the advantage of reducing occurrence of disease and reducing reliance on chemicals or other disease control measures. Improved resistance to worms leads to less reliance on chemical worm drenches, which are becoming increasingly ineffective as worms develop resistance. Flocks that have been selected for lower worm egg counts have reduced the number of worm drenches required. It has taken some 10 to 12 years of selection to significantly

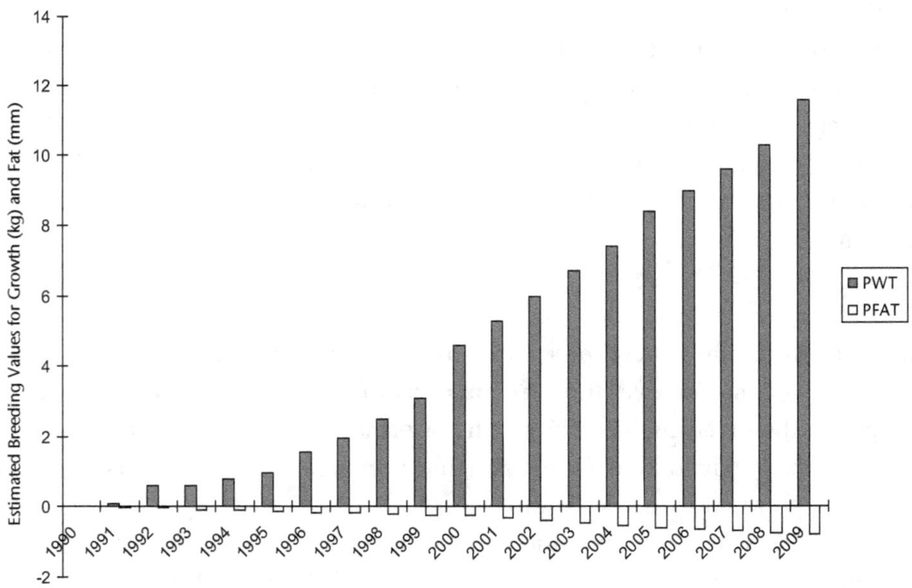

Figure 7.1. Genetic trend for growth (as post weaning weight – PWT) and fat depth (PFAT) in Poll Dorsets.

reduce the number of drenches required in a flock, but, like many outcomes of genetic selection, it takes time but is permanent.

Other traits

There are many traits that can be, and have been, selected for that can change and improve most aspects of livestock production. For example, selection can improve animals' adaptation to the environment (such as resistance to flystrike, body wrinkle and dags), the maternal ability to produce and rear lambs, wool quality (such as colour and comfort) and conformation. Temperament, eating quality and easy care are just a few of the traits that have been, and are being, examined to improve the future sustainability of the sheep industry.

Genetic progress takes time: improvement is not immediate, but is a long-term permanent investment. The key to making 'quick' progress is to be clear about the objective and then to be able to accurately select animals that will help to achieve the objective. Genetic effects are difficult to measure or to see because the impact of the environment is so great. Fortunately, systems have been developed that are capable of reliably predicting genetic merit for the sheep industry.

Genetic principles

Some understanding of genetic principles can help in identifying the potential for genetic improvement, in understanding and questioning breeding strategies, and in using genetic tools for selecting the best animals.

Most traits that can be seen and measured – for example, weaning weight or kilograms of wool – are affected both by the environment and by the animals' genetic make-up (genotype). The environmental influences include nutrition, climate and disease and are usually responsible for 70% or more of an animal's performance. The phenotype is the combined measure of genetic and environmental influences. The phenotype is what is seen or measured.

One half of an animal's genes comes from its dam, the other half from its sire. The environment usually effects how well the animals' genes are expressed (see Figure 7.2).

To illustrate these effects, Table 7.2 shows the performance of three stud groups run at two sites with different environments (differing in geography, rainfall and management). At each site, the difference in clean fleece weight and fibre diameter between each stud is due to the genetic differences between the studs as they were run together as one mob. Stud B cut 200 g more clean wool than Stud C at both sites. However, by improving the environment in which they were run, Stud B

Table 7.2. Clean fleece weight and fibre diameter for three stud groups run at two sites.

	Site 1 High stocking rate High rainfall		Site 2 Low stocking rate Low rainfall	
	Clean fleece weight	**Fibre diameter**	**Clean fleece weight**	**Fibre diameter**
Stud A	4.2 kg	19.1 μ	5.6 kg	20.2 μ
Stud B	4.3 kg	18.9 μ	5.7 kg	20.0 μ
Stud C	4.1 kg	18.8 μ	5.5 kg	19.9 μ

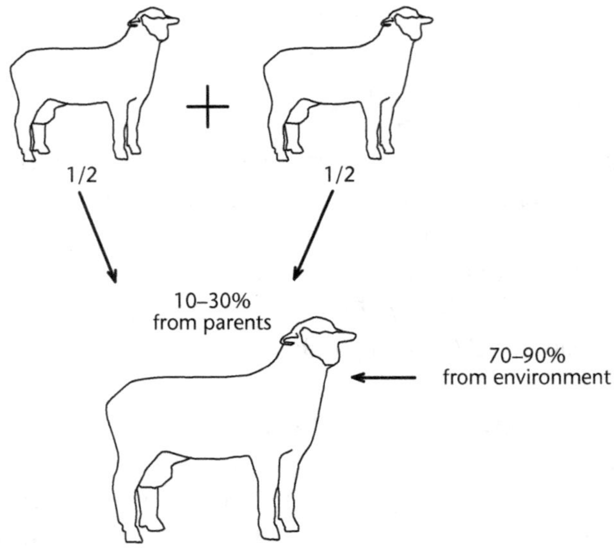

Figure 7.2. The performance of an animal is affected by genes inherited from its parents and the environment in which it is kept.

cut 1.3 kg more clean wool at site 2 than at site 1. Running one thousand animals from Stud B compared with Stud A would result in an extra 100 kg clean wool that is slightly finer.

The things that determine the potential for genetic improvement are:

- how much natural variation exists
- how much of the variation is due to genetics compared with environment (how heritable).

How quickly genetic change can be made will depend on:

- the correlations of traits
- the generation interval.

Natural population variation – how much is there to select from?

Variation or diversity is one of the greatest advantages of Australia's sheep population. Having a large amount of variation for many traits means that there is always potential to identify and select a group of 'superior' animals to breed the next generation. If there are changes in markets, or in the importance of diseases, large and diverse populations will have enough variation to supply even small numbers of animals with the required trait(s).

Even individual sheep flocks show a large amount of variation or diversity. It is quite normal to observe a range in fibre diameter of 6 to 8 μ from finest to strongest, a range in fleece weight of 2 to 3 kg between the lightest and heaviest cutting sheep, and a range in body weight between the lightest and heaviest sheep of up to 10 kg, in a mob of sheep run together. A proportion of this variation is due to the animal's genes. Therefore a large variation in performance means that there is good potential for selection of superior sheep.

Heritability – how much is genetic and how much is environment?

The heritability of a trait is a fundamental concept in genetic improvement. Heritability is how much of the variation within the population for a trait is controlled by genes, and therefore can be passed on to the next generation. Traits that are highly heritable will be easy to improve. For example, carcase weight has a moderate heritability of 20%. This means that 20% of the variation in carcase weight of animals within a population is due to genes (received from their parents) and 80% is due to the environment (such as lactation performance of the dam, other nutrition and disease).

A trait such as fertility has very low heritability of around 5 to 10%, which means that environment has a 90–95% impact on fertility and so progress through genetic selection of fertility will be slower. Although heritability of fertility is very low, there is a large natural variation across and within breeds and so there is variation to select from. It is difficult to measure the fertility trait in rams because it is the repeated lambing performance of his daughters that provides an accurate measure of fertility and it takes a long time to get this information.

Table 7.3 provides heritability estimates of some of the main selection traits for sheep. These heritability values vary between meat and wool breeds (as indicated in some instances by a range). The key message is to select for traits with a high heritability because these will show the highest rates of genetic progress. Traits are considered highly heritable if heritability is greater than 30%, moderately heritable if between 10 and 30% and lowly heritable if less than 10%.

Genetic correlations – changing some traits will affect others

Genetic correlations describe the genetic relationship between traits. If traits are strongly positively correlated, the selection for one trait will also lead to an increase in the other trait. If traits are strongly negatively correlated, then improvement for one trait will lead to a reduction in another. When developing selection criteria to meet a breeding objective, it is important to take genetic correlations into account so that selection for one or more traits does not lead to a detrimental change in others. Often the challenge in genetic improvement is to select animals that defy the natural trends; that is, to identify animals that cut a lot of wool but are fine, or that grow fast but lay down muscle not fat.

Genetic correlation is also an important consideration when evaluating new traits, such as disease resistance, to ensure that the progress made in other key production traits is not impeded.

Tables 7.4 to 7.7 show some genetic correlations between traits. To read these tables, select the trait of interest along the top row and then scroll down that column to see what effects selection for this trait will have on other traits. For example, selection for greasy fleece weight will also increase clean fleece weight, staple length and fibre diameter, but have little impact on coefficient of fibre diameter, staple strength or yield.

Genetic correlations can be extremely useful. Some traits are difficult or expensive to measure, or can only be measured in one sex. If there is a high genetic correlation between two traits, and one is easier to measure than another, then that trait can be used to change the first. For example, measuring staple strength in Merino sheep is expensive, but there is a high genetic correlation between coefficient of variation of fibre diameter (CVFD) and staple strength. CVFD comes with fibre diameter measurements at no additional cost.

Rate of genetic gain – how fast can change be made?

In principle, the rate at which genetic improvement occurs for any particular trait depends upon:

- the heritability of the trait in question – how much of the animal's superiority (or inferiority) is passed onto the next generation

Table 7.3. Heritability estimates for some of the wool, carcase, maternal and disease traits in sheep. (Source: Sheep Cooperative Research Centre)

	Trait	Heritability
Wool traits	Clean fleece weight (kg)	0.36–0.51
	Fibre diameter (μ)	0.59
	Yield (%)	0.48–0.56
	Staple strength (N/ktex)	0.34
	Staple length (mm)	0.47
	Coefficient of variation of fibre diameter (%)	0.52
	Breech cover	0.10[a]
	Breech wrinkle	0.30[a]
Growth traits	Growth to weaning (kg) Growth post weaning (kg)	0.2–0.3 0.3–0.35
Feed intake	Feed intake	0.13
	Efficiency of wool growth	0.25
Carcase traits	Carcase weight (kg)	0.20
	Fat depth (mm)	0.20
	Eye muscle depth (mm)	0.24
	Lean meat yield (%)	0.35
	Meat pH	0.18
Reproduction	Ovulation rate	0.15
	Fertility (ewes lambing/ewes joined)	0.08
	Lambs weaned per ewe joined	0.07
	Lambs born per ewe joined	0.10
	Lamb survival	0.03
	Scrotal circumference (cm)	0.21
Disease	Worm resistance (worm egg count)	0.27
	Fleece rot (incidence)	0.17

[a]limited accuracy as reported in 2008

Table 7.4. Genetic correlations for some key wool traits. (Source: Sheep CRC)

Trait	Greasy fleece weight	Clean fleece weight	Fibre diameter	Coefficient of variation of fibre diameter	Staple length	Staple strength
Clean fleece weight	0.86					
Fibre diameter	0.36	0.28				
Coefficient of variation of fibre diameter	0.09	0.19	−0.10			
Staple length	0.44	0.36	0.19	−0.06		
Staple strength	0.16	0.20	0.37	−0.52	0.05	
Yield	−0.14	0.38	0.04	−0.08	0.25	0.35

Table 7.5. Genetic correlations for some key carcase traits. (Source: Sheep CRC)

Trait	Carcase weight	Carcase fat depth
Carcase fat depth	0.39	
Carcase eye muscle depth	0.54	−0.33

Table 7.6. Genetic correlations for some key growth traits. (Source: Sheep CRC)

Trait	Birth weight	Weaning weight	Post weaning weight	Adult weight
Weaning weight	0.47			
Post weaning weight	0.29	0.85		
Adult weight	0.22	0.75	0.93	
Growth rate	0.27	0.79	0.19	0.78

Table 7.7. Genetic correlations for some key reproduction traits. (Source: Sheep CRC)

	Number lambs born (per ewe joined)	Number of lambs weaned (per ewe joined)	Weight weaned
Number lambs born (per ewe joined)		0.84	
Weight weaned	0.60	0.80	
Number of lambs born (per ewe lambing)	0.89	0.62	0.05
Fertility (as ewes lambed per ewes joined)	0.79	0.73	0.70

- the selection differential – the better the animals selected are from the average then the higher the selection differential and the higher the rate of genetic gain)
- the generation length (the **average** age of the ewes and rams when the **first** offspring are born).

If ewes are joined at 18 months and mated for five seasons, then the average age of ewes when lambs are born is the generation length for ewes. In this example, it is 4 years (2 + 3 + 4 + 5 + 6 years = 20 years ÷ 5 seasons = 4 years). Similarly, if rams are joined 18 months and used for four seasons, then the average age of rams when lambs are born is the generation length for rams and, in this example, is 3.5 years (2 + 3 + 4 + 5 years = 14 years ÷ 4 seasons = 3.5 years). The generation length is an average of ram and ewe generation lengths (4 years + 3.5 years = 7.5 years ÷ 2 = 3.8 years in this example).

$$\text{Rate of genetic gain} = \frac{(\text{Selection differential} \times \text{Heritability})}{\text{Generation length}}$$

The generation length can be reduced to increase the rate of genetic gain by either joining ewes younger or retaining fewer old ewes.

It is not desirable to select on one trait only. Often two or three major traits, and many minor traits, are selected for at the same time. However, selecting for more than one trait is likely to slow the rate of genetic progress in any one trait because the selection differential applied to any one trait is reduced.

Selection should be based on measured traits that are of the greatest importance to the profitability of the enterprise. Objective measurement will maximise the efficiency of selection and therefore result in the greatest rate of genetic gains. Visual assessment should remove undesirable traits, allowing as much selection intensity as possible to be maintained on the economically important traits. In summary, genetic progress is hastened by using animals that are superior for traits that are highly heritable and by reducing the generation length.

Genetic by environment interaction

'Genetic by environment interaction' is a term that is used when the genetic expression of a trait is affected in different ways by different environments. If this occurs, it is not possible to predict accurately how an animal will perform, relative to others, in another environment. An example of this is the incidence of fleece rot, which is only expressed if the environment is conducive to the occurrence of fleece rot. Therefore, from direct measurement, it is not possible to predict how animals will perform in regard to fleece rot in another environment. Table 7.8 shows the example of the three stud groups used previously (Table 7.2) with the incidence of fleece rot reported at each site. At the low rainfall environment (site 2), no incidence of fleece rot was recorded on any of the sheep but in a high rainfall environment (site 1), the incidence of fleece rot was recorded and differed in severity. (i.e. studs B and C had a much higher incidence than stud A).

The effects on sheep of changing their environment can be categorised into two types: scale effects and ranking effects. A **scale effect** occurs when a change in the environment produces an absolute change in a trait but the relative performance of sheep remains unchanged. For example, sheep that are far superior for growth rate compared with others at one property might only perform slightly better for growth rate at another property where the nutrition is very poor. Hence, even though the absolute growth rates are changed by shift in environment, the relative ranking of the two groups of sheep in terms of growth rate is unchanged.

A **ranking effect** occurs when a change in environment produces different changes in a trait such that the relative ranking of different animals is changed. (The best sheep for a trait in one environment would not always be the best sheep for that trait in a different environment).

Ranking effects are present for certain characteristics, such as susceptibility to fleece rot and resistance to parasites, but ranking effects are not significant for most traits. This is why, for most

Table 7.8. Measure of clean fleece weight and incidence of fleece rot reported on three stud groups represented at two environments (Hamilton and Stawell).

	Site 1 High rainfall		Site 2 Low rainfall	
Stud group	**Clean fleece weight (kg)**	**Incidence of fleece rot (percentage)**	**Clean fleece weight (kg)**	**Incidence of fleece rot (percentage)**
Stud A	4.2	24	5.6	0
Stud B	4.3	53	5.7	0
Stud C	4.1	55	5.5	0

production traits that are highly heritable, it is possible to use genetic principles to predict how animals will perform relative to each other in other environments.

When animals are introduced from another district into a commercial flock, scale and ranking effects are difficult to distinguish. Furthermore, these environmental effects are confounded by the differences in management practices and by genetic interactions between the ram and the ewe flock. Ranking effects may be observed on some characteristics (such as susceptibility to fleece rot) in conjunction with scale effects on other characteristics (such as fleece weight) and this makes it difficult to distinguish, in a farm situation, what is real 'genetic environment interaction' for some traits.

Key messages – genetic principles

The principles of genetics help to separate the gene effect from the environment effect.

Natural variation determines the potential to make genetic gain. Having a large amount of variation for many traits mean that there is always potential to identify and select a group of 'superior' animals to breed the next generation.

Heritability is how much of the variation within the population for a trait is controlled by genes and therefore can be passed on to the next generation. Select for traits with a high heritability because these will show the highest rates of genetic progress.

The **genetic correlation** between traits will help determine what traits need to be selected for, as they show which traits will automatically affect others.

Making **fast genetic gain** requires:

- being able to select animals that are far genetically superior to the current animals
- selecting for fewer rather than more traits
- selecting for traits that are heritable
- having short generation lengths (using more young animals).

Breeding objectives

Given the time it may take to see the impact and the permanence of genetic change, it is important to determine what the business and the enterprise are aiming to achieve.

A breeding objective must reflect the business plan and the overall objectives of the enterprise. It is important that:

- the desired changes will result in increased profitability
- the desired changes are achievable (for example, the traits to be selected for are heritable and there is enough variation in the population and that animals with the desired traits can be identified and used)
- selection for the desired traits does not have negative impacts on other traits (for example, will selection for large and fast-growing lambs lead to increase in birth weight and possible lambing difficulty; will selection for large and fast growing lambs lead to large adult ewes, which may lead to a decrease in the number of ewes that can be carried?)
- issues that affect the long-term sustainability of the enterprise, such as disease resistance, are considered.

Setting a clear objective will have an impact on the strategies used to achieve genetic change, such as what rams to purchase and from where. Different objectives will mean that the best rams and breeds will not be the same for everyone.

Developing a breeding objective

A breeding objective is the part of the business objective that can be changed by genetics in a cost-effective manner.

A breeding objective will be based on the view of future demand for the product (whether meat or wool) and the ability to produce the product in the environment (in regard to resistance to disease).

The questions that need to be answered are:

- What product is being produced (e.g. meat, wool, dams for meat or milk)?
- What market is it aimed at (e.g. fine wool apparel, trade lamb or store lambs for heavy export lamb)?
- Which production issues are important to the business profit and management (e.g. disease, conformation, fertility, mature size of dams)?
- Which traits needs improving (do not continue to produce finer wool if the target for fineness has been achieved; concentrate on traits that have the potential for improvement, such as wool cut, staple strength, worms or body weight)?

The product

The product will depend on personal preference, environment and limitations from labour, markets, and so on. The choice between meat, wool or both will also depend on the individual's view of the long-term trend and demand for the product. Farm profitability analyses show that producers who do the job well can be profitable in most market climates and so focusing on producing the chosen product well can be just as, or more, important than the choice of that product.

The market

There are a range of markets and specifications for both wool and meat.

The main consideration in wool production is the future value of finer wool and other factors that influence wool price. The trends in market discounts and premiums can be examined. Figure 7.3 shows the trend in price for 19 and 21 micron over a 10-year period and illustrates the change in premiums for fine wool over time.

Demand for other traits, such as staple length, strength and style also varies between the wool markets. These change over time, but generally most of the wool-quality traits are more important for the fine wool apparel markets.

Similarly for lamb: there are a number of markets based primarily on carcase weight and fat that can be identified and targeted. Figure 7.4 illustrates markets for lamb production that can be used to drive the genetic direction for lamb.

Changing demands from consumers (especially from export markets) will continue to influence breeding and production systems. There is clear evidence of demand for high standards of animal welfare and environmental management and niche markets for organic products. Genetic selection can assist in meeting these market demands through selecting animals that are resistant to some diseases or that display certain temperament traits.

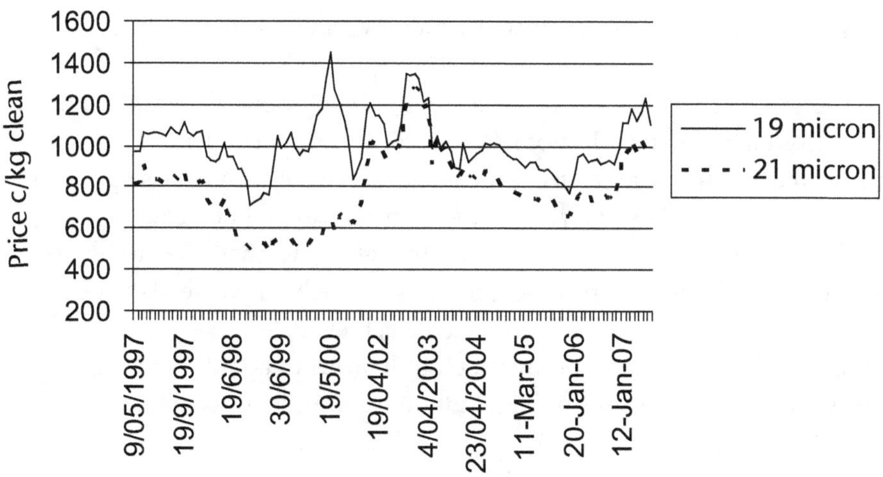

Figure 7.3. Price trend of 19 and 21 micron wool from 1997 to 2007.

Which traits should be selected?

The traits that are used for selection are known as selection criteria. Selection criteria will include traits that are major determinants of profitability. Clean fleece weight and fibre diameter are obvious examples for wool production. Fertility, growth and carcase value are examples for lamb production.

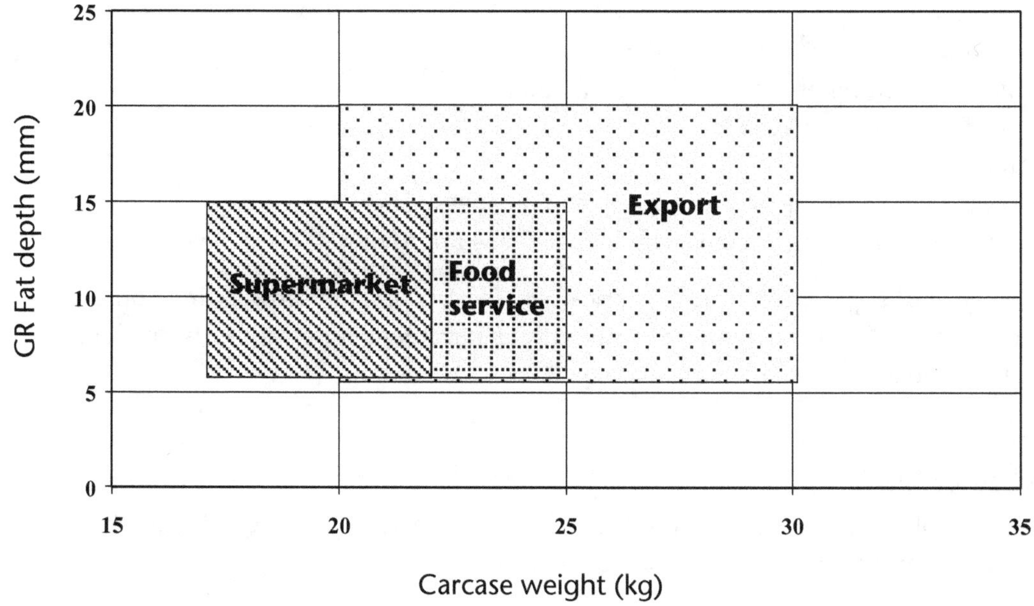

Figure 7.4. Example of market specifications for lamb.

Selection may also include traits that do not have such a direct impact on profit, but are still significant production issues. In a wool enterprise, selection criteria could include live weight, resistance to fleece rot, fibre diameter variation and lambing percentages. In a lamb enterprise and for terminal sires, selection criteria may include worm resistance, birth weight, wool and temperament.

Selection criteria should address what is currently being done well and what needs improvement. Feedback from market specifications (sale records) and farm benchmarking analyses can be used to identify weaknesses or areas for improvement. Visual characteristics, such as the need to improve conformation, may also be targeted if weaknesses are identified.

Progress will be quicker if there is a focus on the key traits that need to be improved, because the more traits selected, the slower the rate of progress in each. There needs to be a balance between what would be nice to achieve and what needs to be achieved.

There will be continual pressure on livestock industries to adapt to changing needs and markets. Forward-thinking stud breeders will include some of these traits in their breeding objective so that these genetics are available when required.

Breeding objectives will describe the direction (and amount) of change required for the traits identified. Examples of broad breeding objectives might be:

- to maintain fibre diameter (at 18 μ), to improve wool cut per head and staple strength with some increase in ewe body weight
- to produce fast-growing heavy export lambs with an emphasis on increasing muscle, reducing fat, but not increasing birth weight.

Once the breeding objective has been set and key selection traits identified, the next step is to find the appropriate animals. This may include selecting the breed, the ram source (or bloodline) and then individual sires. Sire, rather than ewe, selection has the most significant impact on genetic change, because each ram produces far more progeny than an individual ewe. (The impact of culling, mating strategies and ewe selection is discussed towards the end of this chapter.)

Key messages – breeding objectives

In setting the breeding objective, the 'need to knows' are:

- the product being produced
- the market
- the traits that will improve profitability and business goals
- what is currently being done well and what needs improving.

Industry market projections and historical records can assist in setting direction.

Market feedback, benchmarking analysis and management issues can assist in identifying areas for improvement.

Breed selection

The selection of breed and/or ram source will be determined by the enterprise and breeding objective, the environment and ability to source good genetics.

There are many sheep breeds available in Australia that have been selectively bred for different traits for many years. The main sheep breeds in Australia have genetic variation available because of the large numbers of animals.

Sheep meat

Merinos make a significant contribution to the sheep meat industry as a result of crossbreeding with other breeds. The resultant progeny are known as either first or second cross, depending on whether the dam was pure or half-Merino.

Generally, the Merino is a slower growing animal than a first or second cross. A cross with a maternal sire (such as a Border Leicester or Coopworth) provides increased growth rate and reproduction, and terminal sires (like a Poll Dorset or White Suffolk) contribute meat traits of leanness and high growth rates. This crossbreeding system allows each breeder to concentrate on relatively few traits – for fast genetic gain – but also requires the annual purchase of replacement ewes.

Composite breeds that use a range of genetics and breeds have been developed to produce a self-replacing meat enterprise.

More recently, breeds that do not require shearing have been introduced into Australia for use in the meat industry. These may be sought for more challenging environments and where wool production does not contribute to the enterprise due to the costs involved. In a traditional meat specialist enterprise, wool has contributed significantly (25–40% over recent years) to the gross income and so this is a consideration when using alternate non-wool-producing breeds.

Dual purpose – meat and wool

Dual-purpose animals have traditionally been produced by crossing specialist wool-producing animals with a maternal breed. A dual-purpose flock aims to be self-replacing: producing both meat and wool. Examples of dual-purpose breeds are fertile Merinos either mated to a terminal sire, such as a Poll Dorset or to a Border Leicester, to produce prime lamb dams or straight Merinos that are capable of turning off high-quality lambs. Historically, Polwarth and Corriedales have been used and more recently SAMM's (South African Meat Merino) and Dohnes have been introduced.

While selecting a breed to produce prime lambs, direct comparisons of the performance of sires between sheep breeds have indicated that there is more variation for the main traits that drive profit within a breed than between a breed. The significance of this is that for sheep-meat production enterprises, breed selection is secondary to sire selection. The task is to identify and use sheep that best meet the breeding objective.

Similarly, for maternal breeds used in meat production, breed selection is less important than selecting the best sire. A large maternal progeny test trial (the Maternal Central Progeny Test), which was run between 1997 and 2005, compared 91 sires from many maternal breeds including Border Leicester, Coopworth, East Friesian and Finnsheep. This trial measured 3 years drop per sire. Overall profitability was assessed by the number of lambs reared per ewe, market specifications plus the ewe wool income. The results showed that there were large differences in maternal genetics contributing to significant profit variation, but that most of the variation was within the breeds rather than between breeds. Hence the selection of where to source rams should be based on

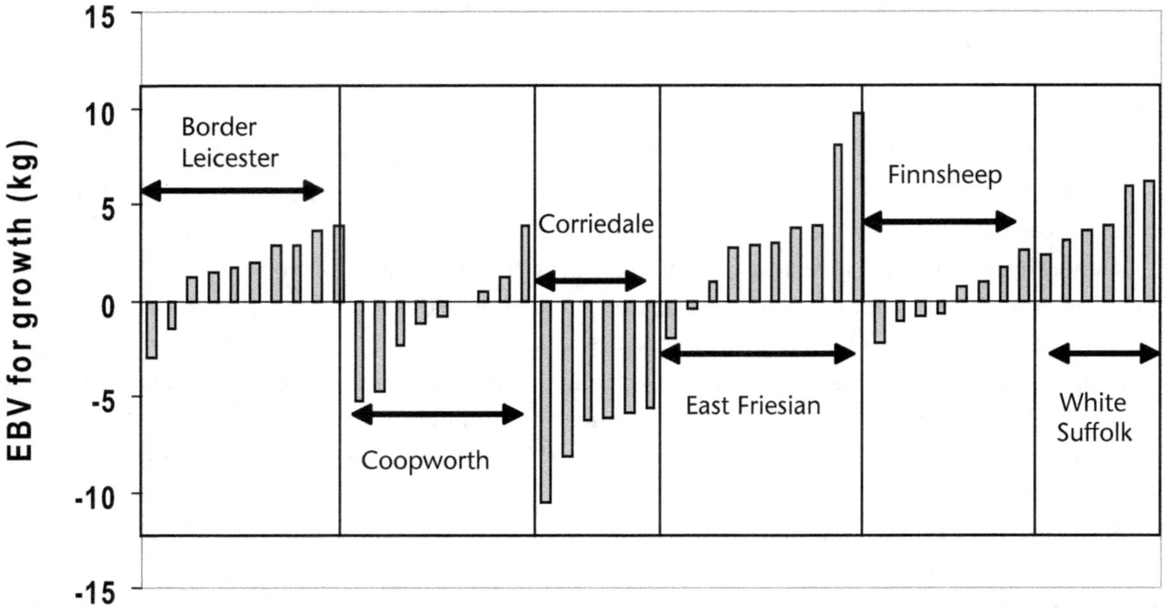

Figure 7.5. From the Maternal Central Progeny Test – highlighting the range in growth rates (measured as estimated breeding values (EBVs) for growth) for sires of the different breeds that were compared.

the stud's ability to provide better than average rams for the traits required. Figure 7.5 shows the variation in genetics for growth (measured as estimated breeding values for growth) between sires for each breed in this trial. This shows the variation for one trait only, but illustrates the overall variation in returns from sires within breeds at the time.

Some breeds will claim particular trait advantages (such as easy care, fleece shedding and hardiness) that may be attractive. The philosophy remains the same – are these traits profitable and can high-quality animals be sourced?

Wool

The Australian sheep industry is dominated by Merinos, which are traditionally selected for wool production, due to the historical importance of the wool industry. Different strains of Merino have been used to develop sheep and wool types for various markets and environments and so it is a diverse breed, providing a large pool of genetic variation. The wide diversity and range in breeding objectives within the industry means that bloodline or stud selection is still an important choice.

Crossbreeding and hybrid vigour (heterosis)

When two breeds are crossed, often the resulting progeny are better than predicted by the performance of the parent breeds. This extra performance is called heterosis or hybrid vigour and has been a feature of both crossbreeding in the cattle industry and also the crossbreeding system traditionally used in the lamb industry as described previously. The maternal heterosis expressed by the Border Leicester cross Merino ewe has been estimated to be as high as 63% extra lambs weaned per ewe mated. Similarly, the effects of heterosis on crossing strains of Merinos are estimated as:

greasy fleece weight	3.0–5.5%
hogget live weight	4.2%
reproductive performance	10%.

Key messages – breed selection

- In the sheep meat breeds, there is more variation between sires than between breeds or studs for the main production traits, and so the selection of the breed and stud should be primarily based on where the best sires can be sourced.
- Merinos have been developed to respond to environment and market differences and so there are significant differences in production traits between some studs.

Bloodline or stud selection

Where to buy rams has been a very important decision for the wool industry because of the range of breeding objectives and sheep types or strains within the Merino breed and the wide range of environments in which they are managed. Bloodline trials illustrate the enormous variation in the profitability of different bloodlines of Merinos owing particularly to variation in wool production and fibre diameter.

The primary considerations when sourcing rams should be that:

- the stud has a similar breeding objective
- the stud is making good genetic progress, relative to other studs, for the traits required to meet the breeding objective
- the stud can provide information to assist in sire selection
- the chosen sires will perform in the relevant environment.

Bloodline comparisons

Well-designed and managed bloodline comparisons provide a valuable tool for Merino breeders, when selecting productive studs. Well-run bloodlines comparisons are those where animals from different sources are managed together in the same environment. Because all of the sheep have access to the same conditions (food, climate, and so on), the differences between the bloodlines are assessed as genetic differences.

Good bloodline comparisons must:

- ensure that the sheep selected do represent the strain or bloodline
- select sheep randomly from the original flock or flocks to ensure that the selection of sheep representing each bloodline is the same
- have adequate numbers of sheep to determine whether or not the differences in the traits that are being measured are real or have occurred by chance (for example, a random selection of 15 sheep would be adequate to detect reasonable differences in fleece weight and fibre diameter, but at least 50 would be required to measure differences in reproduction).

The New South Wales Department of Primary Industries and Australian Wool Innovation have developed methods to combine the information from wether trials across the country to produce bloodline comparisons. These analyses pull together the results of many trials to evaluate genetic differences between Merino bloodlines.

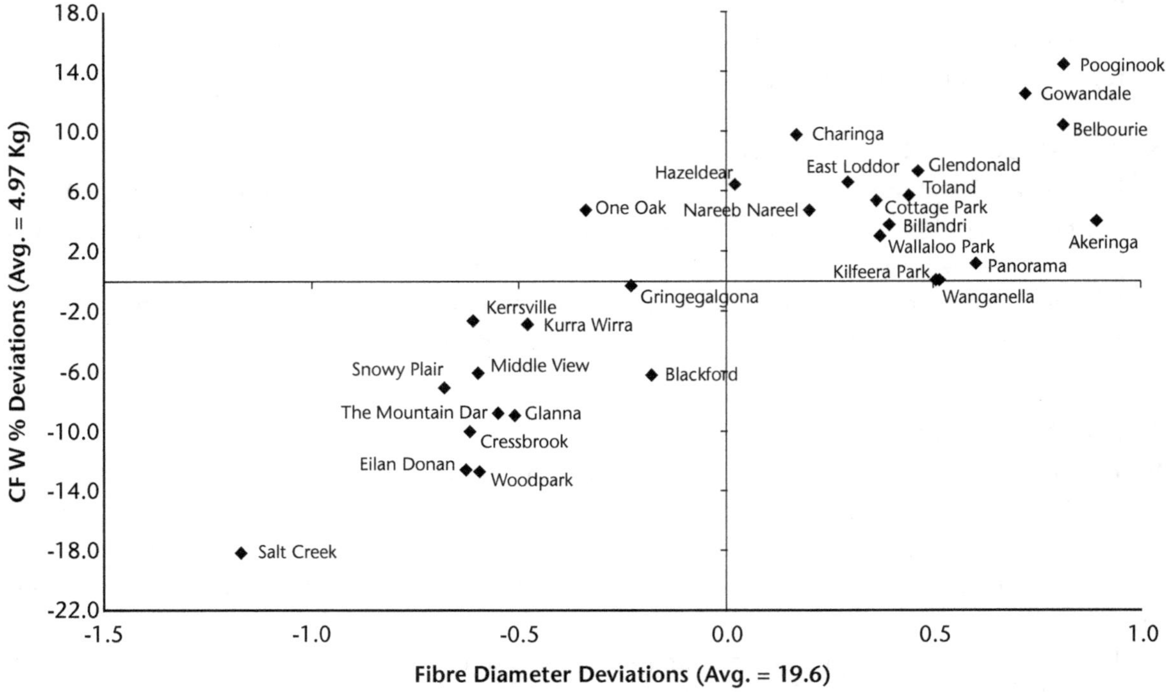

Figure 7.6. Variation in clean fleece weight and fibre diameter for 29 studs in a stud evaluation, Merinos to Match, Victoria 2002.

Figure 7.6 illustrates the ranges in performance in studs, as evaluated in a stud comparison trial run in Victoria (2004 and 2005). The trial was designed specifically to make these comparisons, but represents only a subset of the trials compiled by the New South Wales Department. The Figure illustrates the variation in performance for fleece weight and fibre diameter.

Moving from left to right across the graph shows studs that increase in fibre diameter, so all studs on the right of the central line are above average for fibre diameter in this trial. A bloodline that is to the left of the vertical line is below average in fibre diameter compared with one that is to the right. The vertical axis represents the clean fleece weight of the bloodlines in the comparison. Clean fleece weight is expressed as a percentage deviation from the average. Therefore a bloodline with a clean fleece weight of +10 is one that produced 10% more wool compared with the average. Studs in the top left-hand corner therefore are above average for fleece weight and below for fibre diameter.

Example 1 – using bloodline analysis to select the best choice

Jumbuck Station has been using Hazledean rams for many years and wishes to reduce fibre diameter (FD). Their adult flock currently measures on average about 21 micron wool. The Merinos to Match report shows that Hazledean is average for FD (FD deviation = 0). Therefore they could consider using rams from Cressbrook, Eilan Donan, Kerrsville, Middle View, Salt Creek, Snowy Plain or Wood Park because all are significantly finer (FD deviation of at least –0.6). In taking this course, there will be a likely reduction in wool cut because all the above studs have a significantly lower CFW than Hazledean (difference in CFW deviation of 5.8% or greater).

Alternatively, if Jumbuck had a key breeding objective to increase clean fleece weight (CFW) and was not concerned about a likely increase in FD, they could consider using rams from Pooginook or Gowandale because these cut more than 5.5% more clean wool per head.

But, will these studs perform at Jumbuk Station? Fleece weight and fibre diameter do not generally exhibit genetic by environment interaction, but Jumbuk Station has a high rainfall and is concerned about worms and fleece rot. Therefore, they could either look at trying a small number of rams and making some comparisons or see if there are other flocks locally that use them and make some enquiries.

There will be a significant time lag in the genetics of the sires used in these wether comparisons. In many cases, the wethers measured in these trials are from a stud's commercial clients and can reflect the genetic merit of the sheep stud 5 to 10 years previously. Therefore, if the sheep stud has made major changes to their breeding program – that is, by using new outside sires or by changing what traits they are emphasising in their selection program – this will not be reflected in this type of information.

To further illustrate the diversity in Merino studs, and to illustrate that differences in breeding objectives will change which studs perform 'best', the 29 studs in the Merinos to Match trial were compared using different breeding objectives. As an example, two breeding objectives were:

- reduce fibre diameter, maintain fleece weight body weight and staple strength for a fine wool enterprise
- increase fleece weight and body weight for a dual-purpose enterprise.

These two breeding objectives were estimated for each stud group by placing different economic values on the traits required. So a dual-purpose objective places more economic importance on body weight and fleece weight, whereas the fine wool objective places more importance on reducing fibre diameter, maintaining fleece weight and maintaining staple strength. Table 7.9 shows how four studs were estimated to be quite similar based on a simple gross margin per DSE (hence all breeding objectives are the same) compared with the estimated dual-purpose objective or a fine wool objective. These were measured as \$/DSE/year. Depending on the objective, the choice of stud would be quite different. Note that these were estimations made in 2006, representing genetics that are several years old and so cannot be interpreted as current comparative values for these studs.

Table 7.9. Gross margin (\$DSE) comparison and dual-purpose and fine wool breeding objectives (as \$/DSE/year) for four studs in Merinos to Match comparison 2006.

Stud	Gross margin (\$/DSE/year as a percentage deviation from average)	Dual-purpose objective \$/DSE/year	Fine wool objective \$/DSE/year
Belbourie	+4.2%	\$0.90	–\$0.53
Nareeb Nareeb	+6.2%	\$0.42	\$0.26
Cressbrook	+5.6%	–\$0.88	\$0.87
Snowy Plain	+6.0%	–\$0.60	\$0.56

Source: Merinos to Match 2006

On-farm comparisons

Where current or relevant stud comparisons are not available, some producers have also successfully run their own on-farm comparisons. This requires being able to source wethers (or ewes if maternal and reproductive traits are being assessed) from a good source of sheep that represent the bloodlines to be assessed. A commercial flock that has been on the targeted stud bloodlines for a number of years (more than 5 years) may be suitable. Some studs have a range of grades of rams separated on performance and price, so it may not be fair to compare sheep from a flock that uses low-grade rams to another that uses high-grade rams.

Young animals should be selected at random from the chosen flock and run with a similar number of randomly selected (of the same age) sheep from the home flock. A minimum number of 24 wethers is recommended; 50 ewes are recommended if reproduction rates are to be assessed. At least 50 wethers will be required if worm resistance is of interest (using worm egg counts). Increasing the number of animals per bloodline increases the confidence that real differences between groups of animals can be measured.

Guidelines for random selection

An example of how to get a random selection from a mob of sheep is as follows:

- All sheep in the mob should be presented for selection; if some have already been culled then this should be taken into account. For example, if 20% have been culled, then no culling should be done in the selection.
- Divide the number required into the total number of the mob and add 20%. For example, if 30 sheep are required, it is necessary to draft off 36 sheep, so divide the number in the mob by this number and then draft off at regular counts. If there are 360 sheep in the mob, then 360/36 = 10, so draft off every tenth sheep to get a selection of 36.
- A 20% cull can then be made to ensure no obvious cripples or culls are included, to leave a random selection of 30 sheep; this culling level will vary depending on personal preference. The culling level should not be much higher because the aim is the have a random representation of the original flock.

Ideally sheep should be shorn about 2 months after introduction onto the property to take out wool growth effects that have occurred elsewhere. If this is not possible, a true comparison may not be evaluated until the second shearing. At shearing, fleeces should be weighed individually and sampled for individual wool testing, either from a mid-side sample or a fleece 'core' such as made by some in-shed fleece testing machines.

Alternatively a number of sires can be purchased from each stud of interest and syndicate-mated to a random selection of ewes.

Sire groups – bloodline sources or grades

Comparing bloodlines using sires requires at least five rams per bloodline (preferably 10) and at least 100 progeny per bloodline. This will allow measurements to show differences of more than 1 micron or more than 10% in fleece weight.

Table 7.10. Number of sires per ram source required to detect varying levels (percentage) differences for three traits.

	Reproduction			Greasy Fleece weight			Fibre diameter		
Difference between progeny groups to be detected (%)	5	10	15	2	5	10	2	5	10
No. sires required per ram source	173	45	20	100	16	5	34	5	2

When groups of sires are joined as a syndicate, the number of sires required to represent each group or bloodline will depend on the level of difference in the trait that needs to be detected. Table 7.10 shows the number of sires required (assuming 20 progeny per sire) to detect varying differences (in percentages) between progeny groups for three traits. For example, if confidence is needed to detect very small differences in fleece weight, such as 2% (for example 100 g difference for fleece weights of 5kg), then 100 sires would be required to represent each group. However, if a 10% difference in fleece weight is acceptable (equivalent to 0.5 kg greasy for 5 kg cutting sheep) then only five sires are required. As with most farm trials, there needs to be a compromise between what is practical and what is accurate.

Key messages – bloodline selection

- Good stud comparisons are those where sheep have been randomly selected from flocks that truly represent the stud and have reasonable numbers of sheep per stud to ensure adequate genetic variation is measured.
- There are significant differences between some studs for most production traits and the breeding objective will determine which studs will be the most profitable.
- On-farm comparisons can be made as long as the comparisons are well designed and measured.

Ram selection

The selection and use of rams in a commercial flock is the prime driver of genetic change. This is because one ram can sire between 50 and 80 lambs every year, whereas a ewe will commonly produce only 0–2 lambs per year.

Commercial producers can either purchase rams from studs that can provide the traits required or breed their own rams.

Sires with the highest genetic merit should only be selected if they meet criteria for conformation and health. A useful checklist should include:

- acceptable conformation
- correct and healthy mouth, feet and testicles
- appropriate animal health certification to reduce the risk of importing disease.

This check list applies to all breeds of sheep and may include other key requirements for an animal to be considered.

Estimated breeding values have been developed as a means of showing the genetic merit of an animal independently of the affect of the environment. These values are available for cattle, sheep and goats from studs that participate in performance recording. They are the best tools available for helping to assist in selecting animals on their genetic merit and so it is worthwhile trying to purchase rams from studs that do provide this information (see the section below on 'Estimated breeding values').

If estimated breeding values are not available, objective (measured) data on the important traits is useful, providing that all rams were run together and therefore that the figures and their visually important traits can be compared.

Using raw measurements

Interpreting measurements on sale rams can be difficult if they have been taken at different times, at different ages and even different measurements used for the same traits.

Although average fibre diameter, the spread or variation of fibre diameter and fleece weight are highly heritable, they are also very much affected by nutrition, age and where and when samples were taken for testing. If actual or raw wool measurements are provided on rams for sale, these things need to be taken into account.

Fibre diameter

The fibre diameter of a ram available for sale may not be a good indicator of what he will breed, or even of the average micron of his fleece at shearing. Too often complaints are heard that fibre diameter provided at sale differed from fibre diameter when taken home.

Effect of age

Average fibre diameter increases with age; by how much will differ for ewes and rams. If the nutrition and environment are constant, the fibre diameter generally increases until approximately 2 to 2.5 years of age, at which point it will stabilise (not all animals do this, but this is the general trend). This general trend is illustrated in Figure 7.7.

Effect of nutrition

The quantity and the nutritional value of the pasture or supplementary feed available to the ram at certain times of the year will affect the micron (fibre diameter) of the wool at that time. The micron of the wool will be usually be high in spring when the quantity and quality of the pasture is high.

Figures 7.8 and 7.9 illustrate realistic changes in fibre diameter due to nutrition and hence the impact of taking wool samples for testing at different times of the year.

Hence, the actual measurement of an animal can vary enormously; in this example, the test could have been 20 or 25 micron, but the ram will still breed the same.

Fleece weight

Fleece weight will vary depending on:

- age at shearing
- birth and rearing status (for example, twin or single)
- whether shorn as a lamb or not
- number of months of wool growth
- feed and climatic conditions during wool growth.

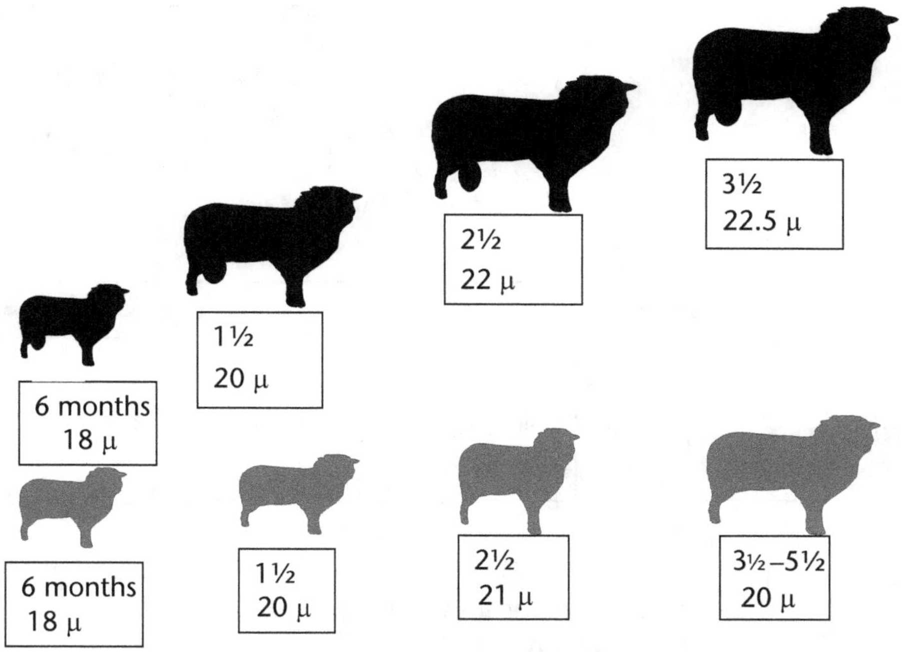

Figure 7.7. Example of the effect of age on rams and ewes on fibre diameter, flock average 20 μ.

To be able to usefully compare fleece weights from animals that have been treated the same way, fleece weights should be taken when animals are older than 10 months, and have at least 6 months wool growth. This approach to comparing fleece weights ensures that some of the early age effects are reduced and provides a good relative ranking of wool weight performance.

Repeatability

The **repeatability** of a trait is the likelihood that a mob of animals that have been managed together will retain their relative rankings of a trait over their lifetime. Generally, the animal with the lowest

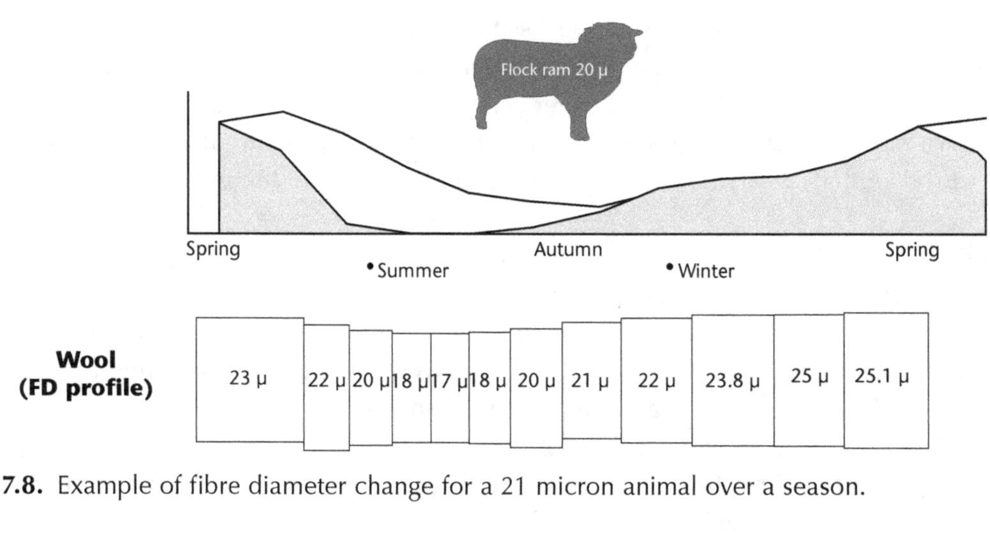

Figure 7.8. Example of fibre diameter change for a 21 micron animal over a season.

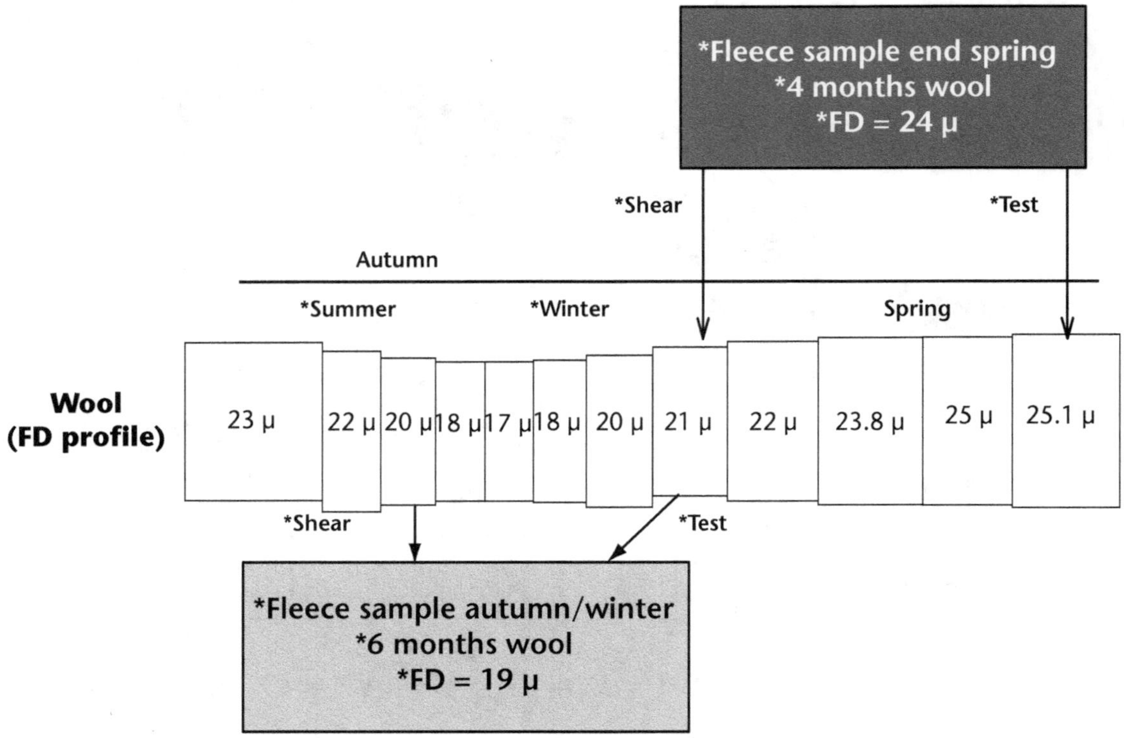

Figure 7.9. Impact of shearing or wool sampling at different times on the average fibre diameter.

body weight at 10 months of age will be the animal with the lowest body weight at 5 years of age. Similarly, for fibre diameter in Merino sheep, although the absolute fibre diameter changes for the mob and the individual over time, if all of the animals are managed together over their lifetime, the relative rankings remain the same. Although individual animals may change their fibre diameter rankings, as a general rule, the top 10% remain the top 10% and the bottom 10% remain the bottom 10%. The same is true for body and fleece weights. In other words, in most cases, the best animals will stay the best and the worst the worst. Table 7.11 shows the results of fibre diameter measurements on ewes from the QPLUS$ selection research flocks at Trangie from the fine and broad wool bloodlines over 5 years. Each group represents a sub-group within the bloodline in which ewes were placed based on the first measurement. Fibre diameter performance was very repeatable across groups (that is the relative rankings between groups and bloodlines stayed constant), illustrating that measurements made on animals when they are young are a good indicator of lifetime performance.

Micron blow-out

Micron blow-out is when some sheep grow broader micron wool as they get older, compared with others. It is a problem for breeders who are concerned about this trait because sheep have to be retained and measured for a long time. Although individuals do vary, bloodline trials that have measured sheep for up to 4 years of age have generally shown that groups or bloodline groups tend to be consistent, as illustrated in Table 7.11.

The examples below show how information on three rams can be used and improved to make valuable selection decisions on how they will breed.

Table 7.11. QPLUS$ fine and medium wool bloodlines divided on fibre diameter groups and measured over 5 years.

Group	1994	1995	1996	1997	1998
Fine wool bloodline					
A	17.7 μ	19.1 μ	19.0 μ	19.0 μ	19.3 μ
B	18.4 μ	19.7 μ	19.6 μ	19.4 μ	19.5 μ
C	19.0 μ	20.5 μ	20.6 μ	20.6 μ	20.9 μ
D	19.7 μ	21.1 μ	21.0 μ	21.0 μ	21.3 μ
E	20.6 μ	21.6 μ	21.6 μ	21.3 μ	22.2 μ
Broad wool bloodline					
A	23.1 μ	25.3 μ	23.9 μ	25.9 μ	26.9 μ
B	23.2 μ	25.7 μ	25.6 μ	25.7 μ	27.0 μ
C	22.6 μ	24.5 μ	25.1 μ	24.9 μ	25.2 μ
D	22.3 μ	24.2 μ	25.2 μ	24.9 μ	26.4 μ
E	21.8 μ	23.3 μ	24.6 μ	24.3 μ	25.7 μ

Example 1a – Three sires are being considered for purchase for a self-replacing Merino flock. Body weight is important for fertility and sale of wether lambs. All rams have acceptable conformation and testicles. The following information is available:

Ram A	Ram B	Ram C
Fleece weight – 8 kg	Fleece weight – 7.6 kg	Fleece weight – 7.8 kg
Fibre diameter – 19 μ	Fibre diameter – 19.2 μ	Fibre diameter – 18.6 μ
Body weight – 90 kg	Body weight – 95 kg	Body weight – 85 kg

The measurements relating to the three rams in Example 1a provides no information on how these rams will breed unless they are known to be all run and managed together and the measurements taken at the same time. If this is so, the information tells the buyer that Ram C is the finest, Ram A cuts the most and Ram B weighs the most. It still does not help decide if any of them will breed finer, cut heavier or produce bigger animals for the buyer.

If these animals were at a central sale and came from different studs then they would have been managed differently, received different treatment and/or may differ in age. The genetic differences between them will be masked by the large differences they have experienced in the way they have been managed. In this scenario, it would be impossible to predict the relative performance expected in the progeny of these rams, based on the ram's measurements alone.

Deviations from average

The actual performance of rams from a stud are not very useful to anyone apart from the stud owner because the actual performance figure will depend upon how the animals are managed, the environment in which they live, how much food they have available and whether or not they have been healthy. In most cases, the relative performance between the progeny of two rams for a particular trait remains the same, so deviations from the average can be a better indication of how an animal will perform, relative to the average. For example, a ram that is reported to cut 10kg greasy wool is unlikely to cut this exact amount on a commercial property, but, knowing that this animal cut 10% more than the average is a more useful indicator of the likely performance of his progeny.

Fleece weight may be expressed as a percentage, rather than as actual deviations. The greasy fleece weight percentage gives the superiority or inferiority for that trait by that animal and provides a better opportunity to predict the relative differences in progeny performance between two rams.

Example 1b – The ram buyer now has more information. Firstly, he knows that at home his wethers cut 5kg of 18µm wool, so is unsure whether these rams will in fact make the progeny broader in fibre diameter. However, the stud animals are fed well and he has been on the bloodline for a while. The stud breeder has tested the entire drop and has provided deviations from the average for these measurements.

Ram A	Ram B	Ram C
Fleece weight +6%	Fleece weight +1%	Fleece weight + 4%
Fibre diameter 0 µ	Fibre diameter +0.2 µ	Fibre diameter –0.4 µ
Body weight +10 kg	Body weight +15 kg	Body weight +5 kg

Now the ram buyer knows that, Ram A cuts 6% more clean wool than the average for the stud ram drop, has the flock average fibre diameter and is 10 kg heavier than the flock average. Now it is easier to assess that if he has been using average rams from this stud then there may be a good chance that Rams A and C will not breed broader micron rams at home, despite being broader themselves, because both rams are average or below average for fibre diameter for that drop.

Maternal affects have an influence on some traits. For example, weaning weight is highly influenced by the mother's milking ability, whether the animal was born early or late in the drop and whether the animal was a twin or a single. Therefore, the repeatability of body weight, based on weaning weight is not high. Similarly, an animal's fleece weight is highly affected by its maternal nutrition, so the ideal time to measure fleece weight is after an animal is 10 months of age or older, was shorn as a weaner and has more than 6 months wool growth. Measurements of fleece weight that meet these requirements are more reflective of the relative lifetime performance compared with those that do not.

Estimated breeding values

The best information available to determine the genetic merit or superiority of animals is provided as **estimated breeding values (EBVs)**. EBVs take into account the principles of genetics using the measured performance of an animal and of its relations, plus the heritability of each trait, to make the best estimate of how an animal is likely to breed (see Figure 7.10).

Studs that provide EBVs are required to run the animals under the same management and/or record different management groups so that these can be accounted for. In this way, an EBV measures only that amount of the trait that the animal can pass on to its progeny.

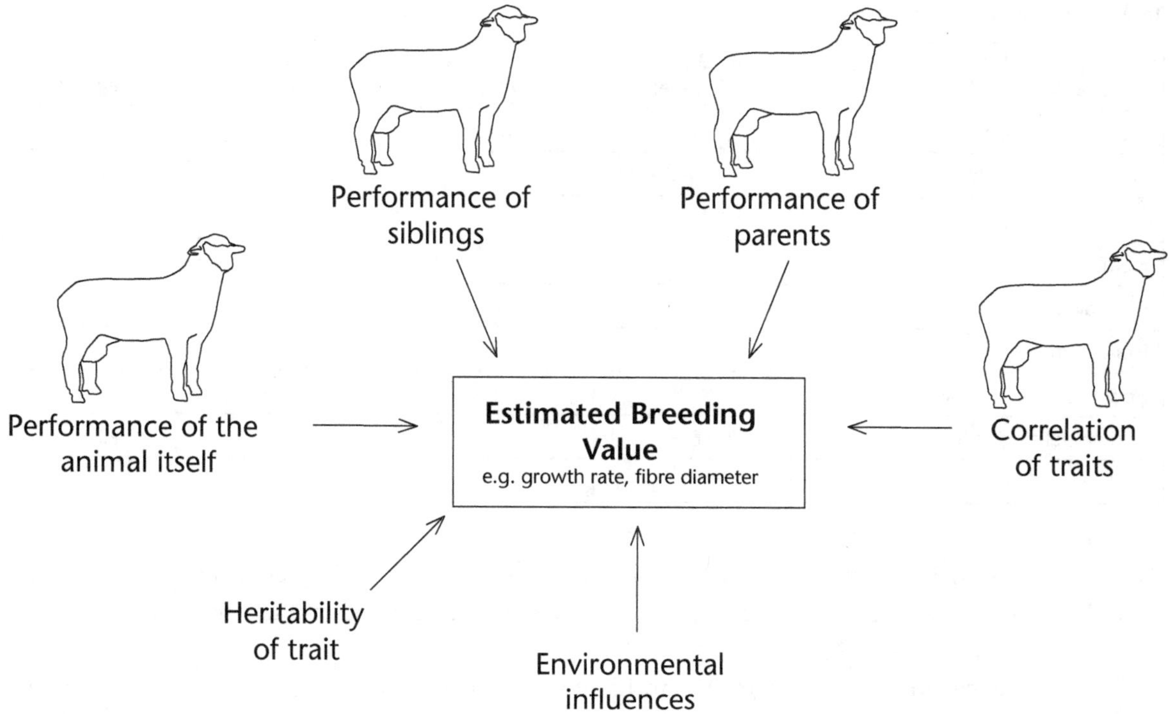

Figure 7.10. Estimating breeding values.

EBVs therefore provide the best estimate available on the likely genetic performance of an animal's progeny by taking into account:

- the maternal background (twin or single)
- the known heritability of the trait
- correlations between traits (how they affect each other)
- the performance of relations (parents, grandparents and siblings).

EBVs estimate how much better or worse one ram's progeny will be compared with another, so are reported as deviations of an animal's performance from the average of the group.

As heritability is a measure of how much of an animal's superiority (or inferiority) will be passed onto its progeny, in its simplest form, an EBV, is calculated by multiplying the animal's deviation for a trait by the heritability for that trait.

If the ram is measured for more than one trait, the genetic correlation between these traits provides further information about the genes the ram is carrying for that trait using the other trait's measurement. This can be included to further refine the estimate of the ram's EBV.

Knowledge of the maternal environment, such as whether the ram was reared as a single or a multiple or by a maiden ewe, can be corrected to better predict the components of genetic performance.

Table 7.12 shows an example of adjustments (as percentages) made for sex, birth and rearing type and ewe on fleece weight and fibre diameter used previously in genetic evaluation (Source: Kevin Atkins, NSW DPI).

Table 7.12. Young animals – fixed effects.

Trait name	Sex (female)	Birth type (multiple)	Rearing type (multiple)	Age of dam (maiden)
Clean fleece weight -yearling	+1%	+9%	+2%	+2%
Clean fleece weight – hogget	+1%	+5%	+1%	+1%
Clean fleece weight – adult		+3%	+1%	
Fibre diameter – yearling		–1%		+2%
Fibre diameter – hogget		–1%		+1%
Fibre diameter – adult		–1%		
Body weight – yearling	+6%	+4%	+2%	+5%
Body weight – hogget	+6%	+3%	+1%	+2%
Body weight – adult		+2%		

These differences have been demonstrated in the Lifetimewool Project where similar differences have been measured from hoggets reared from ewes under good and low nutrition, and these results are shown in Table 7.13.

The differences in the tables are similar. Lambs born as multiples or under poor nutrition are likely to cut 5 to 7% less clean fleece weight as hoggets and test 1 to 2% broader.

So, for example, in the ram buying examples above, if Ram B was reared as a twin and the Rams A and C were singles, then the raw measurements would underestimate the genetic merit of Ram B and so there should be a correction for all three traits in estimating their genetic merit. If there is good pedigree information, performance of relations can also provide information about his genes and, again, further refine or better predict the performance of his progeny.

Breeding values are available for many traits for sheep enterprises. In Australia, Sheep Genetics produces estimated breeding values for meat sheep, maternal breeds and Merino sheep. Some examples of breeding values that are available are provided in Table 7.14. These values are continually refined with research and added to as the industry requires new traits.

EBVs can be provided for a flock with no comparison with other flocks. In Australia, these are called **flock breeding values.** Good linkages across flocks and breeds can provide EBV that compare flocks (and even breeds in some cases) and these are known as **Australian sheep breeding values.**

Flock breeding values

Within-flock breeding values are estimates of the relative performance of animals within that flock. They will provide a better guide to the most productive genetics available in that stud, compared

Table 7.13. Hogget wool production from low and high nutrition in mid pregnancy to weaning.

	Clean fleece weight	Fibre diameter
High nutrition (ewes in good condition and plenty of feed)	3 kg	16.8 μ
Ordinary nutrition (lean ewes and low feed)	2.8 kg (–6%)	17.1 μ (+2%)

Table 7.14. Examples of some estimated breeding values for wool, meat, carcase and disease traits. Most traits are available for different ages (for example, a weaning or yearling age).

Trait	Trait	Estimated breeding value
Growth	Early growth	Weaning weight (Wwt)
	Later growth	Post weaning weight (PWt)
Wool	Wool cut per head (greasy and clean)	Clean fleece weight (CFW), Greasy fleece weight (GFW)
	Fibre diameter	Fibre diameter (FD)
	Coefficient of variation of fibre diameter (CVFD)	Coefficient of variation of fibre diameter (CVFD)
	Staple strength	Staple strength (SSTR)
Maternal	Milk or maternal environment	Maternal weaning weight (Mwwt)
	Fertility	Number of lambs weaned (NLW)
Carcase traits	Fat depth (C site)	Fat (FAT)
	Eye muscle depth	Eye muscle depth (EMD)
Disease	Worm egg counts Breech strike	WEC at different ages Breech wrinkle Breech cover

with individual animal measurements, because they take environmental differences into account. Flock breeding values do not provide any guidance on how those animals compare with others of the same breed outside the flock.

Example 1c – The ram buyer now has within-flock breeding values for these three rams. Knowledge of the parentage, heritability and correlations has enabled a measure of the genetic performance and some of the relativities between the rams have changed:

Flock breeding values

Ram A	Ram B	Ram C
Fleece weight +3%	Fleece weight +8%	Fleece weight + 2%
Fibre diameter +0.2 μ	Fibre diameter 0 μ	Fibre diameter –0.2 μ
Post weaning weight +4 kg	Post weaning weight +8 kg	Post weaning weight +3 kg

Flock breeding values provide better predictive value of each ram's genes and, in this example, Ram B has the genes that will produce heavier cutting progeny than the other two rams, and will be bigger at hogget age (post weaning).

For a buyer, this information can be used to estimate how these rams will perform at home, compared with the average of the ram drop. Rams supply half their genes to their progeny, so only half the breeding value can be used to estimate progeny performance.

- Ram A will have progeny that cut 1.5% more wool, that is 0.1 μ broader and hoggets will be 2 kg heavier.

- Ram B will have progeny that cut 4% more wool, that is average for fibre diameter and hoggets will be 4 kg heavier.
- Ram C will have progeny that cut 1% more wool, that is 0.1 μ finer and hoggets will be 1.5 kg heavier.

What is still not known is whether these rams are better or worse than others available from other studs. If ram buyers do not have good knowledge of how the stud performs at home, then there is no information to assist in knowing what the 'average' is to compare with.

Across-flock breeding values – Australian sheep breeding values

Across-flock breeding values are available as Australian sheep breeding values (ASBVs). Where these are available, animals are being compared with all animals in flocks that were measured, either of the same breeds or even across breeds. For studs to provide ASBVs, they must have good linkage across those flocks.

Many of Australia's ram-breeding flocks have used genetics from one or more flocks or had their genetics used in other people's flocks. Sheep flocks that have common genetics are known to be 'genetically linked'.

When a ram or rams are used in several flocks, these animals have the same genetics and so the relative performance of these animals to others in the flock can be used to compare animals across flocks. The principles of calculating across-flock breeding values are explained in Figure 7.11.

Figure 7.11 illustrates the basis of developing across-flock estimated breeding values, which allow ram breeders to compare the value of animals between different flocks.

Example 1d – The ram buyer now has across-flock breeding values for these three rams. This now provides a measure of how these rams will breed for the three traits compared with all tested animals within the breed.

Across-flock breeding values as Australian sheep breeding values

Ram A	Ram B	Ram C
Fleece weight +9%	Fleece weight +12%	Fleece weight + 4%
Fibre diameter +0.1 μ	Fibre diameter –0.1 μ	Fibre diameter –0.6 μ
Post weaning weight +6 kg	Post weaning weight +7 kg	Post weaning weight +5 kg

In example 1d, the provision of across-flock breeding values provides the buyer with confidence about how these animals will perform compared with many other animals from other flocks. ASBVs from other flocks will provide the buyer with knowledge as to which rams from which stud are better and so, in many cases, the choice of stud is less important than whether the stud has rams that will perform well for the traits required. In the examples provided above, the provision of ASBVs for the three rams would mean that it would not matter if the three rams were from the same stud or different studs. Whether they are above average of all rams tested requires further information because the average may not be zero and will change each year as the breed improves for the trait.

Fred's Flock
Average Body Weight = 40kg

John's Flock
Average Body Weight = 50kg

In the above diagram, each ram represents a group of progeny sired by different rams. Sam has been used as a sire in both flocks. All the other rams are different sires that have been used on each of the properties.

If we measure the body weight of all the progeny of each of these sires in each of the flocks, and make the assumption that each of the sires has been joined to a group of ewes of similar genetic merit of the sires and rams born on each of the two properties.

Body weight of the progeny on each of the two properties

Sam = 45kg
A = 35kg
B = 43kg
C = 37kg
Average = 40kg

Sam = 45kg
D = 55kg
E = 52kg
F = 48kg
Average = 50kg

Sam's genes are the same on both properties, let's compare the difference in progeny performance for the sires on each property to Sam

Sam = 0kg
A = -10kg
B = -2kg
C = -8kg

Sam = 0kg
D = +10kg
E = +7kg
F = +3kg

D = +10kg
E = +7kg
F = +3kg
Sam = 0
A = -10kg
B = -2kg
C = -8kg

We can now compare the sire's performance across the two properties

Figure 7.11. How genetic linkage works.

As animals improve, the average of animals tested will improve compared with when measurements started. The average for each trait can be sourced as well as information on the variation available for any trait measured as an ASBV. This information is provided as deciles. The 50% decile for any trait means that 50% of all animals measured had this breeding value (i.e. the average breeding value is the 50% decile). Animals that have breeding values at or above the 10% decile are in the top 10% of all measured animals for this trait.

Examples of deciles in June 2008 were:

For terminal sires, the 50% decile for post weaning weight ASBV was +9.5 kg.

This illustrates how much the terminal sires have improved because the average in 2008 was 9.5 kg heavier than when first measured. This also means that rams that were better than average for post weaning weight in June 2008 would have had an ASBV greater than +9.5 kg.

To expand on the examples 1a to 1d above for Merinos:

The 50% decile (2008) for clean fleece weight (as yearlings) was 4.7%

(So Rams A and B were well above average. Ram A at +9% would have been in the top 35% of rams and Ram B at 12% would have been in the top 25%)

The 50% decile for fibre diameter (as yearlings) was –1.4 μ

(So all rams were broader than average of all rams tested)

The 50% decile for yearling weaning weight was 1.4 kg.

All the rams in the sale example were well above average for yearling weight and all in the top 15% or better for this trait. These rams would probably not be a choice for a buyer that was aiming to reduce fibre diameter. However, different rams will be best for different producers, depending on their particular breeding objectives. Given most breeding objectives should, and do, include selection for a number of traits, it can be difficult to identify the best animal when comparing more than two traits.

Selection indexes

The selection of sires requires a balanced selection of a number of traits, some of which will be more important than others. It can be difficult to compare rams for three or more traits and to get the right balance to achieve the right product.

Selection indexes have been developed to reflect different breeding objectives by combining a number of traits into one number. These indexes provide a single value to rank animals on the basis of the relative importance of each trait. For example, a balance of fat, muscle and growth will ensure animals are not too lean or too 'muscular', which can impact on eating quality and bone development.

The key to understanding selection indexes is to identify what traits are included in the index and how much emphasis is placed on each of these traits within the index. Their aim is to make the maximum genetic gains for the particular production system. The breeding or selection index should reflect the breeding objective of the stud.

A selection index can help select for a number of traits, without compromising the overall genetic progress. It allows for differences between animals with extreme performance in one trait, with mediocre or low performance in another, by aggregating the overall performance into a score or 'profitability' value. It does this by placing higher emphasis on the most important traits and less on others. The weightings, or emphasis, are done using economic weightings called ***relative***

economic weightings. An economic value is therefore placed on each trait so that most important traits will be worth more money and so have a higher value.

Example 2 – calculating an index on fibre diameter and fleece weight

In this example, assume 1 kg of clean fleece wool is worth \$10, and that reducing fibre diameter by 1 μ will result in an increase in wool price of \$2/kg.

Our sheep cut 4 kg CFW. Therefore, the average sheep produces \$40 worth of wool.

If CFW is increased by 1 kg, then the total fleece value increases by \$10 = \$50 total.

If fibre diameter is reduced by 1 μ, then the fleece value will increase by \$8 (i.e. 4 kg × an extra \$2 per kg).

In this example, fibre diameter is worth –\$8, and fleece weight is worth +\$10. These figures are known as relative economic values.

Values for all the traits in an objective can be calculated, including reproduction, body weight, carcase weight, fertility, staple strength and other traits to be improved.

Index values also take into account the heritability and correlations between traits. This extra information is used to help more accurately estimate the breeding potential of a ram.

In this simple situation, the index that combines fleece weight and fibre diameter is simply:

$$10 \times \text{CFW EBV} - 8 \times \text{FD EBV}$$

Selection indexes and individual breeding values

Selection indexes should be used in combination with individual breeding values. This is because an animal may have a high index because it performs well for the most important traits, but is actually poor for some trait that is required. This is most important for individual sire selection (rather than teams as described below). For example, if high lamb birth weight is becoming an issue for a breeder, a selection of animals with the required indexes is listed, but any ram with an individual breeding value for high birth weight lambs may be rejected. This is often referred to as using 'gates', rather like using a drafting gate to draft off animals from the top 'pen'.

Selecting ram teams

In most commercial flocks, rams are joined as a group of rams, not mated as single sires. Single sire mating is important for studs to identify the performance of different sires and provide ram buyers with good information using the pedigree of each animal.

Single sire mating is generally not practical or worthwhile in a commercial flock (unless breeding rams) so rams are purchased to form part of a team.

Individual animals may have high indexes because they have high performance for one trait, which overrides poor performance in another trait. Most commercial producers purchase and join groups of rams and so this is usually not a concern. Table 7.15 provides an example of seven rams; all rams are above average for the selected index – in this case, the Border \$ index. (The Border \$ index is used for maternal sire selection and includes growth, fat, eye muscle depth, fertility, milk and fleece weight.) Four traits (as across-flock sheep breeding values) are provided for each ram and, as can be seen by the average, these rams will lead to an overall improvement in all traits if joined as a group (assuming all rams join similar numbers of ewes). For a stud producer wishing to

Table 7.15. ASBVS for post weaning weight (Pwt), maternal weaning weight (MWwt), eye muscle depth (EMD) and fertility, and number of lambs weaned (NLW).

Ram	MWwt (kg)	Pwt (kg)	EMD (mm)	NLW (%)	Border $ index
1	–0.2	7	–0.6	6	115
2	0	5.5	0	5	112
3	1.5	7.5	-0.9	4	117
4	0	5.2	0.9	2	113
5	0.3	7.7	0.2	2	117
6.	0	3.4	0	7	111
7.	–0.5	4.7	0.7	4	112
Average rams purchased	0.2	5.9	0	4.3	114
Average all rams tested in breed	–0.1	3.7	0	3	109

improve maternal weaning weight, rams 1 and 7 will not do this if joined singly and, if the emphasis is on eye muscle depth, the stud breeder would not use sire 1 or 3.

If single sire mating, breeders should also look at single trait performance as well as the index that has been selected to reflect their breeding objective.

Progeny testing

Progeny testing is measuring the performance of an animal's progeny to assess his genetic merit, rather than a measurement of the animal itself, if EBVs are not available. The performance of 20 progeny provides a far more accurate assessment of a ram's breeding value than his own performance and has most value when breeding values are not available on the ram of interest. Progeny testing can be used to compare individual sires or groups of sires to enable comparisons of grades of rams from one stud, or to compare studs or bloodlines. Below are guidelines for progeny testing to reduce the environmental and ewe differences, the accuracy required and the numbers of progeny required for the traits to be examined, to improve the value of the information collected.

Guidelines for progeny testing are as follows:

Reduce ewe differences: each ram should be joined to 35 to 50 ewes and it is important that the ewe groups are as similar as possible for each joining group (for example, do not join different sires to different age groups). A random draft of one or a combined mob of ewes is better. Joining groups need to be identified in some way for at least 7 months.

Reduce environment effects: limit the difference in paddock condition for joining and lambing to avoid differences in nutrition (and shelter) that may affect factors such as ovulation rate, lamb survival and milk production. Ewe groups need only be separated for the few weeks of joining and then should all be run together and separated again at lambing. Once lambs are permanently identified to their sire group, then all ewes and lambs should be run together. All weaners should be run together from weaning until shearing.

Shearing: the lambs should have their lamb tip removed before comparing fleece weights to remove the early maternal affects. Fleece comparisons can be made at the next shearing, providing this represents at least 6 months of wool growth.

Table 7.16. Accuracy levels for reproduction, greasy fleece weight and fibre diameter for different numbers of progeny for single sire comparisons.

	Reproduction	Greasy fleece weight	Fibre diameter
No. of progeny		Accuracy	
10	0.45	0.72	0.77
15	0.53	0.79	0.83
20	0.58	0.83	0.86
25	0.63	0.86	0.88
30	0.66	0.88	0.90
40	0.71	0.90	0.82

Minimise the spread of lambing: have ewes in good condition at joining and join for 5 weeks.

The accuracy of estimating the breeding value of individual rams from a progeny test must be reasonable for valid comparisons. In this context, accuracy means the chance that the same result would be achieved if the comparison was repeated.

In general, accuracy values of lower than 0.6 would be regarded as poor, 0.7–0.8 as acceptable and values greater than 0.8 as good to highly accurate. The accuracy will depend on how many progeny are tested per sire and the heritability of the trait. The more progeny, the higher the accuracy, so the usual aim is to balance what is practical with acceptable accuracy. Traits that are low in heritability (such as reproduction) will require more animals (because there is greater likelihood that there will differences between groups of animals that are due to chance and natural variability than due to genetics).

Table 7.16 provides the numbers of animals and likely accuracy for three traits.

How much to pay for rams?

The value of a ram is dependent on the number of progeny that he has and their subsequent value; that is, the value of lambs for sale or of wool cut per head. In deciding how much more to pay for a 'superior' sire, the extra value produced by his progeny should therefore be taken to account.

Meat rams

The number of progeny from meat sires will depend on the number of ewes mated, weaning percentages and the number of years joined.

For example, if a ram is joined to 50 ewes for 4 years, with an average annual weaning percentage of 85% and 95% of weaners sold, then the ram will sire 164 saleable lambs. If ASBVs are provided then a straightforward estimate can be made.

Example

Ram A	Ram B
ASBV for post weaning weight = +8 kg	ASBV for post weaning weight = +14 kg.

The difference in breeding values for weight between Ram A and B is +6 kg (14 – 8 kg), so progeny from Ram B will weigh 3 kg more than those from Ram A (because progeny will get half of the genes from the sire). At slaughter, assuming a 47% carcass dressing percentage, the progeny of Ram B would produce 1.4 kg (3 kg × 0.47), more carcase weight over 164 lambs; this is equivalent to 213 kg. Prices vary greatly, but in a market where carcase weight is $3/kg, Ram B would return $639 (213 kg × $3) more than Ram A, and so this gives a guide to estimating his extra worth.

Merino sheep bred for wool production

The value of a ram in Merino wool production also depends on his number of progeny. However, the value of a Merino ram is more complicated because each progeny that remains in the flock leaves behind fleeces, and the number of fleeces that a ram influences depends on whether or not, and how long, wethers remain in the flock.

The example below assumes that:

- a ram is mated for 4 years to 50 ewes
- the weaning percentage is 75%
- 80% of the progeny are kept and 20% are culled at one year of age, providing one fleece before sale
- wethers are kept for 2 years and ewes are kept for 5 years.

Example

Number of lambs weaned per year = 38 lambs
(joined to 50 ewes to give 75% lambs weaned – 50 × 0.75)
Lambs kept in the flock (per year) = 30 lambs
(80% kept, so 38 lambs weaned × 0.8 = 30 lambs kept and 8 sold)
Total lambs kept from the ram if ram used for 4 years = 120 lambs
(30 lambs × 4 years = 120 lambs total)
Assume half the lambs are ewes and half are wethers, so 60 ewes and 60 wethers
Total number of ewe fleeces (kept 5 years) = 300 fleeces
(60 ewes kept for 5 years = 300 fleeces)
Total number of wether fleeces (kept 2 years) = 120 fleeces
Therefore, if the ram is used for 4 years, his legacy is = 420 fleeces
(300 ewe fleeces and 120 wether fleeces)

Ram A	Ram B
•Clean fleece weight = +12%	Clean fleece weight = +4%
•Progeny get half = +6%	Progeny get half = +2%
(300 g for 5 kg fleece)	(100 g for 5 kg fleece kg)

Ram A will have progeny that will cut 200 g more than Ram B's progeny, which equates to 84 kg more clean wool from 420 fleeces. At $7/kg clean fleece weight: Ram A has returned $588 more than Ram B.

At $10/kg clean fleece weight: Ram A has returned $840 more than Ram B. The same estimation can be done for the value of reducing fibre diameter.

Although this is a very simplistic comparison, it provides guidance for estimating how much extra should be paid for better rams. Obviously if animals are used for longer or joined to more ewes and/or have a number of valuable traits, their value will increase. Tables A4.1 to A4.4 in Appendix 4 tabulate costs for different joining rates and years kept, so the cost can be compared quite easily. The tables show how expensive a ram is if he is used for one year only.

Breeding rams

Most commercial breeders purchase rams annually and rely on the stud or studs to make good genetic progress for the traits that are relevant.

Genetic improvement within a flock enables producers to define their own objectives, control their own enterprise, make their own decisions and to make sure that the available technology is being used. The flock must be big enough to avoid inbreeding (2000 or more breeding ewes) and the manager must be capable of implementing a breeding program.

The most practical program is a ram-breeding nucleus that consists of the best 10 to 15% of breeding ewes mated with the best rams to produce replacement flock and nucleus rams. In this case, artificial insemination can be justified and is often used to rapidly boost the genetic potential of the ram-breeding flock.

The genetic improvement in the flock follows that of the nucleus, but lags a little behind because the best rams would be reserved for 1 years work in the nucleus.

Whether the decision is to breed or buy rams, the confident identification of rams that have superior genetics is essential (not how well he responded to feed or if he was a single lamb from a good ewe). An understanding the principles of genetics shows how important it is to be able to separate the environmental effects from the genetic effects and how to use the information.

Key messages – ram selection

- Estimated breeding values use genetic principles to assess the genetic merit of animals and so take the environmental effects into account.
- Across-flock breeding values use good linkage between flocks to enable comparisons of animals in different flocks.
- Selection indexes are a means of combining a number of traits into one figure to reflect the breeding objective.
- The value of rams will depend on the years used, number of ewes joined and weaning percentages multiplied by the extra value of the product he produces.
- Ensure rams have a good constitution (teeth, testicles toes, tossle and torso), good genetic credentials (good accurate information on traits required) and are well managed (disease and nutrition).

Other selection strategies

Wool traits – visual and measured

Australian Merino breeders have used a range of strategies to select sheep without the use of breeding values. Strategies have evolved around different emphasis on objective and subjective assessments of a range of traits. In 1996, a selection demonstration flock was set up in South Australia to compare the impacts of a number of selection philosophies on achieving a common breeding objective, which was to maintain or improve clean fleece weight and body weight and to improve wool quality through reducing fibre diameter.

Starting with a flock of 1200 South Australian Merino ewes as a base, ewes were drafted randomly into four mobs, each of equal genetic merit. The selection strategies were:

1. control with no selection
2. selection predominantly driven by measurement and performance recording
3. selection by professional classers driven by visual and tactile (feel) appraisal
4. selection based on philosophies of advocates of 'elite wool' or 'soft rolling skin'. (This philosophy uses indirect assessment of skin and wool quality to improve wool cut and quality.)

Each mob was managed by a committee who were advocates of the relevant selection strategy. For the first 2 years, outside rams were selected and then the flocks were closed so that the only sheep that were used in the breeding program were those that were bred within that flock. Although all sheep in the trial were measured at many stages for a variety of traits, the sheep breeders within each group were only provided with the type of information that they would normally collect. For example, the group with a philosophy of using measured performance information had access to all the measurements collected, including EBVs for each of the sheep – both rams and ewes. The group with a philosophy of professional classers had access to a lot less information compared with the others. Figures 7.12 and 7.13 illustrate the changes in fibre diameter and fleece weight for these selections. These trials showed that:

- The use of the measurement and performance recording, provided the best progress (the flock was finer) and returned more dollars in the profit analysis, despite the higher cost.
- Strategies relying heavily on visual selection, whether professional classers or using the 'elite wool method', also produced significant progress. However, a commercial producer would need confidence in both the classer and the strategy to be able to repeat this at home.
- All flocks (except the control) made improvements in wool quality, but had an increase in body wrinkle.
- Major changes can occur rapidly in flock performance through the selection of outside sires.
- Genetic progress can be significant, regardless of the method of selection, as long as the direction is clear.

Ewe selection and culling

Sheep may be culled from a flock at various ages for many different reasons. Normally, the major culling and selection of breeders is done at hogget stage. Often, to retain flock size and lambing rate, the number of ewes that can be culled does not allow for more than removing the obvious culls for a range of reasons including conformation, dry, wool type and susceptibility to breech strike and so on.

The selected or retained portion of the mob will change the average for each trait selected because the culls or undesirables have been removed. The 'new' average of the selected sheep will move according to the percentage of culls removed and the degree of spread of the initial population.

Table 7.17 shows how culling will affect the average for body weight and clean fleece weight depending on the percentage culled for these traits.

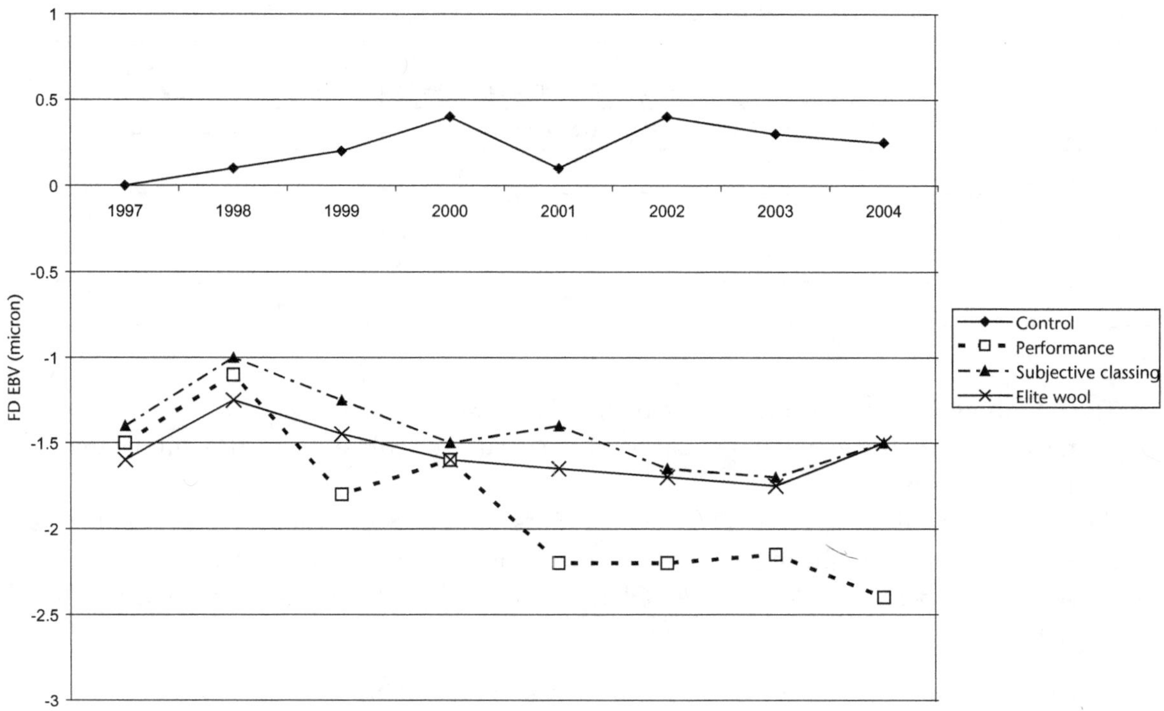

Figure 7.12. Genetic trend for mean fibre diameter (FD) – Merino Selection Demonstration Flocks.

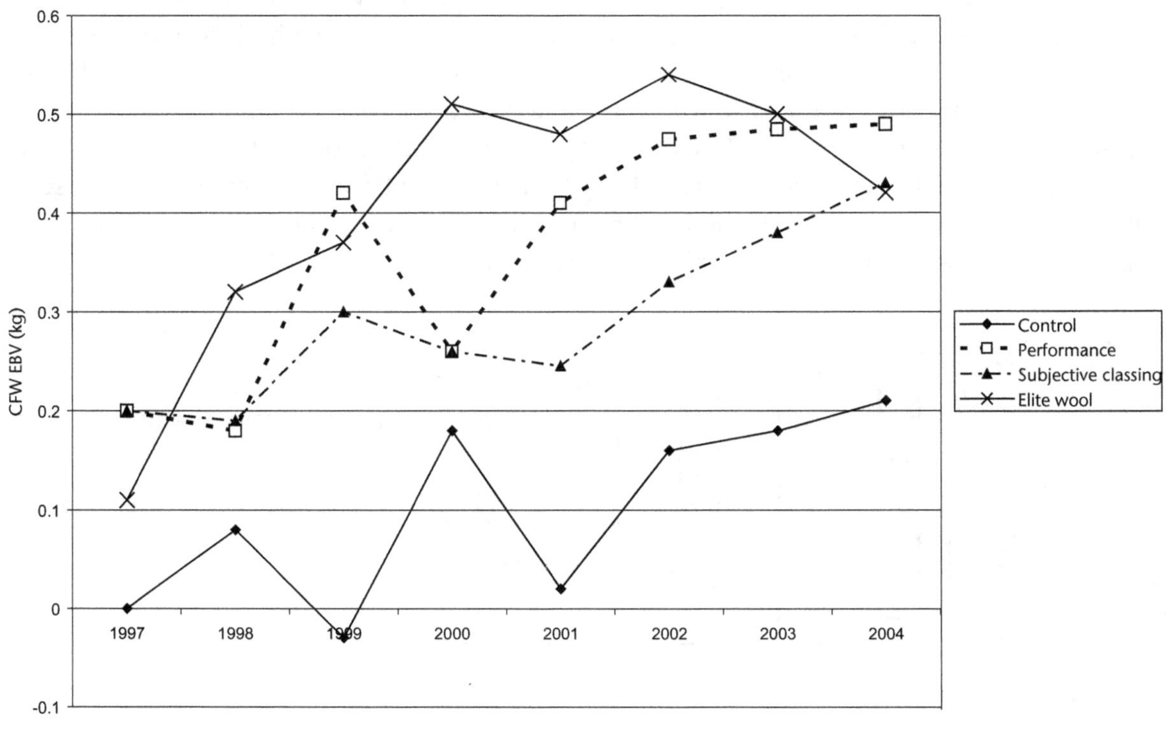

Figure 7.13. Genetic trend for clean fleece weight (CFW) – Merino Selection Demonstration Flocks.

Table 7.17. Effect of culling on the flock average body weight or clean fleece weight.

Percentage culled	Selected average, percentage above initial population average	
	Body weight (%)	Clean fleece weight (%)
10	1.6	2.5
20	2.7	4.4
30	3.9	6.2
40	6.2	10.0

The important points from Table 7.17 and Figure 7.14 are:

- a large proportion of the initial population (mob) need to be culled if a worthwhile change in the new average is to occur
- the response to culling as seen in Figure 7.14 is for selection for one trait only; if several traits are selected for at once, the response to selection in any one trait will be reduced: this is because the proportion culled for one trait alone must reduce if the total proportion of culls stays the same.

On a typical commercial sheep enterprise in southern Australia, rams contribute more than 80% of the overall genetic gain if the weaning percentages are low. In a sheep flock with a weaning percentage of 70 to 75%, at least 80% of the ewe weaners are required as ewe replacements. After the obvious culls are identified and rejected, there is little or no opportunity to place much emphasis on production traits. Unless ewe replacements can be culled heavily (such as following a high weaning percentage), there is little to be gained from placing a heavy emphasis on ewe selection.

Ewe selection plays a more important role in a ram-breeding program because ewes can become the mothers of the future sires. Also, when ram breeders have a commercial flock supporting the ram-breeding flock, measurement of this commercial flock, and selection of the top ewes based on the flock's breeding objective, can lead to a high selection differential and therefore can increase the rate of genetic gain. In meat sheep ram-breeding flocks, these top ewes generally have higher weaning rates, which can allow for higher selection differentials.

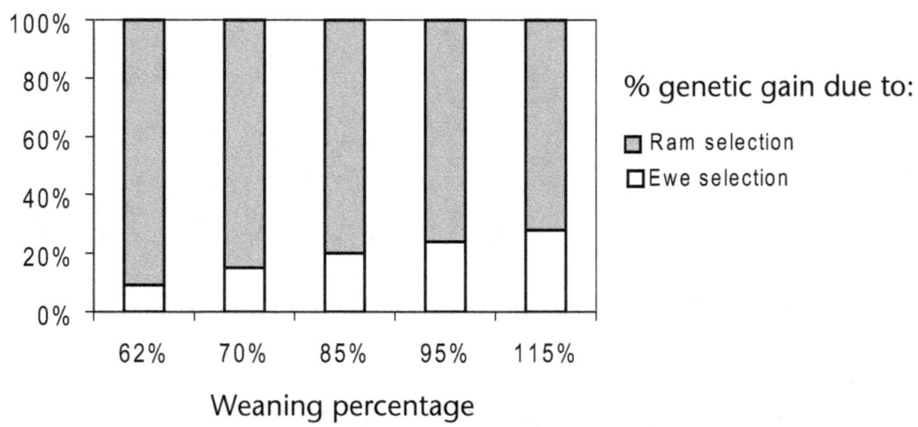

Figure 7.14. The impact of ram selection versus ewe selection.

On-farm fibre measurement

In markets where high micron premiums exist, there can be opportunities to do individual on-farm testing of animals for fibre diameter. Although, initially, the advantage of doing this was to increase returns from opportunistic clip preparation (by forming lines into micron groups aimed at price premium targets), analysis illustrated that the main advantages were in longer term returns from breeding and selection.

On-farm fibre measurement (OFFM) is individual fibre measurement of all sheep at or before shearing. Most commonly, only hoggets are measured given the cost and repeatability of micron in subsequent years. The results of a study into the profitability of this exercise and the accuracy of the measurements are summarised below:

- Two instruments are currently commercially available to test wool samples on farm. They are the optical fibre distribution analyser (OFDA) 2000 (which tests mid-side wool samples) and Fleecescan (which tests whole fleeces). When doing individual measurements in the shearing shed, both instruments gave equivalent results and were only slightly less precise than laboratory tests.
- Midside samples are better than samples taken at the pin bone because the latter did not give results that were as accurate or as precise.
- Most benefits from testing animals individually were from choosing more valuable animals and the consequent cumulative genetic gain.
- The benefits from clip preparation are dependent on micron premiums and also the impacts of other fibre measurements (for example, staple strength and wool quality can also be important determinants of price in fine and superfine markets).
- The benefits from testing ewes will depend on whether a significant proportion are culled (in this study, 42% were culled) or if they stay in the flock and are selectively joined to meat sheep breeds.
- If individual testing for ram breeding or selecting ewe replacements is being considered, fleece weights should also be recorded. Software packages and breeding companies can provide ways of combining the data into useful rankings of animals for selection. A spreadsheet, using 'Example 2 – calculating an index on fibre diameter and fleece weight' (page 251), which places weightings on fleece weight and fibre diameter, can also be a simple means of ranking animals. As a guide, using a 10% micron premium would aim to reduce fibre diameter but not lose fleece weight (10% micron premium means that if 19 μ is 1000 c/kg then 18 μ will be worth 1100 c/kg (10% more).

Mating strategies

In most commercial enterprises, the mating strategy is based on the practicalities of mob and paddock size. Mating strategies that allocate rams to ewes require good information about the ewes as well as the rams, and so selective mating is most used by studs that have the time and capacity to collect this information.

Corrective mating

Corrective mating is the joining of ewes that are poor in a trait to rams that are superior for the trait in question in the hope that the progeny will inherit the good or superior performance of the relevant traits from each parent. The outcome is to reduce the variability in the flock and to bring the flock closer to the average for the traits being corrected. Although this strategy is used widely, genetic progress from corrective mating can be quite slow.

Inbreeding

Inbreeding is the mating of individuals that are more closely related than the average of the population from which they came. When animals are very closely related, there can be an associated reduction in performance of some traits and of characteristics related to physical fitness. Therefore inbreeding tends to decrease fitness.

Inbreeding in the Australian Merino has been found to:

- decrease viability, because inbred animals die younger
- lower fleece weights and body weights
- reduce skin folds.

The size and structure of the Merino sheep population suggest that inbreeding is unlikely to be a major issue, provided closed flocks are greater than 2000 ewes in size. However, in some meat sheep or maternal flocks, the total population size is significantly lower. This means that strategies to reduce or minimise inbreeding are relevant to maximise the rate of genetic gain. As a minimum, some strategies that can be used include using rams from outside the flock that are genetically unrelated and avoid mating close relatives. The degree of inbreeding can be estimated as an inbreeding coefficient and this can be used by studs to manage the level of inbreeding.

With good information, computer models can be used to:

- account for (and avoid) inbreeding
- select superior ewes
- incorporate information where major genes have been identified (for exclusion or inclusion in a flock)
- provide mating options that will maximise genetic progress for an objective.

Line breeding

Line breeding is breeding to produce offspring closely related to some ancestor to produce animals that retain particular traits. It is a form of inbreeding but it is managed to produce beneficial effects. It should only be used by breeders who recognise the potential and limitations and can confidently identify animals of superior genetic worth.

Contract mating

Contract mating can be used to ensure that a lamb producer or finisher can reliably source ewes that are superior for the traits required. The lamb producer contracts a second party to produce ewes for prime lamb mothers and is therefore better able to concentrate on the feeding and management of terminal lambs.

Key messages – other selection strategies

- Genetic progress can be significant, regardless of the method of selection, as long as the direction is clear and there is confidence in the method and measurements used.

- The use of the performance recording and measurement produced better progress than other strategies (visual and 'elite') in a demonstration trial and generated higher profits despite the higher cost.
- Strategies relying heavily on visual selection, whether professional classers or using the 'elite wool method' also produce significant progress. However, a commercial producer would need confidence in both the classer and the strategy to be able to repeat this on the home farm.
- The culling rate of ewes needs to be high to make good genetic progress.
- The selection of ewes to be mated as groups to rams is most use in a ram-breeding situation or where significant number of ewes can be culled.

Technologies for increasing genetic gain

Genetic gain can be optimised by:

- selecting for traits that are highly heritable
- using animals that are highly superior for the traits required
- reducing the generation length by joining animals younger (on the assumption that they are superior).

Ram breeders are most able to take advantage of more expensive technologies to improve the rate of genetic gain. Technologies include artificial insemination and embryo transfer, which both aim to increase the number of progeny from superior animals in a shorter time.

The use of artificial insemination allows the few best rams to be used and also provides breeders with much wider access to sires that they can not buy. It also provides an easy means of linkage to other flocks to allow across-flock comparisons without moving rams.

Multiple ovulation and embryo transfer allow the best ewes to contribute more lambs than they could do if they were mated naturally. The generation length can be decreased by providing multiple lambs from individual young ewes. Selected ewes are treated to produce multiple ova, which are artificially inseminated and removed and implanted into other ewes. It is common to transfer embryos from Merino ewes into first-cross ewes to improve the maternal environment that the recipient ewes will provide for these expensive lambs.

These procedures can be very expensive (and require a high level of technical skill), so only ewes and sires that will perform for the traits required should be used.

Genomics and DNA markers

Genomics is the science of understanding the genetic make-up of an animal at a molecular level and includes the use of deoxyribonucleic acid (DNA) markers. Genes are made up of short lengths of DNA, and modern gene technology is able to make changes at the level of individual genes.

DNA markers are pieces of DNA that are used in diagnostic testing. If major genes are responsible for certain traits, then a DNA marker can identify at a very early age whether or not animals have the desired gene.

DNA makers are useful for assessing traits that are:

- difficult or expensive to measure (e.g. staple strength and parasite resistance)
- require animals to be slaughtered to measure the traits (e.g. meat eating quality)
- where animals have to reach adulthood to measure the trait (e.g. fertility or fecundity).

Much genetic research focuses on identifying major genes that have an affect on important traits. For humans, the identification of major genes that impact on heritable diseases or susceptibility to disease, means that susceptibility to disease can be determined at an early age (before symptoms appear), allowing for early intervention to improve quality of life and life span.

In sheep, there are some major genes that are responsible for deleterious effects. Such genes include the spider syndrome (a lethal condition in Suffolk sheep), black wool gene, the Callipyge gene (which causes very high muscling, with tough meat quality) and the Inverdale gene (which causes high fecundity but also total infertility if the gene is inherited by a ewe from both parents).

If a gene is found which is responsible for a trait, then a blood test, or piece of DNA from an animal, can be tested very early to determine whether it carries the gene.

There are currently gene markers available that have been commercialised such as Inverdale™ and LoinMax™ (for muscling). Other production genes will increase over time and will improve the ability of geneticists and ram breeders to identify, with confidence, highly superior animals very early and so increase the genetic gain in the sheep industries.

Other technologies identify more complex genetic interactions, and all these technologies will allow early, and more confident, selection of superior animals.

Key messages – increasing genetic gain

- Genetic progress can be hastened by
  - selecting for traits that are heritable
  - using superior animals
  - shortening the generation length.
- Genetic technologies such as artificial insemination and embryo transfer increase the number of superior animals that can be joined, so it is important that superior animals are clearly identified.
- Genetic technologies such as DNA markers will improve the confidence and ability to detect both superior and inferior animals much earlier.

CHAPTER 8

Animal welfare

Animal welfare has always concerned sheep producers given that the cornerstone of animal agriculture is good husbandry, including the provision of the right quality and quantity of feed, managing reproduction to ensure good ewe and lamb performance, good handling practices and protecting sheep from disease, predators and extremes of climate. The objective of the first seven chapters of this book is to provide advice on protecting and promoting animal welfare through appropriate nutrition, disease control, reproductive management and genetic improvement. However, animal welfare requires some discussion in its own right because consumers generally do not think in terms of feed, disease control, reproductive management, stockmanship or good husbandry, but in terms of what they believe to be good welfare for the animals whose products they, or others, may consume.

Animal welfare and quality standards

The wellbeing of farm animals now concerns a large proportion of the general population and can no longer be regarded as of concern only to extremists. Members of the public want animal products, but and they also want assurance that the animals from which those products were derived were fit and happy. Increasingly, consumers see animal welfare as a quality characteristic and markets are looking at ways of measuring it and passing this information on to consumers. If welfare standards do not meet consumer expectations, markets can be lost. Consumers and citizens want assurance that animals had a life that was worth living (a good life) and a good death.

What is animal welfare?

'The welfare of an animal is its state as regards its attempts to cope with its environment' (Broom). Approaches to animal welfare can consider animal feelings (mental state), animal bodies (their biological functioning) and animal nature (natural conditions and behaviour). Good welfare includes both fitness and sense of wellbeing and goes beyond preventing unnecessary pain and suffering. Animals have essential needs that must be met to enable individuals to maintain their health and happiness. The 'five freedoms' have been proposed to evaluate and safeguard animal welfare.

The five freedoms are:

- freedom from hunger and thirst
- freedom from discomfort
- freedom from pain, injury and disease
- freedom to express normal behaviour
- freedom from fear and distress.

The five freedoms require good management, good stockmanship, an appropriate environment, considerate handling and transport, and humane slaughter.

When an animal is in a situation that does not provide all its physical needs or is fearful and distressed, it compensates by diverting energy away from non-essential functions, such as growth, reproduction and production, to maintain its internal environment. The lack of adequate food, water, shelter or social contact over a long period of time, result in poor growth, production and reproduction. Fulfilling animal welfare requirements is complementary with running a productive and profitable sheep farm.

Animal welfare is a science and is underpinned by research that informs the development of animal welfare standards. There is a large and increasing body of research, in which Australian research is very well represented. Our welfare standards compare favourably with standards in the developed world. Taking the opportunity to embrace animal welfare as a quality standard will ensure continuing access to markets.

Legislation and codes of practice

Failure to provide sufficient food or water, or failure to control disease or to provide appropriate treatment for a sick or injured animal, and so on, is considered to be cruel – as a result, prevention-of-cruelty laws have been developed.

The emphasis in prevention-of-cruelty legislation is the avoidance of unnecessary suffering of animals (which also serves to re-enforce the negative image of farming and food and fibre production). Animal industries and governments have developed minimum standards to focus on a good quality of life rather than suffering. This approach puts livestock industries 'on the front foot' and promotes a positive image.

The vision of the Australian Animal Welfare Strategy is as follows: 'Welfare of all animals in Australia is protected and promoted by development and adoption of sound animal welfare standards and practices'. The Australian Animal Welfare Strategy will guide the development of new, nationally consistent policies and will enhance existing animal welfare arrangements.

Codes of accepted farming practice for the various livestock species, including sheep, have been developed. These will soon be replaced by Australian Standards and Guidelines for the Welfare of Animals, which describe the minimum standards to ensure the physical and mental comfort and health of livestock on farm, during transport, and at saleyards and slaughter. These standards 'provide the basis for developing and implementing consistent legislation and enforcement across Australia. They reflect available scientific knowledge, current practice and community expectations'.

The scope of prevention of cruelty laws is deviant behaviour towards animals, not standard agricultural practice (as described in the codes of accepted farming practice or the Australian Standards and Guidelines for the Welfare of Animals). Standard agricultural practice is exempt from prevention of cruelty laws. Given that the vast majority of the general population have never visited a working farm, it is not surprising that some farm practices, which are carried out in good faith by those that are very familiar with the requirements of managing a sheep enterprise, can shock the general population. The public has grown increasingly concerned about animal suffering, even when the source of suffering is not intended cruelty, but has grown out of goals such as increased productivity. Some standard agricultural practices are now the subject of public and consumer interest and rigorous scientific research. As a result, new practices and techniques will be developed, such as providing pain relief, employing genetics to increase resistance to disease, and the old practices will be phased out. For example, mulesing has been widely adopted for many years as the most important strategy to prevent breech strike in sheep. However, this procedure will be phased out in 2010, in part due to increasing consumer concerns about the practice and to maintain market access. Alternatives are been developed including clips and intradermal injections, although at this stage they are not as effective as mulesing and management changes will be necessary including additional chemical control and crutching. In the longer term, genetic solutions will be beneficial, although for most producers this will have little impact for the next 5 to 10 years.

Consumers and citizens are now able to influence farming practice, they want continuous improvement in quality (animal welfare) standards, and some want particular husbandry practices and procedures to cease.

Pain

Pain is an unpleasant sensory and emotional experience associated with actual or potential tissue damage. Pain is protective, because it acts as a warning that tissues are being damaged and it induces the sufferer to move from the harmful situation and helps to adjust behaviour to increase avoidance of that particular harmful situation in the future.

Livestock feel pain. Sheep do not express pain in the same way as humans do: they have evolved not to attract attention to themselves when in pain so they do not to alert predators of their more vulnerable state.

Invasive husbandry procedures

Invasive husbandry procedures, such as tail docking, castration, pizzle dropping and surgical mulesing (to be phased out in 2010) cause pain, distress and fear. The codes of practice and the Australian Standards and Guidelines for the Welfare of Animals provide information on the appropriate age and minimum requirements for ensuring the comfort of animals undergoing invasive husbandry procedures.

Performing an aversive treatment at a location not used for more routine handling or by either an unfamiliar operator, or a familiar operator wearing a style of clothing not usually worn, can help to prevent animals from associating the negative aspects of husbandry procedures and handling with their regular carer.

Tail docking of lambs

The public debate and interest surrounding mulesing highlights the need for good animal welfare management in all aspects of sheep meat and wool production. Tail docking is a less invasive procedure than mulesing, and research has shown that it ranks relatively low on the scale of pain and stress for sheep, if done by particular methods. Tail docking represents the removal of living innervated tissue. Some methods of tail removal cause less pain and suffering than others. Research conducted at New Zealand's Massey University found that on a scale of 1 to 6, tail docking with a knife produced a level five pain response, while use of a heated docking iron or rubber rings produced only a mild level one response. The release of stress hormones was used to measure animal responses. The reason for the difference is that a cut with a knife tends to preserve and irritate the sensitive nerve endings that remain in the tail stump. In contrast, the cauterising action of the docking iron is thought to damage these nerve endings and minimise the pain signals.

Tail docking pain in sheep is short lived, and the need for the procedure is broadly recognised in Australia and overseas. However, producers can boost their animal welfare credentials and optimise the management of their sheep by adopting methods that are the most welfare-friendly.

The docked tail should be just long enough to cover the vulva in female sheep and be of similar length in the male.

Castration

Castration is unnecessary if all lambs are to be marketed for slaughter before puberty, which generally occurs at 3 to 6 months of age. Where castration is required, follow the codes of practice and the Australian Standards and Guidelines for the Welfare of Animals.

Pizzle dropping

Pizzle dropping is sometimes performed on Merino wethers to reduce urine stain of belly wool and flystrike around the pizzle. The procedure also reduces pizzle rot. The risk of pizzle strike is greatest when wethers have more than 6 to 8 months wool growth during the fly season. Wethers shorn in late spring (that were crutched and given a pizzle ring in early spring) will be at low risk of pizzle strike. In contrast, wethers shorn in winter and not crutched or given a pizzle ring until mid autumn will be at risk of pizzle strike in late summer and early autumn. Although the procedure reduces urine stain of the belly wool, the need for this operation should be considered according to the risk of pizzle rot and pizzle strike, and information should be sought on the correct procedure from the State Department of Primary Industries or a sheep health adviser. The procedure should not be performed on wether lambs at marking because the length of the pizzle is too variable, but is normally performed when wethers are less than 1 year old in their first winter.

Teeth grinding/trimming

Grinding, clipping or trimming of teeth must not be performed as a routine flock measure because these procedures do not have any beneficial effect on health, wellbeing or productivity of the animals, but do have the potential to cause significant pain and infection. Corrective dental procedures on individual sheep may be undertaken to alleviate a specific dental problem if it is considered to be beneficial to its health and wellbeing.

Manage sheep to satisfy nutritional targets and complete all important husbandry procedures

All professional livestock managers understand the importance of nutrition and adopting management procedures to prevent disease. These two factors are perhaps the most important criteria for a sheep manager to fulfil to be able not only to run a productive enterprise but also minimise the risk of welfare problems. For example, ewes that fail to achieve body condition benchmarks before lambing are more likely to have excessive lamb losses, which reduce flock productivity and profitability. Ewes that fail to achieve body condition benchmarks before lambing and their progeny will also be at much greater risk of developing internal parasite problems. These issues will be less important in flocks where managers have adopted preventive worm control programs and where ewes have access to adequate pasture or are supplemented with feed if seasonal conditions are poor. Successful managers have the skills to assess pasture and stock condition and so make rational decisions to feed adequate pasture to ensure productivity is optimised and profitability is maximised, while ensuring sheep meet welfare targets.

Human–animal interaction

Sheep, like humans and some other animals, are sentient beings. A sentient being has ability to evaluate the actions of others in relation to itself and third parties, to remember some of its own actions and their consequences, to assess risk, to have feelings and emotions – including fear, happiness, pain and distress – and to have a degree of awareness.

Most stock people understand, and can identify, the effects that poor feed, water and husbandry have on the welfare of animals, but they are unaware of the impact that humans can have. Poor animal handling can result in livestock that are highly fearful of humans. Fear of humans in animals that are handled poorly can decrease welfare, production, reproduction and growth as a result of higher stress levels. Stress responses burn up energy that could be otherwise used for maintaining the internal environment of the sheep or for production of meat, wool or lambs.

Animal behaviour and low-stress stock handling

Research has shown that inappropriate handling can result in up to a 20% reduction in productivity, reproduction and product quality, especially in more intensively handled livestock such as pigs, poultry and dairy cattle. Inappropriate handling can lead to fear of humans and, in turn, depending on the regularity of contact and handling, chronic stress responses that burn up energy for no good purpose.

Understanding the behaviour of stock and how they perceive (understand) humans is a key factor in ensuring stock handlers behave in ways that will not initiate a fear response. Although fear is a normal and adaptive response, which is designed to prevent injury, chronic fear can cause long-term stress responses, decreasing the productivity and welfare of animals and making routine handling very difficult.

Enabling stock to become familiar with humans in a neutral setting, such as in the paddock, may also reduce the impact of negative procedures that are necessary as part of normal sheep husbandry. Properly designed sheep yards that allow efficient movement of sheep are beneficial for livestock wellbeing, less stressful for stock handlers and are more labour efficient.

Sheep behaviour

Producers can use knowledge of sheep behaviour to their advantage in all aspects of sheep production and management.

Sheep have evolved as social animals living in groups as protection from predators. Sheep will flock together and follow other sheep. A sheep that has been separated from its companions will try hard to get back to the mob. Sheep have a good long-term memory, especially with respect to unpleasant experiences and will try hard to avoid places where they have had bad experiences, even if there are no operators present. However, they also learn from, and remember, good experiences: lambs fed supplementary feed with their mothers will consistently come to a feed trail later in life.

Sheep have good hearing and are fearful of loud noises, yelling and barking. Sheep see dogs and humans as predators and will generally try to get away from them. In sheep-handling facilities oncoming sheep should not be able to see the stockperson. Likewise, sheep like to see what is pressuring them and will cut back if forced too hard. When moving sheep, the stockperson should be positioned so the lead animals will move rather than forcing from behind.

Sheep have good vision and, with eyes placed on the side of their head and a slight movement of the neck, they can almost see right around their body. Sheep prefer moving on flat ground and will move up an incline in preference to moving down an incline. Sheep move better through facilities if the same flow directions are maintained. They do not like strong contrasts in light and prefer to move from a darkened area to a lighted one.

Finding the time to bring stock into yards and run them through races without performing a procedure can help to familiarise stock with handlers and improve the ease and speed with which husbandry procedures can be carried out. This is particularly important for young or new stock. Young sheep move through facilities better if, on their first time, they are accompanied by well-trained older sheep.

Face recognition in sheep

The human brain has evolved neural mechanisms for visual recognition of faces, which allow us to distinguish between, remember and think about many hundreds of different individuals. Research has shown that 'sheep also recognize and are attracted to individual sheep and humans by their faces, as they possess similar specialized neural systems in the temporal and frontal lobes for assisting in this important social task … individual sheep can remember 50 other different sheep faces for over two years, and … the specialised neural circuits involved maintain selective encoding of individual sheep and human faces even after long periods of separation.' Sheep can distinguish between familiar and unfamiliar handlers and don't forget a face, so be nice to them!

Plan for adverse situations such as drought, fire, cold conditions or flood

On any farm, adverse situations and disasters, such as drought, fire, cold wet windy conditions or flood can occur. These are times of severe stress, and the potential for significant welfare problems is high. The farm management plan should include a plan for prompt action when adverse events occur, whether this involves a slow build up to drought, where drought feedlots are necessary to ensure adequate nutrition and maintain the farm soil and pasture resources, or a sudden event such as fire or flood. Sheep welfare and nutrition is paramount not only to protect sheep welfare and farm resources and to ensure the quick recovery of the farm business.

Another important event is the risk of off-shears losses due to exposure to cold wet windy conditions. This can be minimised by ensuring adequate shelter for all sheep shorn for between 2 and 4 weeks off-shears if severe weather is forecast.

Humane destruction of sheep

The aim of humane destruction is to render animals instantaneously unconscious (and insensible to pain), and for them to remain unconscious until cardiac and/or respiratory arrest lead to complete loss of brain function and death.

Humane destruction can be a two-part process: initially a 'stun' to render the animal instantaneously unconscious (and insensible to pain), followed by a procedure to ensure the animal dies without regaining consciousness; however, in some cases, the 'stun' alone will lead to death without the animal regaining consciousness.

The three key killing methods that result in loss of brain function and death are:

- the physical destruction of the brain
- stopping the blood supply to the brain, resulting in lack of oxygen to the brain
- chemical depression of the brain.

These methods are often used in combination.

Physical destruction of the brain

The physical destruction of the brain by a penetrating captive bolt or shooting is often sufficient to kill sheep. A successful shot depends on the position of the shot, the energy of impact and the area of the brain destroyed.

However, sometimes more shots, pithing or bleeding out are required to ensure death without the animal regaining consciousness. Pithing can be achieved by inserting a rod through the hole in the skull created by a captive bolt and manoeuvring the rod to cause massive brain damage (pithing does not mean severing the spinal cord).

Stopping the blood supply to the brain

Bleeding out (exsanguination) following stunning, deprives the brain of oxygenated blood. Following stunning, animals can be bled out by cutting major vessels of the neck, chest or axilla (armpit).

Cutting the neck vessels involves cutting the throat. With the neck of the unconscious animal extended and the cutting edge facing outwards, thrust a long, sharp strong boning knife into the neck below the ear, behind the angle of the jaw, through to the other side. Then cut outwards, severing all neck tissues including trachea and skin. The carotid arteries and jugular veins are usually cut in the initial thrust, but this should be checked once the throat cut is completed. Turn the knife around and cut down onto the bones of the neck to ensure these major blood vessels are cut.

When cutting the throat of a conscious animal, some operators also attempt to sever the spinal cord or dislocate the atlanto-occipital joint to minimise animal movement and hasten death. A word of caution: killing conscious sheep by cutting the neck vessels should only be done in an emergency and stunning is not possible, such as when a sheep is severely injured and equipment to stun the sheep or kill it by other methods described, is not available. The spinal cord must not be severed or broken until the animal is unconscious. Done prematurely, this procedure is painful, produces only paralysis and adds to the potential for distress.

Cutting the chest vessels means cutting the aorta and carotids arteries in the chest and is sometimes called chest sticking. Stab a long sharp knife into the thoracic inlet just in front of the first rib at the start of the jugular groove and manipulate the knife to cut the vessels at the top of the heart. This should only be carried out by a trained person and must never be used in a conscious animal.

Cutting the axillary vessels means cutting the major blood vessels of the armpit in an unconscious (stunned) sheep. This is simply done by slightly elevating a forelimb and stabbing into the armpit cutting the major vessels. Bleeding out is slower than by the other two methods, but will achieve the same result. The armpit stick must never be used in a conscious animal.

Bleeding out should be done using a sharp boning knife at least 15 cm in length, so the full incision can be made with one swift stroke.

Chemical depression of the brain

High doses of sedative or anaesthetic drugs and other chemicals cause the depression of the brain sufficient to cause unconsciousness and death. Most sedatives and anaesthetics can only be obtained from, and administered by, a veterinarian. Most are injected intravenously, or directly into the heart, which usually necessitates veterinary involvement.

The carcase should not be fed to dogs (or other animals), because the tissue of the carcase will contain the anaesthetic drugs or other chemicals, which will be absorbed by the dog causing depression of its central nervous system, and is likely to be followed by unconsciousness and death.

The use of intravenous agents may replace the need for bullets, bolts or knives to be used, thereby enhancing safety and aesthetics.

Signs of death

The sheep is dead when:

- the eye reflexes are absent and the pupils are dilated (test by touching the eyelids and cornea and shine a light into the eye)
- rhythmical breathing has stopped (stand back and observe)
- jaw and tongue tone are absent (open the jaw and grab the tongue).

If the above three indicators of death are not present, then the procedure must be repeated or additional procedures such as pithing or bleeding out should be performed.

The heart may continue to beat for a short period and some limb movements and respiratory gasps may occur, but if the three indicators are present, the animal is dead.

Using firearms to kill sheep on farm

Aiming sites

These depend on the position of the brain and brainstem and their relationship to external features such as eyes, ears and horns (see Figure 8.1).

The most appropriate point of penetration for polled sheep is in the midline on the highest point of the head aiming down towards the angle of the jaw.

For horned sheep, aim in the midline just behind the bony ridge between the horns, aiming towards the back of the throat. It is important to bleed sheep promptly when using this site because they have been known to regain consciousness.

Licensing and proficiency

Firearms must only be used by licensed and proficient operators in accordance with recognised safety procedures. Firearms can only be used in residential or urban areas by a person authorised

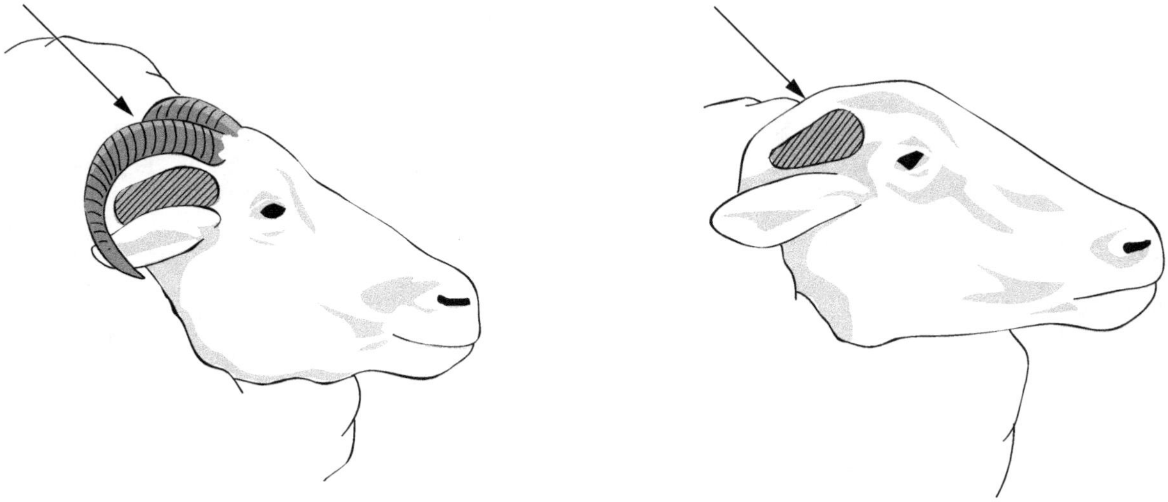

Figure 8.1. Sheep target sites for shooting and stunning.

to use firearms in these areas. Hand guns should only be used at close range owing to their inaccuracy over longer distances.

Captive bolts are not firearms so licensing is not required. Only discharge captive bolts with the end of the barrel in contact with the head. In contrast, never hold the barrel of a firearm against the target otherwise blow-back and rupture of the barrel will occur. Guard against ricochet should the bullet pass through the animal.

Size of projectile

Animal and aim site	Rifle projectile – minimum	Captive bolt – minimum
Sheep – ewe head	.22 long rifle/hollow point	2.5 grain (purple)
Sheep – lamb head	.22 long rifle/hollow point	1.25 grain (pink)
Sheep – ram head	.22 magnum	3 grain (green)

CHAPTER 9

Working dog health

This chapter deals with the basics of health, housing and transport of working dogs. The health and welfare of working dogs can be overlooked, and yet they are an integral part of the labour force of the farm. Working dogs contribute many hours of labour, so care should be taken to ensure that dogs can work to the best of their ability.

Working dogs require routine care including adequate housing and shelter, nutrition, vaccination, regular worm control (including heartworm prevention in some areas), control of fleas and desexing if dogs are not intended to be used for breeding.

Desexing

Dogs of all ages can be desexed. It is now commonplace for pups to be desexed as they leave the litter. The procedure usually only involves 1 day at the veterinary clinic and the recovery period is short (10 to 14 days). Early desexing prevents roaming in males and the problem of bitches coming on heat. Desexing before sexual maturity also ensures that diseases that may occur later in the dog's life – including mammary cancer, uterine pyometra and prostate disease – are much less likely to occur.

Bitches begin cycling at 6 months of age; if they are to have pups, they should not be bred until after their second heat (that is, until 18 months of age). Male dogs are generally sexually mature from 7 months of age or sometimes younger.

Vaccination

Vaccination protects dogs against fatal diseases such as parvovirus, distemper and hepatitis. A veterinarian can advise on appropriate vaccination schedules.

Viruses can spread within the dog population as quickly as the human cold spreads from person to person. Wild dogs and foxes can spread viruses to working dogs. Isolation on the farm is no protection. The spread may be by flies, car tyres, or by the clothes and shoes of people. Some viruses persist in the environment for many years.

Dogs can be vaccinated from 6 weeks of age for protection against parvovirus, hepatitis, distemper and canine cough. All dogs should be fully vaccinated.

Parvovirus

Parvovirus is common in unvaccinated working dogs, is highly contagious and is fatal in more than 80% of cases. This virus attacks the intestine, causing blood stained diarrhoea, vomiting, severe abdominal pain and weakness. Immediate veterinary attention is required because the disease progresses quickly and is usually fatal unless treated aggressively early in the course of the disease. In very young puppies, the virus can invade the heart muscles, causing death within a few hours.

Contact with other dogs is not necessary for this disease to spread. This virus persists in the ground for many years. The environment around an infected dog needs to be cleaned with potent disinfectants to prevent spread to other dogs. Outbreaks occur regularly throughout Australia, especially in summer.

Hepatitis

Hepatitis is a highly contagious disease that spreads to unvaccinated dogs of any age. However, canine hepatitis is more severe in dogs less than 2 years of age. Symptoms include high fever, depression and acute abdominal pain. Death can occur within 24 to 36 hours. Dogs that recover may develop long-term liver and kidney problems and can act as carriers, spreading this disease to other working dogs for many months.

Distemper

Distemper is extremely contagious to unvaccinated dogs of any age and is often fatal. Symptoms can include coughing and sneezing, nasal and eye discharge and depression. Muscle tremors, fits and paralysis can occur later in the course of the disease, as the virus attacks the nervous system. Recovery rates, even with treatment, can be low, and recovered dogs may have permanent brain damage.

Canine cough

Canine cough is caused by several highly infectious disease agents, and is spread wherever dogs congregate, such as on dog trials, in saleyards and in boarding kennels. It is a major problem for working dogs. Affected dogs have a dry hacking cough that persists for weeks and can result in pneumonia. Effective vaccines are available, including those that are given intra-nasally.

All working dogs should visit a veterinarian once a year for a routine health check, vaccination and parasite control program.

Parasite control

Hydatid tapeworm

Hydatids pose a serious health risk to humans. The life cycle of the hydatid tapeworm (*Echinococcus granulosis*) can involve a number of species, including humans, but the most important are sheep, kangaroos and dogs. The hydatid life cycle involves two host animals: an intermediate host (usually sheep, but also cattle, pigs, kangaroos, wallabies and occasionally humans) and a definitive host (dogs, dingoes and, to a lesser extent, foxes).

The hydatid tapeworm can live in the intestine of the dog (and other definitive hosts). It is very small – only half a centimetre long – and is made up of segments, which contain hydatid eggs. When eggs in the last segment mature, the segment is shed and passed in faeces and contaminates pasture. Tapeworm eggs can survive for several months on pasture in cool moist conditions.

The intermediate hosts, usually sheep, become infected by swallowing eggs from contaminated pasture. Humans generally become infected by handling dogs carrying hydatid tapeworms and then not washing hands before putting something in their mouth (humans can occasionally become infected by eating raw vegetables contaminated by dog faeces). Hydatid eggs are sticky and adhere to the coat of the dog. When hydatid eggs are swallowed, they migrate through the wall of the stomach into the bloodstream and can be carried to the liver, lungs, brain or other internal organs, where a hydatid cyst forms. (In humans, major, and sometimes multiple, surgery is required to remove hydatid cysts from internal organs.)

The life cycle is completed when the dog or other definitive host eats offal of sheep or other intermediate hosts containing hydatid cysts.

Control of hydatids is achieved by interfering with the life cycle by:

- ensuring that dogs are regularly treated with effective anthelmintics to kill the hydatid tapeworm (not all dog worm tablets kill tapeworm)
- preventing access of dogs to offal from intermediate hosts.

In addition the following precautions can be taken:

- always wash hands after handling dogs and before eating or drinking
- do not allow dogs to defecate near vegetable gardens or play areas
- do not allow dogs to roam and scavenge
- collect all carcasses and offal and dispose of them appropriately
- feed only commercially prepared dog foods.

Roundworm, whipworm, hookworm and common flea tapeworm

Around 80% of Australian dogs carry intestinal worms. Some adult worms can produce 20 000 eggs a day and some eggs can survive up to 5 years on the ground. Dogs are infected via uncooked meat, rodents and also through the skin, from larvae on grass (whipworm), or from fleas. Even ingestion of material in contact with the ground can be enough to cause infection. Worm larvae can infect people, particularly children who are in close contact with dogs. Dogs with worms may show signs including a poor or dull coat, anaemia (pale gums), diarrhoea, a pot-bellied appearance, weight loss and tail skidding. White tapeworm segments or worms may be seen in the droppings.

All working dogs need regular worming. Pups should be wormed according to their body weight every 2 weeks from birth until 3 months of age, then monthly until 6 months of age. All dogs older than 6 months require worming every 3 months with a product that removes all worms including the hydatid tapeworm. It is advisable to develop a worm-control program in consultation with a veterinarian.

Other worm-control strategies include:

- flea control (see below)
- clean kennels and sleeping areas regularly
- remove faeces from dog yards and around kennel areas
- worm and flea control for farm cats
- ensure dogs and cats are fed quality pet foods and not forced to hunt and scavenge for feed. Always wash hands before eating and after touching dogs and cats.

Fleas

Fleas are a great source of irritation to working dogs, resulting in scratching, chewing and hair pulling. Some dogs develop an allergy to fleas. Although fleas breed rapidly in warmer months, they can be a problem all year round.

Fleas can be killed and re-infestation prevented for a month with 'spot on' and oral treatments that are applied to all working dogs, pet dogs and cats on the property. Fleas are a continual threat, so ongoing control measures are necessary. Veterinarians can advise on effective flea control programs.

Ticks

The paralysis tick is found on the eastern seaboard, ranging from North Queensland to coastal East Gippsland and can live on a wide variety of animals, including humans and dogs. The paralysis tick is grey in appearance, like the common bush tick. It affects the dog by injecting a deadly toxin that causes paralysis progressing to respiratory and heart failure. Signs of paralysis are seen from 3 to 6 days after the tick has attached to the dog.

Early detection increases chances of recovery and veterinary treatment with paralysis tick antiserum is essential. Signs of tick paralysis are loss of coordination, beginning with the hind limbs (drunken appearance) and progressing to the forelimbs, change of sound of bark, retching, coughing or vomiting and difficulty breathing. Some dogs can die very quickly, so seek veterinary help immediately and urgently. Do not offer food or water. If one tick is found on a dog, always go over the dog with a fine toothcomb to look for others: a second or third missed tick can kill the dog.

Help prevent tick attachment by using veterinary insecticidal washes weekly, tick-collars, 'spot on' treatments, sprays or tablets. These may be used in combination under veterinary advice. No single prevention is 100% effective and there is no substitute for regular examination of the dogs at risk.

If possible, avoid taking dogs to tick-prone areas. In summer in tick prone areas, dogs should be thoroughly checked for ticks at least once a day. Look in the ears, under the collar and around the lower neck, between the dog's toes and around the tail. More commonly, ticks are found on the lower areas of the front half of the dog's body. Ticks can be hard to see, although after 2 to 3 days of sucking the dog's blood, the tick will be about the size of a pea, with a flattened oval shape.

Remove ticks immediately and save them to show the veterinarian. To remove the tick, scrape it off the skin with a sharp blade. Alternatively, a loop of cotton can be pulled tight around the tick at skin level, and flick the tick out with a sudden movement. The toxin is contained in the body of the tick, and so once the body of the tick is removed so is the source of toxin.

Mange

This is a skin disease caused by tiny parasitic mites, and can be common in working dogs. There are two forms of mange: sarcoptic mange (also known as 'scabies') and demodectic mange.

Sarcoptic mange mites are transmitted by dogs coming into close contact with infected dogs or wild animals such as foxes and wombats. They can also be transferred to humans. Mites prefer to live on lightly haired areas such as around the dog's face, around the elbow, under the armpits and in the groin area. The mites burrow into the skin and cause intense itching. Dogs will scratch and rub the infected skin until it becomes red and hair falls out. Continued scratching will cause skin to become thickened and crusty. Consult a veterinarian regarding treatment with insecticides, and prevent re-infection by ensuring that all dogs on the property are treated at the same time. Simple topical 'spot on' parasite control can be applied to eliminate this parasite.

Demodectic mange mites live in the hair follicles of dogs. The skin of infected dogs will **not** be intensely itchy and there will be few signs of infection, other than hair loss resulting in large bald patches all over the dog. Veterinarians can provide appropriate 'spot on' or injectable treatment that is highly effective in controlling this parasite and allows the hair to regrow. Mild cases will generally clear up within 4 to 6 weeks. However, in the event of secondary bacterial or fungal infection, further veterinary treatment will be required.

Heartworm

Adult heartworms live in the pulmonary artery. Affected dogs have a chronic cough, weight loss and, ultimately, congestive heart failure. Mosquitoes spread this blood-borne parasite.

Heartworm is relatively easy to prevent, but difficult and costly to treat. There are many preventative treatments including daily tablets, monthly tablets, 'spot on' and yearly injections. Consult a veterinarian for advice on heartworm prevention.

Other problem areas

Grass seeds

Grass seeds can end up in the ear canal, up the nose or prepuce and can penetrate the skin (particularly between the toes). They can also lodge in the eye under eyelids or in hair mats on the coat (particularly in the groin and armpit). Seeds that have awns, such as barley grass and spear grass, cause the most problems.

A dog with a grass seed in its ear will shake its head, scratch its ear and often walk with its head tilted to the side. Specialist equipment is needed to remove grass seeds from dogs ears, so seek prompt veterinary attention. Dogs with floppy ears should have the hair clipped short on the inside of the ear flap and at the entrance to the ear canal to reduce the risk of grass seeds entering the ear canal.

A dog with a grass seed up its nose will sneeze persistently, so seek prompt veterinary treatment.

During spring and summer, the hair on the paws should be clipped short to reduce the chance of catching grass seeds. Grass seeds that have penetrated the skin must always be removed (this may have to be done by a vet); if they are ignored, they can lead to severe infection and sinus formation causing pain and lameness.

Heat stroke

Heat stroke occurs when the body temperature rises above normal levels due to production of excessive heat, exposure to high temperatures or failure to loose heat. Dogs pant to loose heat and reduce their body temperature. Overheating typically occurs on days over 35°C, especially when dogs are left in vehicles or tied up in areas with no shade or water. Working dogs can also suffer heat stroke if they exercise excessively when they already have an elevated body temperature from another cause, such as an infection.

Initial warning signs are open mouth rapid panting and the tongue hanging out, followed by restlessness, dullness, weakness, reluctance to move, recumbency, difficult breathing and excessive drooling, and the dog may vomit or fit. If untreated, the dog will eventually become comatose.

Dogs showing signs of heat stroke require urgent veterinary treatment. Immediate first aid involves wetting the dog all over with water and moving it to a well-ventilated shady place with plenty of good-quality drinking water.

To prevent heat stroke, ensure dogs always have access to shade and water. Never leave a dog tied up in the sun or in a vehicle (even when windows are wound down). Avoid work during extreme temperatures. Limit work, and/or allow regular rest and/or swim breaks, in high temperatures.

Arthritis

Working dogs can suffer joint wear and tear and can suffer in silence. Dogs may limp on and off, reduce their activity and 'stiffen up'. Arthritis can be managed, and many working dogs can return to full function with advanced, affordable easily administered veterinary treatments.

Snake bite

Dogs bitten by venomous snakes will initially be hyper-excitable, may have dilated pupils, develop muscle twitches and may stagger, salivate or vomit. However, symptoms change quickly as the poisoning progresses: seek immediate and urgent veterinary care for cases of known or suspected snake bite. Snake antivenines are available.

Poisoning

Many working dogs are accidentally poisoned. Dogs are particularly susceptible to 1080 poisoning, for which there is no treatment. Care should be taken to confine dogs and keep them away from areas where poisoning programs for pest animals are occurring.

Other poisoning risks include pasture sprays, sheep jetting fluids and dips, snail bait, and rat and mice poisons.

Raising young dogs

New pups need a full vaccination and a parasite -control program.

For the first 6 to 12 months of their life, young dogs need **puppy** formula food designed for optimum bone growth. Supermarkets, veterinarians and farm trade stores can supply specially formulated puppy food in bulk. The total daily food requirement needs to be split into two meals per day until the pup is 6 months old.

Exercise must be restricted because the young dog's bones are forming and too much work, including jumping onto and off vehicles, before the bones stop growing at 1 year of age can lead to poor conformation. Poor conformation can lead to early onset arthritis in working dogs and potentially limit their active working life.

Dogs on moving vehicles

Dogs can suffer terrible injuries if not secured properly while on moving vehicles. Unrestrained dogs that fall from moving vehicles can be injured when they hit the road or be hit by other cars. Restrained dogs can suffer strangulation when their leash is long enough to allow them to fall off the side of the vehicle.

In some states, it is a legal requirement to secure dogs on moving vehicles (e.g. utes, trays and trailers) so they cannot fall or jump from the vehicle.

When securing dogs, ensure that the restraint is long enough to allow the dog to stand and lie down, but not too long to enable the dog's front or hind legs to reach the side of the tray. Attach the restraint by a swivel to an anchor point against the vehicle cabin. Fix the other end of the chain to the dog's harness or leather collar with another swivel to stop the chain from tangling.

Housing and tethering

Ideally, dogs should be kept in securely fenced areas or, where adequate fencing is not available, in pens. Tethering should only be considered as a last resort, and only under certain conditions. Regardless of how they are housed, dogs must always have ready access to a kennel, shed or other protection from the elements and for sleeping. Kennels should be of an appropriate size for the particular animal, have adequate ventilation and, if made of metal, should be kept out of direct sunlight or should be effectively insulated.

Dogs also require daily exercise appropriate to the age, health, working status and breed of the individual animal. Tethered dogs should be allowed off for a minimum of 2 hours daily.

Pens should provide enough space for each dog to comfortably move about and lie down, and provide areas for urinating and defecating that are well away from feeding and sleeping areas. Pens need to be drained appropriately to allow water to run off, and be kept hygienic with regular cleaning (it may also be necessary to seal the surface of pens). A separate pen should be used for any whelping bitches.

Tethering is not recommended, but if dogs are tethered, the following guidelines should be observed:

- The tethering site should be reasonably flat, free of obstructions (including rocks), and not be situated in a waterlogged or flood-prone area.
- Dogs must not be tethered adjacent to a fence or other structure that places them in danger of death by hanging.
- The site must provide a minimum tether radius of 3 metres, allowing 6 metres of run.
- A suitable kennel must be provided for shelter from the elements and for sleeping.

- Dogs must be able to easily reach their kennel, food and water supply.
- The location of the kennel must not cause a threat of entanglement.
- The acceptable material for a tether is a light metal chain.
- Fixed and running tethers require the fitting of an appropriate collar, with a swivel to which the tether is attached.
- Fixed tethers must be fixed to an anchor point (preferably with another swivel) that allows 360 degrees of movement at ground level, and allows the dog to cover the area without tangling.
- Running tethers must have a strong wire that is secured at either end to trees, fences or posts, but must have stops at either end to ensure the tether cannot become entangled or injure the dog.
- Tethered dogs must be inspected regularly (at least twice during daylight hours in each 24 hour period, and increased to three times or more in very hot weather).
- Tethered dogs should be let off tethers at least 2 hours per day during daylight hours.
- Collars, tether chains, swivels, wires and anchor points must be regularly inspected for signs of wear.
- Dogs must not be tethered closely together.
- Dogs less than 4 months old should not be tethered.
- Dogs must be closely monitored when tethered for the first time, until it is certain they have adapted to tethering.
- Bitches in season must not be tethered where entire males may have access to them.
- Bitches about to give birth must not be tethered.

Appendix 1 – Monthly management reminders

The following monthly management reminders can be found in this appendix:

- autumn lambing Merino ewes joined to Border Leicester rams
- a spring lambing wool-producing flock
- winter lambing crossbred ewes joined to terminal sires.

These reminders are preceded by general monthly management reminders that are relevant to all sheep enterprises.

General monthly management reminders on sheep property

January

Pastures

Plan fertiliser application for the year.

General

Maintain flock quarantine: sheep of unknown disease status may introduce footrot, drench resistant worms, ovine Johne's disease or lice.

Assess stock at least weekly.

Check water regularly.

Farm maintenance, including:

- farm machinery
- land and improvements
- infrastructure.

February

Pastures

Order and apply fertiliser before the autumn break.

Manage pastures and avoid overgrazing – destock pasture once ground cover falls below 80% (20% bare ground).

General

Maintain flock quarantine: sheep of unknown disease status may introduce footrot, drench resistant worms, ovine Johne's disease or lice.

Assess stock at least weekly.

Check water regularly.

Review farm budgets and farm plan.

Farm maintenance, including:

- farm machinery
- land and improvements
- infrastructure.

March

Pastures

Prepare paddocks for sowing at or after the autumn break.

Apply fertiliser before the autumn break to maximise autumn pasture growth. Apply on the basis of soil tests or fertiliser test strips.

Manage pastures and avoid overgrazing – destock pasture once ground cover falls below 80% (20% bare ground).

General

Maintain flock quarantine: sheep of unknown disease status may introduce footrot, drench resistant worms, ovine Johne's disease or lice.

Assess stock at least weekly.

Check water regularly.

Review farm budgets and farm plan.

Carry out farm maintenance, including:

- farm machinery
- land and improvements
- infrastructure.

April

Pastures

Manage pastures and avoid overgrazing – destock pasture once ground cover falls below 80% (20% bare ground).

Carry out sod seeding at the autumn break as soon as soil conditions are suitable (a delay will allow competition from established species).

Prepare for pasture renovation once the autumn break arrives.

General

Maintain flock quarantine, sheep of unknown disease status may introduce footrot, drench resistant worms, ovine Johne's disease or lice.

Assess stock at least weekly.

Check water regularly.

Review the farm budgets and the farm plan.

Carry out farm maintenance. including:

- farm machinery
- land and improvements
- infrastructure.

May

Pastures

Control capeweed and other broad leaf weeds in pastures.

Plan winter allocation of stock to pastures to ensure that ewes are allocated the best feed. Complete winter feed budgeting to determine if extra supplementation or strategic fertiliser to boost winter pasture growth is required.

General

Maintain flock quarantine: sheep of unknown disease status may introduce footrot, drench resistant worms, ovine Johne's disease or lice.

Assess stock at least weekly.

Check water regularly.

Review the farm budgets and the farm plan.

Carry out farm maintenance, including:

- farm machinery
- land and improvements
- infrastructure.

June

Pastures

Check for pasture pests such as earth mite or black headed cockchafers to determine if spraying is necessary, especially in new pastures.

General

Maintain flock quarantine: sheep of unknown disease status may introduce footrot, drench resistant worms, ovine Johne's disease or lice.

Assess stock at least weekly.

Review the farm budgets and the farm plan.

Carry out farm maintenance, including:

- farm machinery
- land and improvements
- infrastructure.

JULY

Pastures

Broad-acre spray thistles if they are a problem.

Assess each paddock to determine paddock carrying capacity – this helps to plan future stock allocation, and sets priorities for pasture renovation and fertiliser application.

General

Maintain flock quarantine: sheep of unknown disease status may introduce footrot, drench resistant worms, ovine Johne's disease or lice.

Assess stock at least weekly.

Collect financial and physical information to undertake a farm business analysis.

Carry out farm maintenance, including:

- farm machinery
- land and improvements
- infrastructure.

August

Pastures

Plan fodder conservation (hay/silage) and top-dress pastures to boost fodder production. Prepare summer fodder crop program if applicable.

General

Maintain flock quarantine: sheep of unknown disease status may introduce footrot, drench resistant worms, ovine Johne's disease or lice.

Assess stock at least weekly.

Review the farm budgets and the farm plan.

Carry out farm maintenance, including:

- farm machinery
- land and improvements
- infrastructure.

September

Pastures

Check pasture quality and quantity.

General

Maintain flock quarantine: sheep of unknown disease status may introduce footrot, drench resistant worms or lice.

Assess stock at least weekly.

Check water regularly.

Carry out farm maintenance, including:

- farm machinery
- fencing and gates

- yards, shearing shed.

October

Pastures

Assess pasture species to assist in planning future pasture renovation.

Plan next year's fertiliser program.

Consider soil tests.

Spray top paddocks to be renovated next year.

Conserve fodder.

General

Check disease status of introduced sheep that have been quarantined from home-bred sheep.

Assess stock at least weekly.

Review the farm business analysis and the farm plan.

Check water regularly.

Carry out farm maintenance, including:

- farm machinery
- land and improvements
- infrastructure.

November

Pastures

Eliminate barley, brome and silver grasses from pastures.

Plan fertiliser application for the next 12 months.

Consider soil tests.

Undertake fodder conservation.

Review feed purchase requirements for grain direct from harvest.

General

Review disease status of quarantine sheep with unknown disease status.

Assess stock at least weekly.

Check water regularly.

Carry out farm maintenance, including:

- farm machinery
- land and improvements
- infrastructure.

December

Pastures

Plan fertiliser application for the next year, if not already done.

General

Maintain flock quarantine: sheep of unknown disease status may introduce footrot, drench resistant worms, ovine Johne's disease or lice.

Assess stock at least weekly.

Check water regularly.

Carry out farm maintenance, including:

- farm machinery
- land and improvements
- infrastructure.

Monthly management reminders for autumn lambing Merino ewes joined to Border Leicester rams

January

All sheep

A second summer drench is due from mid January, so do a worm egg count (WEC) to determine if a drench is needed.

Use an effective drench at the correct dose rate; do not under-dose; and set dose rate to the heaviest sheep in the mob.

Watch for blowflies – jet weaners if in a high-risk situation or if supervision is poor.

Ewes

Ensure ewes maintain body condition at an average of condition score 3 over late summer and autumn.

Rams

Remove the rams from the ewes after a 6-week joining period.

Weaners

Carry over lambs (ewe lambs) should be allocated to good-quality feed to ensure they continue to grow.

February

All sheep

Watch for blowflies.

If liver fluke is present, a drench may be required: monitor to determine if a drench is required.

Ryegrass staggers may occur in late summer/autumn so ensure that all livestock management procedures are completed before ryegrass staggers is likely to become a problem.

A second summer drench is due, if not done in January, and should be completed by mid February; do a WEC to determine if a drench is needed.

Decide on a weaning paddock for next season's lambs – graze this paddock only with cattle or sheep that have low egg outputs, such as adult wethers or dry ewes.

Ewes

Ensure that ewes maintain body condition at an average of condition score 3.

Weaners

Ensure that weaners continue to grow to meet market specifications.

March

All sheep

Watch for blowflies.

Carry over lambs

Ensure that lambs are given an appropriate ration supplement or allocated pasture to meet target weights and market specifications.

Monitor WECs about 4 to 6 weeks after the last drench to ensure that worm burdens are not causing production losses.

Ewes

Vaccinate with 5 in 1 or 6 in 1 before lambing.

Increase energy intake in last 6 weeks of pregnancy to meet additional requirements.

Avoid pregnancy toxaemia by preventing under nutrition and stress in late pregnancy.

Monitor body condition and ensure that ewes maintain a condition score above 2.5.

Avoid under nutrition of the ewes, which results in low birth weight lambs that have a very low survival rate.

Scan for pregnancy, especially if supplementary feeding is required due to a poor season or if twin lambing ewes can be identified and allocated very good feed and shelter to ensure high lamb survival rates.

April

All sheep

Watch for blowflies.

Ewes

Monitor body condition and ensure that ewes maintain a condition score above 2.5.

Ensure that pasture availability is greater than 1100 kg DM/ha at lambing – this is difficult to achieve with early lambing and so supplementary feeding is often necessary.

Crutch ewes pre-lambing.

A pre-lambing drench is probably of little value unless there are enough 'worm safe' paddocks to allow a move after drenching or if ewes are in poor condition and grazing short pasture (to get best value out of the drench, ewes should be drenched as close to possible to lambing, this must be balanced with the risk of pregnancy toxaemia).

Carry over lambs

Ensure lambs are supplemented appropriate ration or allocated pasture to meet target weights and market specifications.

Monitor WECs about 4 to 6 weeks after last drench to ensure that worm burdens are not causing production losses.

May

Ewes

Lambing

Ensure good shelter for ewes to increase lamb survival.

A dead lamb pick up is useful to determine the cause of death to aid future management decisions.

The energy intake requirements should meet demands of late pregnancy and lactation.

All sheep

Monitor potential fluke affected mobs and drench if fluke eggs are detected.

Weaners

Ensure lambs are supplemented with an appropriate ration or allocated pasture to meet target weights and market specifications.

Monitor WECs about 4 to 6 weeks after last drench to ensure that worm burdens are not causing production losses.

June

Ewes

Ewes require high levels of energy for lactation, so ensure that pastures have more than 1200 kg DM/ha or supplementary feed as necessary.

Mark lambs.

Separate dry ewes by wet drying ewes at marking if feed conditions are poor.

A post-lambing drench for ewes is not warranted in most seasons; assess the need based on pasture availability (less than 900 kg DM/ha, ewe condition score less than 2.5 and WEC result).

Determine marking percentages; if they are low, determine why.

The lamb marking percentages of the maidens should be within 10% of the adult flock: if not, check the cause.

Lambs

Mark lambs at the end of the month.

Dock the tails at a length below the tip of the vulva, or equivalent in wethers.

Provide the first 5 in 1 or 6 in 1 vaccination for all lambs and provide trace elements supplements as appropriate for known deficiencies.

A worm drench is not necessary for lambs at marking.

July

All sheep

If footrot is present, plan the control strategy for spring.

If seasonal conditions are poor, additional supplementary feeding may be necessary.

Ewe weaners

Monitor WECs of mobs in poor condition or scoring to determine if additional drenches are necessary; weaners should have WECs monitored every 4 to 6 weeks.

Crutch ewe weaners.

August

All sheep

Examine pre-shearing for lice; if lice are found, control lice if fleece damage is occurring and eradicate off shears.

Ewes

Lambs should be weaned 12–14 weeks post lambing.

Lambs

Wean lambs onto a 'clean paddock'; that is, one relatively free of worm eggs and larvae such as a cattle paddock, wether paddock or a hay aftermath.

Do not wean lambs onto a paddock that has been grazed by lambs or lambing ewes since the autumn break.

Drench lambs with an effective drench at the correct dose rate: drench for the heaviest animal in the flock.

Provide additional trace elements as necessary, based on known deficiencies.

Carry out a drench-resistance trial every 2 to 3 years to determine what drenches to use in an annual rotation.

Provide the second 5 in 1 or 6 in 1 vaccination.

Check body weights: weight at weaning should be in the range of 22 to 25 kg of higher.

Ewe hoggets

Monitor WECs to determine if additional drenching is necessary.

September

All sheep

Examine all flocks for footrot; if footrot is present, minimise spread in spring.

Clean up shed before shearing.

October

All sheep

Undertake shearing.

Lambs to be sold within 8 weeks may be better not shorn.

Avoid later shearing if lice are present, unless the lambs can be kept separate.

Do not dip or use a pour-on unless a lice infestation exists or there is a reasonable suspicion that one exists.

Watch for blowflies.

Ewes

Draft ewes into condition score groups; light ewes should get preferential feed before mating.

Rams

Carry out a pre-joining examination: ensure rams are in condition score 3.5 to 4.

Examine the rams' feet, testicles and penis, and assess their general health.

Supplement rams to ensure they are in adequate condition for mating.

Lupins can be fed to increase testicular size.

Ewe hoggets and lambs

Prepare ewe hoggets for sale.
Prepare wether and cull ewe lambs for sale.

November

All sheep

First summer drench can be given after the middle of the month, especially if seasonal conditions are dry.

Ewes

Ewes should be at least condition score 3.5 as they have had ample time to recover body condition after weaning.

Teasers may be necessary to ensure ewes are cycling at start of joining.

Rams

Carry out a pre-joining examination: ensure rams are in condition score 3.5 to 4.

Examine the rams' feet, testicles and penis, and assess their general health.

Supplement rams to ensure they are in adequate condition for mating.

Lupins can be fed to increase testicular size.

Exclude infertile rams from the ram team (the lambing percentage can be reduced by running infertile rams with fertile rams, infertile rams serve ewes and these ewes may not be serviced by fertile rams).

Weaners

Sell store wether lambs.

Give remaining wethers and ewe lambs the best available feed in terms of both energy and protein.

Remaining weaners can be given their first summer drench after the middle of the month and should be jetted to prevent flystrike if wool is still long (beware of withholding periods).

Repeat WEC 4 to 5 weeks following each drench to determine if a further drench is required.

December

Ewes and rams

Join maidens separately at 2% rams for 6 weeks.

Use experienced rams with maiden ewes.

Join mature ewes at 1 to 1.5% for 6 weeks.

Use smaller paddocks with a single watering point and with no geographical barriers such as timber belts or gullies.

Ensure that ewes are grazing good-quality feed and the lightest ewes have access to the best feed.

All sheep

The first summer drench should be completed. Use an effective drench at the correct dose rate.

If footrot is present, tip sheep, and pare and examine all feet of all sheep and cull any infected animals (if spring footrot control was effective, there should be fewer than 5% of sheep infected in summer).

Watch for blowflies.

If liver fluke occurs on the property, a drench may be required to prevent acute fluke disease.

Monthly management reminders for a spring lambing wool-producing flock

January

All sheep

A second summer drench due from mid January, so do a worm egg count (WEC) to determine if a drench is needed.

Use an effective drench at the correct dose rate; do not under-dose; and set the dose rate to the heaviest sheep in the mob.

Examine pre-shearing for lice; if lice are found eradicate after shearing.

Watch for blowflies: jet weaners if high-risk situation or if supervision is poor.

Rams

Rams should be shorn at least 2 months before the start of mating.

Examine the rams' feet, testicles and penis, and assess their general health.

Carry out a pre-joining examination: ensure rams are in condition score 3.5–4.

Supplement rams to ensure they are in adequate condition for mating.

Lupins can be fed to increase testicular size.

Ewes

Ensure ewes maintain condition score over 2.5 and an average of CS 3 at joining.

Weaners

Weigh about 30 to 50 weaners; separate the light weaners from the rest of the mob and give them the best available feed in terms of both energy and protein.

To ensure survival, weaners should grow at least 0.5 to 1 kg per month.

February

All sheep

Watch for blowflies.

If liver fluke is present, a drench may be required; monitor for fluke eggs to determine if a drench is required.

Ryegrass staggers may occur in late summer/autumn so ensure all livestock management procedures are completed before ryegrass staggers is likely to become a problem.

A second summer drench is due, if not done in January, and should be completed by mid February; do a WEC to determine if a drench is needed.

Prepare worm safe pastures for weaners – 'smart graze' preparation.

Ewes

Ensure that ewes maintain a condition score over 2.5 and an average of CS 3 at joining.

Rams

Carry out a pre-joining examination: ensure rams are in condition score 3.5–4.

Examine the rams' feet, testicles and penis, and assess their general health.

Supplement rams to ensure they are in adequate condition for mating.

Lupins can be fed to increase testicular size.

Weaners

Monitor weaners' body weight to assess weight changes to fine tune supplementary feeding strategies.

Weaners less than 20 kg should grow at 1 kg per month.

Weaners greater than 21 kg can be maintained at that weight for 1 to 2 months, although slow growth is still preferable.

Decide on a weaning paddock for next season's lambs – graze this paddock with only cattle or sheep that have low egg outputs such as adult wethers or dry ewes.

March

All sheep

Shear sheep.

Sheep affected with lumpy wool should be clean shorn and dipped last. Lumpy wool infection can spread by dipping and handling wet sheep.

Do not dip or use a pour-on unless a lice infestation exists or there is a reasonable suspicion that one exists.

Watch for blowflies before shearing.

Ewes

Maintain a pre joining condition score above 2.5, with an average condition score of 3.

Maiden ewe weights should be about 75% of their mature body weight.

Rams

Prepare rams for joining.

Exclude infertile rams from the ram team (the lambing percentage can be reduced by running infertile rams with fertile rams, infertile rams serve ewes and these ewes may not be serviced by fertile rams).

Weaners

Monitor the WEC about 6 weeks after the last drench.

April

Ewes and rams

Join maidens separately at 2% rams for 5 weeks.

Use experienced rams with maiden ewes.

Join mature ewes at 1 to 1.5% rams for 5 weeks.

No need for teasers from February onwards.

Use smaller paddocks with a single watering point and with no geographical barriers such as timber belts or gullies.

Weaners

Monitor WEC 4 to 6 weeks after the last drench or drench onto a smart graze paddock.

May

All sheep

Monitor body condition to assess the feeding strategy if seasonal conditions are poor.

Monitor potential fluke-affected mobs and drench if fluke eggs are present.

Allocate sheep to their winter paddocks; take into account the paddock carrying capacity and stock feed requirement.

Rams

Remove rams from ewes after a 4 to 6 week joining period.

Weaners

Monitor WEC 4 to 6 weeks after last drench or drench onto a smart graze paddock.

June

All sheep

Monitor the WEC of weaners and other mobs in poor condition.

Monitor body condition to assess the feeding strategy if seasonal conditions are poor.

Ewes

Scan for pregnancy, especially if supplementary feeding is required due to a poor season or if twin lambing ewes can be identified and allocated very good feed and shelter to ensure high lamb survival rates.

July

All sheep

If footrot is present, plan the control strategy for spring.

Monitor the WEC of weaners or other mobs in poor condition.

Monitor the condition score to assess the feeding strategy if seasonal conditions are poor.

Monitor potential fluke-affected mobs and drench if fluke eggs are detected; a drench in late winter will remove adult fluke and hence remove the source of fluke eggs before the intermediate host snail becomes active in spring.

Ewes

Vaccinate with 5 in 1 or 6 in 1 pre lambing.

Increase the ewes' energy intake in the last 6 weeks of pregnancy.

Ensure that the ewes' condition score does not fall below an average of 2.7 or 3 if ewes cannot gain body condition in late pregnancy (under-nutrition of the ewe results in low birth weight lambs, which have a very low survival rate).

Crutching pre-lambing has no effect on lamb survival.

Control foxes pre-lambing.

Wethers

Pizzle rot can occur in wethers grazing clover-dominant pasture. Testosterone treatment protects against pizzle rot for about 2 to 3 months and should be timed to coincide with growth of clover-dominant pasture.

Young sheep

Assess the trace element status of sheep; trace elements are more likely to be deficient in late winter/ early spring.

August

All sheep

Monitor the WEC of weaners or other mobs in poor condition.

Ewes

A pre lambing drench is probably of little value **unless**

- there are enough 'worm safe' paddocks to allow a move after drenching or
- if ewes are in condition score less than 3 and grazing pasture less than 800 kg DM/ha in high rainfall regions or
- the property history indicates drenching is necessary in young ewes or
- if WECs are high.

Safe paddocks should be reserved for weaners.

September

Ewes

Lambing.

Good shelter can increase lamb survival; a dead lamb pick up is useful to determine cause of death to aid future management decisions.

Ensure that the ewes' energy intake meets the demands of late pregnancy and lactation.

Ensure that ewes are grazing greater than 1000–1200 kg DM/ha.

All sheep

Examine all flocks for footrot; if footrot is present, minimise the spread in spring.

October

All sheep

Examine pre-shearing for lice; if lice are found, control lice if fleece damage is occurring and eradicate off shears.

Ewes

Ensure that ewes receive high levels of energy and protein for lactation.

A post-lambing drench for ewes is not warranted in most seasons.

Determine which ewes have lambed and lost their lambs.

It is possible to improve the reproductive performance of the group by removing failed-to-rear maiden ewes, but it is not worthwhile selecting for rearing ability unless the ram source is doing the same.

Wet-drying ewes is probably not worthwhile unless investigating an infertility problem; if more than 8% of the flock is dry, there is a problem so determine the cause.

The lamb marking percentages of the maidens should be within 10% of the adult flock; if not, determine the cause.

Lambs

Mark at the end of the month.

Dock tails at a length 1 cm below the tip of the vulva or equivalent in wethers; tails cut too short result in dirty breeches, flystrike and cancers of the tail and vulva.

Carry out the first 5 in 1 vaccination for all lambs.

Feed trace elements supplement if deficiency exists; monitor the trace element status of lambs every few years.

A worm drench for lambs at marking is not necessary in most seasons.

All sheep

Crutch or jet ewes to prevent breech strike, if not already crutched.

November

All sheep

Watch for blowflies: a long-term chemical treatment may need to be administered later this month for protection through summer, especially in high-risk situations or when nutrition is poor.

Commence first summer drench; prepare smart grazing paddocks for next winter.

Complete crutching, if not done earlier.

Carry out final inspection of suspect new sheep for release from quarantine for footrot.

December

All sheep

Watch for blowflies: a long-term chemical treatment may need to be administered later this month for protection through summer, especially in high-risk situations or when supervision is poor.

The first summer drench should be completed; prepare smart grazing paddocks for next winter.

If footrot is present, tip sheep and pare and examine all the feet of all sheep. Cull any infected animals. If spring footrot control was effective, there should be fewer than 5% infected sheep in summer.

If liver fluke occurs on the property, a drench may be required to prevent acute fluke disease, but usually only in high risk situations.

Ewes

Wean lambs 12–14 weeks post lambing.

Crutch ewes, if not crutched earlier.

Draft ewes post weaning and provide the lightest ewes with the best feed to ensure they have an opportunity to increase weight before the next mating period.

Lambs

Wean onto a 'clean paddock'; that is, one relatively free of worm eggs and larvae such as a cattle paddock, wether paddock or a hay aftermath (do not wean onto a paddock that has been grazed by lambs and ewes since the autumn break).

Drench with an effective drench at the correct dose rate; drench for the heaviest animal in the flock.

Monitor the effectiveness of the drench by carrying out a WEC 10 days after drenching.

Complete a drench-resistance trial every 2 to 3 years to determine drench rotations.

Give the second 5 in 1 vaccination: a single vaccination protects sheep from the clostridial diseases for only a short time, so a second dose is essential.

Check body weights: the weight at weaning should be in the range of 18 to 22 kg and aim to grow at about 1 kg per month.

Winter lambing crossbred ewes joined to terminal sires

January

All sheep

A second summer drench is due from mid January, so carry out a worm egg count (WEC) to determine if drench needed. Use an effective drench at the correct dose rate; do not under-dose; and set dose rate to the heaviest in the mob.

Rams

Carry out a pre-joining examination: ensure that rams are in condition score 3.5 to 4.

Examine the rams' feet, testicles and penis, and assess their general health, if not done earlier.

Supplement rams to ensure they are in adequate condition for mating.

Lupins can be fed to increase testicular size.

Exclude infertile rams from the ram team (the lambing percentage can be reduced by running infertile rams with fertile rams, infertile rams serve ewes and these ewes may not be serviced by fertile rams).

Ewes

Ensure that ewes maintain a body condition of an average of condition score 3 at joining to ensure adequate conception rates.

Carry over lambs

Ensure lambs are supplemented with an appropriate ration or allocated pasture to meet target weights and market specifications.

February

All sheep

Watch for blowflies.

If liver fluke is present, a drench may be required; monitor for fluke eggs to determine if a drench is required.

Ryegrass staggers may occur in late summer/autumn so ensure all livestock management procedures are completed before ryegrass staggers is likely to become a problem.

A second summer drench is due if not done in January and should be completed by mid February, so do a WEC to determine if drench is needed.

Ewes and rams

Join maidens separately at 2% for 6 weeks.

Use experienced rams with maiden ewes.

Join mature ewes at 1 to 1.5% for 6 weeks.

There is no need for teasers from February onwards.

Use smaller paddocks with single watering point and with no geographical barriers such as timber belts or gullies.

Monitor body condition and ensure that ewes and rams maintain a condition score above 2.5.

Carry over lambs

Ensure that lambs are supplemented an appropriate ration or allocated pasture to meet target weights and market specifications.

March

All sheep

Watch for blowflies.

Ewes

Monitor the body condition and ensure that ewes maintain a condition score above 2.5.

Rams

Remove the rams from the ewes after a 6-week joining period.

Carry over lambs

Ensure that lambs are supplemented with an appropriate ration or allocated pasture to meet target weights and market specifications.

Monitor WECs about 4 to 6 weeks after last drench to ensure worm burdens are not causing production losses.

April

All sheep

Watch for blowflies.

Ewes

Monitor body condition and ensure that ewes maintain a condition score above 2.5.

Scan for pregnancy, especially if supplementary feeding is required due to a poor season or if twin lambing ewes can be identified and allocated very good feed and shelter to ensure high lamb survival rates.

Carry over lambs

Ensure that lambs are supplemented an appropriate ration or allocated pasture to meet target weights and market specifications.

Monitor WECs about 4 to 6 weeks after the last drench to ensure worm burdens are not causing production losses.

May

All sheep

Monitor body condition to assess the feeding strategy if seasonal conditions are poor.

Monitor potential fluke-affected mobs and drench if fluke eggs are detected.

Ewes

Carry out a pre-lambing crutch (an alternative is to crutch post lambing so no clean up dagging is required before shearing).

Vaccinate ewes with 5 in 1 or 6 in 1.

Increase energy intake of ewes in the last 6 weeks of pregnancy.

Avoid pregnancy toxaemia by preventing under-nutrition and stress in late pregnancy.

Ensure that ewe body condition does not fall below an average of condition score 3.

Avoid under-nutrition of the ewes, which results in low birth weight lambs that have a low survival rate.

Allocate the ewes to lambing paddocks: ewes carrying multiple lambs should have the best paddocks.

Carryover lambs

Ensure that lambs are supplemented with an appropriate ration or allocated pasture to meet target weights and market specifications.

Monitor the WECs about 4 to 6 weeks after last drench to ensure that worm burdens are not causing production losses.

June

Ewes

A pre-lambing drench is probably of little value unless:

- there are enough 'worm safe' paddocks to allow a move after drenching or
- if seasonal conditions are poor and ewes are under nutritional stress and grazing pastures are less than 900 kg DM/ha or
- pre lamb WECs are high.

Monitor ewe condition score and pasture availability: it should be greater than 1100 kg DM/ha at lambing.

Carry over lambs

Ensure that lambs are supplemented with an appropriate ration or allocated pasture to meet target weights and market specifications.

Monitor WECs about 4 to 6 weeks after the last drench to ensure that worm burdens are not causing production losses.

Ensure the last carry over lambs are sold before they cut their first two permanent incisors.

July

Ewes

Lamb ewes.

Provide good shelter to increase lamb survival.

A dead lamb pick up and investigation is useful to determine cause of death to aid future management decisions.

Ensure that energy intake meets the demands of late pregnancy and lactation.

August

Ewes

Ewes require high levels of energy for lactation, so ensure that pastures have more than 1200 kg DM/ha.

Mark lambs.

Separate dry ewes by wet drying ewes at marking if the feed conditions are poor.

A post-lambing drench for ewes is not warranted in most seasons: assess the need based on pasture availability (less than 900 kg DM/ha, ewe condition score less than 2.5 and WEC result).

The lamb marking percentages of the maidens should be within 10% of the adult flock; if not, determine the cause.

Determine the marking percentages; if low, determine why.

Lambs

Mark lambs at the end of the month.

Dock tails at a length below the tip of the vulva or equivalent in wethers.

Give the first 5 in 1 vaccination for all lambs.

Provide trace elements supplements as appropriate for property deficiencies.

A worm drench is not necessary for lambs at marking time.

September

All sheep

Examine pre-shearing for lice; if lice are found, control lice if fleece damage is occurring and eradicate off shears.

Examine all flocks for footrot; if footrot is present, minimise the spread in spring.

October

Ewes

Crutch, if not done earlier, before fly season or jet breech if crutching is not possible; observe chemical withholding periods.

Lambs

Drench lambs when 12 to 13 weeks of age and provide trace element supplements as appropriate for property deficiencies.

Consider a breech jet to prevent flystrike, especially if the lambs are daggy.

Complete a drench resistance trial every 2 to 3 years to determine what drenches to use in an annual rotation.

Start to prepare lambs for sale, including dagging.

November

Ewes

Wean lambs 16 weeks post lambing.

Commence the first summer drench from mid month, especially in dry seasons.

Draft ewes into condition score groups: lighter ewes should be allocated the best feed to ensure that they gain the most weight before pasture dries off and weight gain is no longer possible without supplements.

Lambs

First draft lambs: sell as suckers those lambs that meet specifications according to marketing strategy for enterprise.

Wean remaining lambs onto a 'clean paddock'; that is, one relatively free of worm eggs and larvae, such as a cattle paddock or a hay aftermath; avoid weaning onto a paddock that has been grazed by carry over lambs or lambing ewes.

Drench with an effective drench at the correct dose rate.

Drench for the heaviest lamb in the mob, and be aware of withholding periods.

Give the second 5 in 1 vaccination: a single vaccination does not protect against clostridial disease, so a second dose is essential for carry over lambs.

Lambs that will not be sold for at least 8 weeks may be shorn, although they must be kept completely separate if lice eradication is required.

All adult sheep

Shear sheep

Do not dip or use a pour-on unless a lice infestation exists or there is a reasonable suspicion that one exists.

Watch for blowflies in full-wool sheep before shearing.

Carry out a final inspection of suspect new sheep (purchased last year) for release from quarantine for footrot.

Rams

Carry out a pre-joining examination, and ensure that rams are in condition score 3.5 to 4.

Examine the rams' feet, testicles and penis, and assess their general health.

Supplement the rams to ensure they are in adequate condition for mating.

Lupins can be fed to increase testicular size.

Replacement ewes

Plan replacement ewe purchases: purchased ewes should be placed in quarantine until next spring to minimise the risk of the spread of footrot and drenched with a triple combination drench to minimise the risk of introducing resistant worms onto the property.

December

All sheep

The first summer drench should be completed: use an effective drench at the correct dose rate.

If broad-spectrum drench groups are still working, rotate drench groups yearly with a change at the time of the first summer drench.

If liver fluke occurs on the property, a drench may be required to prevent acute fluke disease.

Rams

Carry out a pre-joining examination, and ensure that rams are in condition score 3.5 to 4.

Examine the rams' feet, testicles and penis, and asses their general health.

Supplement rams to ensure they are in adequate condition for mating.

Lupins can be fed to increase testicular size.

Weaned lambs

Give the weaned lambs the best available feed in terms of both energy and protein to assist in finishing for market.

Watch for blowflies if carry over lambs are not shorn.

Repeat the WECs 4 to 6 weeks following each drench to determine if a further drench is required.

Appendix 2 – Recording sheet for estimating average pasture height

Ensure measurements are taken randomly (do not choose an easy spot to measure) and measure only green leaf, but ignore weeds such as onion grass and thistles that sheep will not eat (if the stick lands on these then the pasture height is zero).

Aim to get about 50 random height measures across the paddock and place a tick or number 1 beside the appropriate height (see example below) until finished. Multiply each height by the number recorded at that height, then add up these totals and divide by the number of measurements to get the average height (cm). The blank recording sheet below can be photocopied and used to record pasture heights.

Height (cm) A	Place a mark or tick beside the height whenever a measurement of this height is recorded	Number of recordings B	A × B
0			
1			
2			
3			
4			
5			
6			
7			
8			
9			
10			
11			
12			
	Totals		
	Average height = total (A × B)/total B		

Example of filled in sheet:

Height (cm) A	Place a mark or tick beside the height whenever a measurement of this height is recorded	Number of recordings B	A × B
0	11111	5	0
1	111111111111	12	12
2	11111111111111	14	24
3	1111111111	10	30
4	11111	5	20
5	11	2	10
6	11	2	12
7			
8			
9			
10			
11			
12			
	Totals	**50**	**108**
	Average height = total (A × B)/total B (e.g. 108/50 = 2.16)	**2.16 cm**	

Another simple assessment can be by comparing the height of green leaf to the side of your boot. Note that not all boots will be the same so you will need to make own measurement of the height of toe, heel, and so on (you need to calibrate your boot).

Appendix 3 – Cutting and drying pasture quadrants for estimating pasture availability

1. Make a 25 cm square quadrant out of timber, weldmesh or other suitable material.
2. Place the quadrant on a representative section of pasture.
3. Cut the pasture in the quadrant close to the ground and avoid collecting dirt.
4. Weigh the pasture sample using kitchen scales (e.g. 500 g).

Divide the sample into 4 and discard 2 quarters

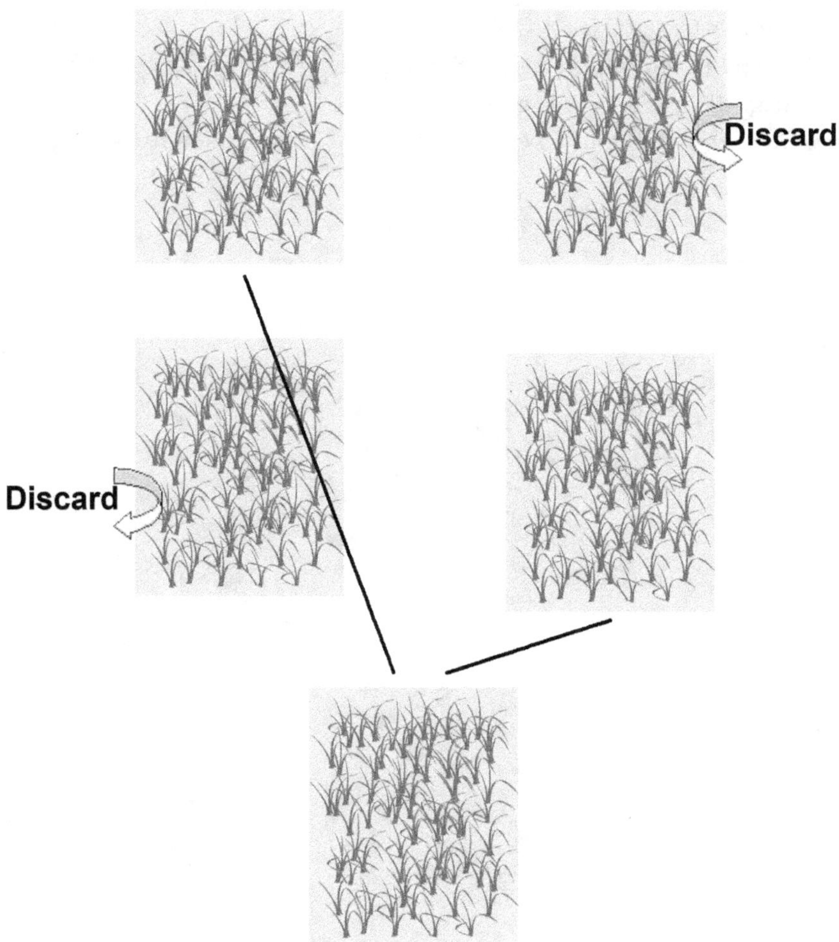

Divide into quarters and start again

Sub-sampling to get a sample size small enough to dry.

5. Sub-sample to get a small enough sample to dry in the microwave (see the instructions above on how to get a representative pasture sub-sample to dry).
6. Weigh the pasture sub-sample, exclude the weight of any sample container (100 g is a workable amount of pasture).
7. Dry the pasture sub-sample in the microwave at 80% of microwave power (include a glass of water in the microwave oven to ensure that the pasture does not catch fire).
8. Microwave the pasture sub-sample for short periods of time and check the sample between bouts in the microwave.
9. When the pasture sub-sample appears to be dry, weigh it and give it another short period of time in the microwave, weigh it again. When the weight of the sub-sample remains constant, it is dry.
10. Weigh the dry pasture sub-sample (e.g. 80 g).
11. Divide the dry weight of the sub-sample by the wet weight of the sub-sample and multiply by 100 to get the dry matter (DM)% (e.g. 80/100 × 100 = 80% DM)
12. Multiply the weight of the original quadrant pasture sample by the DM %/100 to get the DM weight of the sample (e.g. 500 g × 80%/100 = 400 g).
13. Convert DM weight of the sample to g DM per m^2. There are 16 (25 cm × 25 cm) quadrants in a square metre (e.g. 400 g × 16 = 6400 g/m^2).
14. Convert to g DM per m^2 to kg DM per ha. (6400 g/m^2 is equivalent to 6400 kg/ha dry matter).

Appendix 4 – Cost per weaner for different ram prices, joining rates and weaning percentages

Tables A4.1 to A4.4 show the cost per weaner for different ram prices, joining rates and weaning percentages. These can be used to place genetic costs (ram costs) into context with other variable costs, but importantly to identify how to get better value from good sires and make some simple comparisons on ram costs.

Table A4.1. Cost per weaner for different ram prices at variable joining rates when rams are used for 4 years and the weaning percentage is 75%.

	Cost per weaner – ram used for 4 years, weaning percentage is 75%					
Joining	**$150**	**$250**	**$350**	**$500**	**$750**	**$1000**
1%	$0.50	$0.83	$1.17	$1.67	$2.50	$3.33
2%	$1.00	$1.67	$2.33	$3.33	$5.00	$6.67
3%	$1.52	$2.53	$3.54	$5.05	$7.58	$10.10

Table A4.2. Cost per weaner for different ram prices – rams used for 1 to 4 years when the joining rate is 2% and the weaning percentage is 75%.

	Cost per weaner – ram joined at 2% and weaning percentage is 75%					
No. years joined	**$150**	**$250**	**$350**	**$500**	**$750**	**$1000**
1	$4.00	$6.67	$9.33	$13.33	$20.00	$26.67
2	$2.00	$3.33	$4.67	$6.67	$10.00	$13.33
3	$1.33	$2.22	$3.11	$4.44	$6.67	$8.89
4	$1.00	$1.67	$2.33	$3.33	$5.00	$6.67

Table A4.3. Cost per weaner for different ram prices at variable joining rates when rams are used for 4 years and the weaning percentage is 100%.

	Cost per weaner – rams joined for 4 years and weaning percentage is 100%					
Joining	**$150**	**$250**	**$350**	**$500**	**$750**	**$1000**
1%	$0.38	$0.63	$0.88	$1.25	$1.88	$2.50
2%	$0.75	$1.25	$1.75	$2.50	$3.75	$5.00
3%	$1.14	$1.89	$2.65	$3.79	$5.68	$7.58

Table A4.4. Cost per weaner for different ram prices, when rams are used for 1 to 4 years when the joining rate is 2% and the weaning percentage is 100%.

	Cost per weaner – rams joined at 2% and weaning percentage is 100%					
No. years joined	**$150**	**$250**	**$350**	**$500**	**$750**	**$1000**
1	$3.00	$5.00	$7.00	$10.00	$15.00	$20.00
2	$1.50	$2.50	$3.50	$5.00	$7.50	$10.00
3	$1.00	$1.67	$2.33	$3.33	$5.00	$6.67
4	$0.75	$1.25	$1.75	$2.50	$3.75	$5.00

So, in this example, if Ram A costs $500, Ram B costs $700 and Ram C costs $1000.

All rams are equally likely to produce 420 fleeces if used for 4 years, so the cost per fleece is:

Ram A costs $1.19/ fleece ($500/420)

Ram B costs $1.67/fleece ($700/420) 48c/ fleece more than Ram A

Ram C costs $2.38/fleece ($1000/420) $1.19 fleece more than Ram A

So, in this example, the justification for paying $1000 rather than $500 is that each fleece produced from the progeny of Ram C must return more than $1.19 than Ram A, or paying $200 more for Ram B needs to return more than 48c cents/fleece.

If Ram A has an estimated breeding value for clean fleece weight of 0% (average clean fleece weight) and

Ram B has a breeding value for clean fleece weight of +5%

Ram C has a breeding value of +10%

The average clean fleece weight on the home farm is 4 kg.

As the rams contribute half the genes, then:

Progeny from Ram A will cut the same – 4 kg clean fleece weight

Progeny from Ram B will cut 2.5% more (half of 5%) an extra 100 g/fleece

(4 kg fleece × 2.5% or 0.025 = 0.1 kg or 100 g more)

So if Ram B costs 48c more per fleece to provide 100g more clean wool – is the cost justified?

The same applies to comparing the value of Ram C:

Progeny from Ram C will cut 5% more (half of 10%) than A, so an extra 200 g/fleece

(4 kg fleece × 5% or 0.05 = 0.2 kg or 200 g more)

So, if Ram C costs $1.19 more per fleece to provide 200 g more clean wool – is the cost justified?

The same calculation can be done to compare the value of reducing fibre diameter or improving some other trait.

References and further reading

Pasture growth and management

Cayley J & Quigley P (2005) *Phosphorus for Sheep and Beef Pastures*. Victorian Department of Primary Industries: Hamilton.

Glendinning JS (1999) *Australian Soil Fertility Manual – Fertiliser Federation of Australia Inc*. CSIRO Publishing: Melbourne.

Gourley CJP & Victorian Department of Primary Industries (2007) *Making Better Fertiliser Decisions for Grazed Pastures in Australia*. Victorian Department of Primary Industries: Ellinbank.

Kieran Ransom K, Trapnell L, Clune T, Hirth J, Whale J, Bate N & Riad N (2006) *Making Lucerne Pay – Integrating Crops and Lucerne on Mixed Farms*. Victorian Department of Primary Industries.

Lamp C & Collet F (1979) *A Field Guide to Weeds in Australia*. Inkata Press: Melbourne.

McCarthy M & Stapleton P (1996) *Better Pasture Better Crops – Pastures for Cropping/Livestock Farms*. Victorian Department of Natural Resources and Environment.

Mitchell M (1994) *Identification Handbook for Native Grasses in Victoria*. Speciality Press.

Nie Z & Saul G (2006) *Greener Pastures for South West Victoria*. 2nd edn. Victorian Department of Primary Industries: Hamilton.

Peverill KI, Sparrow LA & Reuter DJ (1999) *Soil Analysis – An Interpretation Manual*. CSIRO Publishing: Melbourne.

Slavich P (2001) Soil health: a systems approach to soils. In *Soil Health: Foundation of Sustainable Agriculture Workshop Proceedings*. Wollongbar, New South Wales.

Victorian Department of Primary Industries (2003) *Grazing Management for Productive Hill Country Pastures – The Broadford Grazing Experiment 1998–2002*. Victorian Department of Primary Industries.

Websites

Australian Soil and Plant Analysis Council, <http://www.aspac-australasia.com/>

Australian Soils Resource Information System, Fertiliser Nutrient Loss Index tool, <http://www.asris.csiro.au>

Grain and Graze, <http://grainandgraze.com.au/>

Grasslands Society of Southern Australia Inc, <http://www.grasslands.org.au/>

Nitrogen Best Management Practices, <http://www.nitrogen.unimelb.edu.au>

The University of Melbourne's Faculty of Veterinary Science – The Mackinnon Project, <http://www.mackinnonproject.com.au>

Nutrition and feed management

Cottle DJ (Ed.) (2010) *International Sheep and Wool Handbook*. Nottingham University Press: Nottingham, UK.

Davis J (Ed.) (2003) *Feedlotting Lambs: A Producer's Guide.* Landlinks Press: Melbourne.

Freer M & Dove H (Eds) (2002) *Sheep Nutrition.* CSIRO Publishing: Melbourne.

Nutrient Requirements of Domesticated Ruminants (2007) CSIRO Publishing: Melbourne.

Websites

Lifetime Wool Project, <http://www.lifetimewool.com.au/>

Sheep CRC – Practical Wisdom, <http://www.sheepcrc.org.au/industry-tools-and-information/practical-wisdom-notes.php>

The University of Melbourne's Faculty of Veterinary Science – The Mackinnon Project, <http://www.mackinnonproject.com.au>

Drought management

Court J (Ed.) (2006) *Drought Feeding and Management of Sheep – A Guide for Farmers and Land Managers.* Department of Primary Industries: Melbourne.

McFarland I, Curnow M, Hyder M, Ashton B & Roberts D (2006) *Feeding and Managing Sheep in Dry Times.* WA Bulletin 4697. Department of Agriculture Western Australia.

Websites

The University of Melbourne's Faculty of Veterinary Science – The Mackinnon Project, <http://www.mackinnonproject.com.au>

Disease management

Brightling A (2006) *Livestock Diseases in Australia – Diseases of Cattle, Sheep, Goats and Farm Dogs.* CH Jerram Science Publishers: Mt Waverley, Victoria.

Boray JC (1999) 'Liver fluke disease in sheep and cattle'. Primefact 446, revised by Dr GW Hutchinson and Stephen Love (March 2007). NSW Department of Primary Industries. <http://www.dpi.nsw.gov.au/agriculture/livestock/sheep/health>

Campbell NJ & Miller G (2007) 'Control of liver fluke'. Agnote AG0072. Department of Primary Industries Victoria. <http://www.dpi.vic.gov.au/dpi/index.htm>

Lloyd JB, Boray JC & Campbell NJ (2003) 'Identifying liver fluke snails'. Primefact 476. NSW Department of Primary Industries. <http://www.dpi.nsw.gov.au/agriculture/livestock/sheep/health>

Websites

Australian Pesticide and Veterinary Medicine Authority APVMA, <http://www.apvma.gov.au/>

Australian Wool Innovation, <http://www.wool.com/>

Flystrike prevention, <http://www.wool.com/Grow_Animal-Health_Flystrike-prevention.htm>

Integrated Pest Management sheep, <http://www.wool.com/Grow_Animal-Health_Integrated-Parasite-Management_Integrated-Parasite-Management-sheep.htm>

LiceBoss, <http://www.wool.com/Liceboss>

Making more from sheep: Module 11 Healthy and contented sheep, Australian Wool Innovation and Meat & Livestock Australia, <http://www.makingmorefromsheep.com.au/>

Managing flystike, <http://www.sheepcrc.org.au/industry-tools-and-information/managing-flystrike.php>

Meat & Livestock Australia, <http://www.mla.com.au/>

The University of Melbourne's Faculty of Veterinary Science – The Mackinnon Project, <http://www.mackinnonproject.com.au>

WormBoss, <http://www.wool.com/Wormboss>

Genetic improvement

Court J & Victorian Department of Primary Industries (2006) 'Merinos to match: the right sheep for your environment, across site report 2006'. Victorian Department of Primary Industries: Seymour, Victoria. < http://www.dpi.vic.gov.au/DPI/nrenfa.nsf/93a98744f6ec41bd4a256c8e00013aa9/80b2130a54318b5eca257528000dae3c/$FILE/FinalReportNov06.pdf>

Dun RB & Eastoe RD (1970) *Science and the Merino Breeder.* New South Wales Department of Agriculture.

Safari A & Fogarty N M (2003) *Genetic Parameters for Sheep Production Traits: Estimates from the Literature.* NSW Agriculture: Orange, NSW.

Websites

Meat and Livestock Australia, Australian Wool Innovation Limited Sheep genetics, <www.sheepgenetics.org.au>

Merino Bloodline Performance – NSW Department of Primary Industries and Australian Wool Innovation, <http://www.dpi.nsw.gov.au/agriculture/merino-bloodline-performance>

Merino breeding for profit, <http://www.sheepgenetics.org.au/Document/70/SDF%20brochure.pdf>

WormBoss, <http://www.wool.com/Wormboss>

Results of the South Australian Selection Demonstration Trials??? [AQ: delete?]

Results from the Trangie QPLus project can't find this [AQ: delete?]

Merino Selection Demonstration Flocks – Final report (SARDI) can't find this [AQ: delete?]

Science and the Breeder – can't find this] [AQ: delete?]

The University of Melbourne's Faculty of Veterinary Science – The Mackinnon Project, <www.mackinnonproject.com.au>

Animal welfare

Kendrick KM *et al.* (2001) Sheep don't forget a face. *Nature* **414**: 165–166.

Websites

Meat and Livestock Australia (MLA), <http:// www.mla.com.au>

Australian Wool Innovation (AWI), <http://www.awi.com.au>

Animal Health Australia, <http://www.animalhealthaustralia.com.au>

Animal Welfare Science Centre, <http://www.animal-welfare.net.au>

Royal Society for the Prevention of Cruelty to Animals, <http://www.rspca.org.au>

The Australian Animal Welfare Strategy, <http://www.daff.gov.au/aaws>
The Department of Primary Industries Victoria, <http://www.dpi.vic.gov.au/animalwelfare>

Working dogs

Brightling A (2006) *Livestock Diseases in Australia – Diseases of Cattle, Sheep, Goats and Farm Dogs*. CH Jerram Science Publishers: Mt Waverley, Victoria.

Index

www.ingramcontent.com/pod-product-compliance
Lightning Source LLC
LaVergne TN
LVHW080023210926
848481LV00011B/3094